Statistics
for Economics

Statistics
for Economics
An Intuitive Approach

Alan S. Caniglia

Franklin and Marshall College

HarperCollins*Publishers*

Dedication

To my daughters, Annelies and Ellie:
That they may never be intimidated by numbers.

Sponsoring Editor: John Greenman
Project Editor: David Nickol
Design Supervisor: Lucy Krikorian
Text Design Adaptation: North 7 Atelier Ltd
Cover Design: Heather A. Ziegler
Production Administrator: Jeffrey Taub
Compositor: TAPSCO, Inc.
Printer and Binder: R. R. Donnelley & Sons Company
Cover Printer: New England Book Components, Inc.

STATISTICS FOR ECONOMICS: An Intuitive Approach

Library of Congress Cataloging-in-Publication Data

Caniglia, Alan Scott.
    Statistics for economics / Alan S. Scott.
        p.   cm.
    Includes index.
    ISBN 0-06-041168-6
    1. Statistics.   I. Title.
HA29.C273   1991
519.5′024339—dc20

91-20505
CIP

91 92 93 94   9 8 7 6 5 4 3 2 1

Contents

Chapter 14   Multiple Regression   340

Epilogue   377

Tables   381

Preface

This book is intended primarily for a one-semester statistics course for undergraduate economics students. Two simple beliefs led me to write this book. First, students of economics need a statistics book designed for *them,* and not designed primarily for students of business administration or other disciplines. And second, statistics *can* be presented in a way which brings out the intuition behind the material; this subject does not have to be viewed by students as meaningless sequences of formulas devoid of any connection to the way we naturally think about things.

There are a large number of textbooks in statistics for "business and economics" that cover a broad range of topics. Typically, however, they are not focused on the material most relevant to applied research in economics, which is regression analysis and its variations. If the goal of a statistics course for economics students is to prepare them to be able to read such work and to do such work on their own, the aim of its text should be to prepare such students to understand and perform basic regression techniques. This means that some of the material in a standard business and economics statistics textbook is unnecessary, while other material should receive more emphasis. This text was written to prepare readers to understand and perform basic regression analysis, both of the two-variable and multiple-variable forms.

In this text, statistical concepts and techniques are presented in a mathematically rigorous context, but without using calculus and other forms of advanced mathematics. Verbal explanations as well as numerical examples are provided in an effort to present the intuition behind the procedures because such intuition enables the reader to understand and properly interpret the results of such procedures. (Calculus is referred to in a few places in the text, in situations where a student familiar with it can easily gain a heightened understanding of the concept under discussion. But verbal explanations which make the calculus unnecessary for the discussion are also presented. Additionally, some calculus results are derived in optional appendices.) A student who completes a course based upon this book should be ready to take an undergraduate course in econometrics.

DESIGN OF THE BOOK

There are four major sections in this book. In the first section (Chapters 1 through 4), the fundamental issues in statistics are introduced, including the distinctions between populations and samples and between the processes of description and inference. Simple methods of inference are suggested, and the limitations of these methods are explored. In the second section (Chapters 5 through 9), probability theory and its extensions are presented as material which will provide a bridge from the simple inferences discussed in the first section to the more sophisticated inferences discussed in the third section. In the third section (Chapters 10 through 12), the simple inferences from Section I and their limitations are examined in a more detailed manner, and more sophisticated inferences which attempt to deal with these limitations (in general by bringing in the notion of a "likely margin of error") are presented. Finally, the fourth section (Chapters 13 and 14) takes the foundations discussed in the first three sections and uses them to present, develop, and interpret regression analysis.

   Throughout the text, two criteria have been used in selecting examples and problems: (1) those chosen should be truly useful in conveying to students an understanding of the material; and (2) they should show, whenever possible, either the application of techniques to economics or the types of problems economists typically work with.

   Several real-world data sets from economics are included in examples or in problems. Most notably, Appendix 2 to Chapter 11 and Appendices 3 and 4 to Chapter 13 contain data sets which can form the basis for additional problems of an instructor's own design.

It is assumed that students have access to statistical software on either a mainframe or personal computer. Virtually any statistical software package will be able to handle the examples and problems presented in this text. There are no references to specific statistical software; given the wide availability and ease of use of such software, it is left to the instructor to choose what is most appropriate for his or her course and institution.

Practice problems are, of course, a key tool for students in developing their understanding of the material. Sets of practice problems are therefore interspersed throughout the text; these problems are based upon the concepts and techniques just discussed. At the end of each chapter are review problems which integrate and extend the material from that chapter. Problems marked with a "*" are more difficult, but such problems also have high benefits. Each chapter closes with a set of discussion questions meant to elicit verbal explanations of what has been done and what might be done in future work. Short answers to odd-numbered problems are in the back of the text; a separate instructor's manual contains detailed answers for all problems and discussion questions in the text.

ACKNOWLEDGMENTS

I would like to thank the following people who read various drafts of my text manuscript and offered innumerable useful suggestions:

Roger R. Betancourt, University of Maryland at College Park
William S. Dawes, State University of New York at Stony Brook
E. Philip Howrey, University of Michigan
Wallace Huffman, Iowa State University
C. John Kurien, McGill University
Byung-Joo Lee, University of Colorado
Ram Mudambi, Lehigh University
Pierre Perron, Princeton University
Terry Seaks, University of North Carolina, Greensboro
Charles W. Sinclair, Portland State University
J. Wixon Smith, Rochester Institute of Technology
Edward McKenna, Connecticut College.

I am grateful to John Greenman at HarperCollins for supporting and believing in this project from the very beginning, and to all those who have worked at producing this final product. And finally, I am grateful to my wife, Janneke van Beusekom, for her comments on this project and for her encouragement.

I welcome criticisms and suggestions for improvement from students and instructors who use this book.

Alan S. Caniglia

Section I

Introduction and Simple Inference

Chapter *1*

Introduction

We begin this text with some introductory remarks about the nature of statistical analysis, the role of statistics in economics, and the motivating theme behind the text. We shall also discuss some preliminary material that sets the stage for much of what follows.

STATISTICS IN ECONOMICS

Statistical analysis gives us a way of trying to link economic theory with the real world. By using statistical methods to learn more about what actually occurs, we can evaluate the appropriateness of the theory and perhaps gain insights into how the theory might be further developed or changed. Statistical analysis has historically played a role in economics, but following the development of modern computers in the 1950s there has been a virtual explosion in its use and importance. It is therefore important for students of economics to acquire a basic understanding of statistical techniques, the proper interpretation of the results from these techniques, and their strengths as well as their limitations.

The theoretical side of economic analysis is not always easily separable from the statistical side, inasmuch as many economic theories are informed and motivated by observation or empirical results. To provide a classic example, in macroeconomic theory we learn about the mechanisms by which monetary expansion leads to inflation, growth in gross national product (GNP), or some combination of the two. But this theory is motivated by a large body of statistical information suggesting links (and perhaps very close links) in the historical data between these variables. In this case, the statistical information has played an important role in the development of the theory we learn today.

Given the importance of statistical analysis, we must then ask how to best present it for students of economics. We now turn to that issue.

THE MOTIVATION BEHIND THIS TEXT

The first motivating theme behind this text is that an understanding of statistical analysis is most likely to be acquired *and retained* if the material is presented in a context that emphasizes the intuition underlying it. As we shall see, often what is involved in a statistical procedure or result is quite consistent with our intuition if we only give that intuition a chance to come out. A seemingly complicated procedure in the problem at hand might be quite similar to what we would do instinctively in the context of other situations; presentation of such situations then facilitates our seeing the originally hidden intuition. In general, if the intuition behind a method of analysis is allowed to come to the surface, we wind up with a much stronger understanding of that analysis. And that stronger intuitive understanding should lead to better interpretive ability and a greater awareness of the strengths as well as the possible weaknesses of the statistical results.

The second motivating theme is that a statistics text specifically designed for students of economics, as opposed to a generic statistics text or one designed for students of "business and economics," is more conducive to the acquisition of the techniques and comprehension necessary to understand and practice the statistical methods commonly used in economics. For one thing, such a text allows us to focus as much as possible on examples from economics, so that students are regularly presented with the links between economic questions and statistical methodology and results.[1] But also, focusing on the needs of economists allows us to pursue a series of topics as they form a foundation for a study of regression analysis, which is the context for the vast majority of statistical work in economics, without the distractions that are necessarily a part of a more general text. Upon completion of this text a student will be well prepared for a first course in econometrics, in which the regression framework is applied and expanded to investigate a wide variety of economic problems.

To begin our work, we shall now present some preliminary material to set the stage for the remainder of this text.

SOME BASIC DISTINCTIONS IN STATISTICAL ANALYSIS

There are two fundamentally different types of statistical analyses, and their distinction is a basis for much of the work we shall undertake in this text. In *descriptive statistics* we summarize (describe) a particular set of information; for example, we might find that the average yield on money market funds at ten major banks during the past week was 8.0%. This statistic (the average) summarizes the information on the ten banks. With a descriptive statistic, however, we make no effort to go beyond the information we have available. For example, in the above analysis we made no effort to predict the yields for the current week for these banks or to say anything about the yields at other banks.

In many cases, however, we are interested in going beyond the available data to make statements about some alternative or larger group. For example, we might try to predict next year's inflation rate based on what the inflation rate has been in the recent past. Or we may only have partial information about some group of persons or things that we want to make a statement about; perhaps we are interested in the money market yields at banks other than the ten considered in the above example.

---

[1] Of course, there will be some situations in which the examples are not from economics because an alternative context most clearly brings out the point under discussion. In general, however, a substantial effort has been made to keep the examples as close to economics as possible.

Let us put this into the context of a new example. Suppose that you are a financial aid officer at a college and are interested in learning more about the summer income that students at the college typically earn. Suppose that you go to a class in which there are 20 students, ask everyone what their income was last summer, and find that the average was $2000. You are of course really interested not so much in this particular class but rather in the students at the college as a whole. Your knowledge concerning this class, however, gives you some ideas about what is true for the college as a whole, and you might then attempt to *infer* the characteristics of students' summer incomes in the college as a whole from the information you have for this class.

When we go beyond our available information to make statements about some larger (or alternative) group, we are engaging in *inferential statistics.* Clearly, this process is more complicated than simple description because we are dealing with an unknown. This approach makes the study of inferential statistics more interesting from an academic perspective. But, more important, it is typical in economics (and in many other areas as well) that we do not have complete information, and what we really want to do is to use our incomplete information to make statements about the larger group in which we are ultimately interested.

To pursue this matter further, it is often useful to make a distinction between a *population* and a *sample.* For any problem, the population is the group of individuals or things that we are ultimately interested in making a statement about. If we were interested in the money market yields during the past week at the ten banks considered above, then the ten banks would be our population; if we were interested in the summer incomes of the students at the college discussed above, then all the students at that college would be our population.[2] A population may be observed as a whole (as with the money market example), but often is not (as with the college example).

A sample, on the other hand, is any subset of the population. In our earlier example, the students in the class were a sample from the population of all the college's students. The process of inference

---

[2] Whether a set of observations is a population or a sample depends on the problem at hand. If we were only interested in the performance of the ten banks during the past week, then our data would describe the entire population. On the other hand, if we were really interested in the money market yields of all banks, then our information would be but a part of this population.

generally involves using the information in a sample to infer the characteristics of the larger population, which is not completely observed. As we shall be seeing, we can often make quite sophisticated inferences about a population even though it is not observed in its entirety. (For example, in the college student summer income example, in addition to estimating the average summer income for students at the college as a whole we might also be able to determine a "likely margin of error" for that estimate.) We shall discuss in more detail the process of sampling (and in particular the concept of a *random sample*) in Chapter 4.

Why might we not observe the entire population? This typically occurs because the population is too large to observe given the costs (both in time and money) involved in doing so. It is far easier to observe the incomes of a class of 20 students than it is to observe the incomes of all students at a college. And if we were trying to determine the summer incomes of all college students in the United States, it would be virtually impossible in the time permitted to interview all these students. The point is that inferential statistics as a field of study allows us to address a far wider range of problems than would otherwise be manageable, and it therefore contributes significantly to our ability to advance our knowledge about the world around us.

We shall be spending most of our time considering issues relevant to the problem of inference, although we shall of course develop some descriptive tools as well. The distinction between description and inference—and the distinction between a population and a sample—is, however, fundamental to much of our work.

WHERE DO WE GO FROM HERE?

In the remainder of Section I (Chapters 2 through 4) we shall continue to lay some foundations to prepare ourselves for pursuing the process of inference. In addition to a discussion of description, we shall discuss some basic ways of making inferences. But we shall also discuss the limitations of these basic methods so as to motivate future work. In Section II (Chapters 5 through 9) we shall discuss probability theory and its extensions and applications; this work is a foundation for making more sophisticated types of inferences in which we can often get very precise knowledge concerning a population even though only a small part of it is observed. In Section

III (Chapters 10 through 12) we shall discuss some ways of making such sophisticated inferences. Finally, in Section IV (Chapters 13 and 14) we shall study regression analysis and its applications. Regression analysis builds upon our earlier work, and it is the form of statistical analysis that is most commonly used in economics.

DISCUSSION QUESTIONS FOR CHAPTER 1

1. Suppose that we are interested in the average income of all families living in a particular geographical area (these families form our population). Consider two different ways of observing a sample from this population: (1) We interview the first 25 individuals who come to a new car dealer's showroom one day; and (2) we open the phone book to an arbitrary page and call the first 25 families (omitting the businesses) that are listed there. Which sample would you expect to give a more accurate picture of the typical income within this geographical area? Carefully explain why.

2. In this chapter we mentioned the relationship between monetary growth and macroeconomic performance as an example in which statistical results had motivated the development of the theory— that is, one in which theory and observation were linked quite closely. Can you think of another such example in economics?

Chapter 2

Technical Background on Random Variables and Linear Functions

In this chapter we discuss some concepts that are a foundation for much of our future work. In particular, we'll discuss (1) random variables, (2) the Σ notation, and (3) linear functions.

VARIABLES AND RANDOM VARIABLES

A *variable* represents a quantity that takes on different values for different persons or things; height, for example, is a quantity that varies across individuals. (A *constant,* on the other hand, is a quantity that does not so vary. The number of heads that a human has is a constant equal to one.) A *random variable* is a special type of variable that has its value determined at least in part by an element of chance. If, for example, we were to roll a normal die, the number showing would be a random variable that can take on a value of 1, 2, 3, 4, 5, or 6. It is a random variable because the number that turns up is determined by chance. Or, if we were to arbitrarily pick a person from the population as a whole, his or her income would be a random variable because the person actually chosen is determined by chance.

What we might refer to as a *nonrandom variable,* then, is a quantity that varies across persons or things but through a process that does not include an element of chance. If, for example, we were doing a controlled experiment on a group of individuals in which we imposed price increases of 1%, 2%, 3%, 1%, 2%, 3%, . . . on commodity X as we moved from person 1 to person 2 and so on (perhaps to observe how their quantity demanded changed), the series of price increases is due to a process containing no element of chance. The rate of price increase is then a nonrandom variable; the price increase for the second person is 2% as a result of the experimental design.[1] The responses of quantity demanded to the price changes, however, are undoubtedly random, because different persons will undoubtedly react in different ways to the price increase; persons 1 and 4, who both experience a 1% price increase, are likely to not change their quantity demanded by identical amounts.

We shall here focus on random variables. In the die example, suppose that we refer to the number showing on the die as the random variable X; the possible values of this random variable are $X = 1$, $X = 2$, $X = 3$, $X = 4$, $X = 5$, and $X = 6$. We often observe a set of outcomes (from rolling the die several times), in which case X_1 is the result of the first roll, X_2 is the result of the second roll, etc., up to X_n, which is the result of the nth (or last) roll. In general, X_i is the result of the ith roll of the die.

This X is an example of what we refer to as a *discrete* random variable. For such a variable the possible values are specific individual points along a number line. For this X we can illustrate the possible values as

where we have noted the six possible values of the random variables along the number line. The possible values are single points, and points in between are not possible (e.g., a value of 1.5 is not possible).

In contrast, for a *continuous* random variable any value within some range of values is a possibility. For example, if Y represents the heights of individuals, Y can take on any value within some range of biological plausibility (for adults, say from 2 feet to 9 feet). On the number line, the possible values of Y are

[1] If, however, we were to arbitrarily choose one of the individuals participating in the experiment, an element of chance would then be introduced.

and any point along the number line between 2 and 9 feet is a possibility (thus we've shaded that range of the number line). (The presumption, of course, is that we are able to measure height *exactly*—that is, to an infinitesimal degree. Even if we cannot, this variable is reasonably modeled as being continuous.)[2]

As we shall see in Chapter 3, the distinction between discrete and continuous random variables is an important one in that it leads to different ways of characterizing their behavior.

We often observe a set of outcomes for a random variable—for example, the incomes of the individuals appearing in a sample. We are often interested in combining these observations numerically. We now turn to a notation that simplifies such matters.

PROBLEMS

2.1 Give at least two examples of economic variables that are discrete, and then give at least two examples of economic variables that are continuous.

2.2 The Federal Reserve Board is conducting an experiment wherein on the first day of each of a set of successive months, money is injected into the economy as follows: $1 million the first month, $2 million the second month, $3 million the third month, $1 million the fourth month, $2 million the fifth month, $3 million the sixth month, and so on. Is the injection of money into the economy on the first day of a month a nonrandom variable, a random variable, or a constant? Do you expect that the response of the price level to these changes is a nonrandom variable, a random variable, or a constant? Fully explain your reasoning.

THE Σ NOTATION

Suppose that we have a set of observations on the variable X. That is, we observe $X_1, X_2, \ldots, X_n$ (or X_i, $i = 1, \ldots, n$). Often we need to combine these observations in the process of computing some value. We define

---

[2] A continuous random variable may have possible values over the entire number line—that is, from $-\infty$ to ∞. Models of such variables are of some importance, as we shall see in subsequent chapters.

$$\sum_{i=1}^{n} X_i = X_1 + X_2 + \cdots + X_n$$

The Σ (read "sigma") sign is a mathematical operator that tells us to sum the quantities that follow. What follows? The X_i's, allowing i to vary from 1 (under the Σ sign) to n (above the Σ sign).

It will typically be the case that the summation is over all the observations we have; in such a case the $i = 1$ and n are often omitted, and we simply write ΣX_i. Note that ΣX_i is much easier to write than is $X_1 + X_2 + \cdots + X_n$.

For an example, consider the arithmetic average (or mean) of the X's. How do we calculate this? We add up the X's and then divide by how many of them there are (n in this case). We can write the formula for this average as

$$\frac{1}{n} \sum_{i=1}^{n} X_i \quad \text{or} \quad \frac{1}{n} \sum X_i$$

The Σ sign is a mathematical operator; these operators have properties that can be usefully exploited. The plus sign ("+") has the property, for example, that $a + b = b + a$ (i.e., it has the property of commutativity). The minus sign ("−"), however, does not have this property. What properties does Σ have? We now turn to this problem, and in our discussion we will develop the idea of a linear function and suggest its importance in statistical analysis.

PROPERTIES OF THE Σ OPERATOR

In this section we shall develop and motivate the major properties of the Σ sign that will be used throughout this text.

Property 1: $\displaystyle\sum_{i=1}^{n} a = na$ where a is some constant.

This property suggests that the sum (over a set of observations from i equals 1 to n) of a constant number a (note that since a is a constant, it does not need an i subscript) is the number of observations times that constant number (a). To demonstrate this result, remember that the definition of the Σ sign tells us that

$$\sum_{i=1}^{n} a = a + a + \cdots + a \qquad (n \text{ times})$$

$$= na$$

Property 2: $\Sigma a X_i = a \Sigma X_i$ where a is a constant
and X_i is a variable.

This property tells us that a multiplicative constant a can be written either inside our outside of the Σ sign without changing the value of the expression. It is here of particular importance to reiterate that a is a constant, not a variable. To obtain this result, let us note that

$$\Sigma a X_i = a X_1 + a X_2 + \cdots + a X_n$$

which we know from the definition of the Σ sign. Note that on the right side of this expression there is a common factor of a. We can then write

$$\Sigma a X_i = a(X_1 + X_2 + \cdots + X_n)$$
$$= a \Sigma X_i$$

which proves our result. (The term in parentheses in the next to last line is, by definition, ΣX_i.)

For example, suppose that X_i is the monthly rent paid by family i, and that $a = 12$ (translating monthly rent into annual rent). The quantity $a X_i$ then represents the annual rent paid by family i. With $\Sigma a X_i$ we add up the annual rents of a number of families. With $a \Sigma X_i$ we first add up the monthly rents of the families (ΣX_i), and we then multiply this sum by 12 (the value of a). You should convince yourself intuitively that these two processes lead to the same answer.

Property 3: $\Sigma(a + b X_i) = na + b \Sigma X_i$
                        where a and b are constants
and X_i is a variable.

This property is the *linearity* property of the Σ sign in the case of a function of one variable, and it is a generalization of the first two properties. A linear function of one variable is any function that can be written in the form $Y_i = a + b X_i$, where a and b are constants. (We call this a linear function since its geometric representation is a line.) Linear functions are very important in statistical analysis as we will see again and again in this text, and this property is an important tool to have available for simplifying expressions.

To prove this property, note that

$$\Sigma(a + bX_i) = (a + bX_1) + (a + bX_2) + \cdots + (a + bX_n)$$

which comes from the definition of the Σ operator. The equation may be rewritten as

$$\Sigma(a + bX_i) = (a + a + \cdots + a) + (bX_1 + bX_2 + \cdots + bX_n)$$

$$= (a + a + \cdots + a) + b(X_1 + X_2 + \cdots + X_n)$$

$$= na + b\Sigma X_i$$

which proves the result. We often say, then, that the Σ sign can be "pushed through" linear functions.

As an example, suppose that $Y_i = 5 + 4X_i$ and we know that $\Sigma X_i = 50$ in a group of ten observations on X. Then $\Sigma Y_i = 10(5) + 4(50) = 250$.

We now consider properties of the Σ sign for the case of more than one variable.

Property 4: $\Sigma(X_i + Y_i) = \Sigma X_i + \Sigma Y_i$ where X_i and Y_i are variables.

If we have two variables X and Y, we might be interested in their sum. For example, suppose that X_i represents the monthly rent paid by family i and that Y_i represents the monthly utilities paid by this family. $X_i + Y_i$ is then a measure of total housing expenditures for this family. This proposition simply says that to find the sum of total housing expenditures for a group of families we may add X and Y for each family and then add these sums (the left side) or, equivalently, we may add the X's across families, add the Y's across families, and then add these two sums (the right side).

To prove this result, note that from the definition of Σ we know that

$$\Sigma(X_i + Y_i) = (X_1 + Y_1) + (X_2 + Y_2) + \cdots + (X_n + Y_n)$$

The X's on the right side may be grouped together, and the Y's on the right side may also be grouped together, giving us

$$\Sigma(X_i + Y_i) = (X_1 + X_2 + \cdots + X_n) + (Y_1 + Y_2 + \cdots + Y_n)$$
$$= \Sigma X_i + \Sigma Y_i$$

Property 5 generalizes this result to the case of a linear function of two variables.

Property 5: $\Sigma(aX_i + bY_i + c) = a\Sigma X_i + b\Sigma Y_i + nc$

where a, b, and c are constants and X_i and Y_i are variables.

This expression represents an extension of the linearity property of Σ to the case of two variables. A linear function of two variables is any function that can be written in the form $aX_i + bY_i + c$. (This equation is that of a plane, which is the extension of a line into a higher dimension.)

To show this property, note that

$$\Sigma(aX_i + bY_i + c)$$
$$= (aX_1 + bY_1 + c) + (aX_2 + bY_2 + c)$$
$$+ \cdots + (aX_n + bY_n + c)$$
$$= (aX_1 + aX_2 + \cdots + aX_n) + (bY_1 + bY_2 + \cdots + bY_n)$$
$$+ (c + c + \cdots + c)$$
$$= a(X_1 + X_2 + \cdots + X_n) + b(Y_1 + Y_2 + \cdots + Y_n) + nc$$
$$= a\Sigma X_i + b\Sigma Y_i + nc$$

which proves the result.

As an example, suppose that we have 10 observations on X_i and Y_i; we find $\Sigma X_i = 20$ and $\Sigma Y_i = 30$. If $W_i = 2X_i + 3Y_i + 2$, then $\Sigma W_i = 2(20) + 3(30) + 10(2) = 150$.

In general, a linear function of the k variables $X_1, \ldots, X_k$ would be $Y_i = a_0 + a_1 X_{1i} + a_2 X_{2i} + \cdots + a_k X_{ki}$, where (for example) X_{1i} is the value of the variable X_1 for observation i, and the a's are a set of constants. The linearity property extends directly to this case.

It is important to note that although the Σ sign can be pushed through linear functions, in general it *cannot* be pushed through nonlinear functions. To illustrate this, let's ask whether $\Sigma(1/X_i)$

equals $1/\Sigma X_i$. (Note that the function $1/X_i$ is not a linear function; if we were to graph this function its representation would not be a line.) The best way to see that these two expressions are not equal is to create a simple numerical example. Suppose that $n = 2$ and that the two values of X are $X_1 = 1$ and $X_2 = 2$. Then

$$\Sigma \frac{1}{X_i} = \frac{1}{1} + \frac{1}{2} = \frac{3}{2}$$

However,

$$\frac{1}{\Sigma X_i} = \frac{1}{1+2} = \frac{1}{3}$$

Clearly these two quantities are not equal. This one counterexample is sufficient to show that $\Sigma(1/X_i) \neq 1/\Sigma X_i$, and to suggest in general that the Σ sign, though a linear operator, is not a nonlinear operator.

Let us illustrate now that it is also true that nonlinear applications of Σ are "not allowed" in the case of functions of more than one variable. Suppose that $n = 2$, $X_1 = 1$ and $X_2 = 2$, and $Y_1 = 1$ and $Y_2 = 2$. Let us determine whether $\Sigma X_i Y_i$ equals $(\Sigma X_i)(\Sigma Y_i)$:

$$\Sigma X_i Y_i = X_1 Y_1 + X_2 Y_2 = 1(1) + 2(2) = 5$$

$$(\Sigma X_i)(\Sigma Y_i) = (1+2)(1+2) = 9$$

Clearly these results are not equal. It makes a difference whether you multiply first and then add, or add first and then multiply.

The Σ sign is a very useful tool; remember, however, that it can be "pushed through" expressions only when these expressions are in fact linear. This observation provides a first hint of the key role played by linear functions in statistical analysis.

PROBLEMS

2.3 Suppose that we have the following four observations on X and Y.

X	Y
1	2
2	2
3	4
4	4

For each of the following, demonstrate whether the equation is true or false using the above data. Provide a general explanation for your finding based on the properties of the Σ sign.

(a) $\Sigma 2X_i = 2\Sigma X_i$

(b) $\Sigma 2X_i = 2(4)\Sigma X_i$

(c) $\Sigma(X_i + Y_i) = \Sigma X_i + \Sigma Y_i$

(d) $\Sigma 2X_i Y_i = 2\Sigma X_i Y_i$

(e) $\Sigma 2X_i Y_i = 2(\Sigma X_i)(\Sigma Y_i)$

(f) $\Sigma 2X_i^2 = 2\Sigma X_i^2$

(g) $\Sigma 2X_i^2 = 2(\Sigma X_i)^2$

(h) $\Sigma(2 + 3X_i + 4Y_i) = 2 + 3\Sigma X_i + 4\Sigma Y_i$

(i) $\Sigma(2 + 3X_i + 4Y_i) = 2(4) + 3\Sigma X_i + 4\Sigma Y_i$

(j) $\Sigma(Y_i + 3) = \Sigma Y_i + 3$

2.4 Suppose that the tax (T_i) paid by individual i is the following function of his or her income (X_i):

$$T_i = .1(X_i - 2000) = -200 + .1X_i$$

(i.e., there is a flat tax of 10% on all income over $2000). You know that total income in a geographical area containing 500 individuals is $5,000,000. What will total tax revenue be?

2.5 There are three wage earners in a family (Mom, Dad, and one child). Their earnings are $20,000, $15,000, and $3000 per year, respectively. Taxes for each person are

$$T_i = -200 + .1X_i + .00001X_i^2$$

Find the total taxes paid by this family per year.

*2.6 In a particular geographical area, the demand function for each person i for a particular good is

$$q_i^d = 100 + .2I_i - 9P$$

where q_i^d is quantity demanded for person i, I_i is person i's income, and P is the commodity's price. (Note that P is constant across individuals, since all individuals face the same price. Therefore no i subscript is needed.) For the firms in this area the supply function for this good is

$$q_j^s = -50 + 20P$$

where q_j^s is the quantity supplied of the good by firm j. (Note that a different subscript is needed. Why?) There are 5 firms and 100 consumers. Total income of the consumers is $100,000. Equilibrium occurs where

$$\sum_{i=1}^{100} q_i^d = \sum_{j=1}^{5} q_j^s$$

(i.e., market demand equals market supply). Find the equilibrium price.

SUMMARY REMARKS

In this chapter we have laid some foundations for our subsequent work. We now turn to a discussion of the problem of description in the context of populations, which is the simplest forum for statistical analysis. (Since we observe the population, no inference is necessary, just description.)

REVIEW PROBLEMS FOR CHAPTER 2

2.7 X_i represents the outcome of the ith roll of a die. We roll the die five times, and form

$$Y = 2 + \Sigma X_i$$

Is Y a random variable? Fully explain your answer.

2.8 Suppose that X_i is a random variable but Y_i is a nonrandom variable. If

$$W_i = X_i + Y_i$$

is W_i a random variable or a nonrandom variable? Fully explain your answer, and suggest its implications.

DISCUSSION QUESTIONS FOR CHAPTER 2

1. If X is the height of a person chosen from your class, is X a random variable, a nonrandom variable, or a constant? Once this person is chosen, is his or her height a random variable, a nonrandom variable, or a constant?

2. Suppose that you roll a die several times and consider the average value that shows up on these rolls. As the number of rolls increases, what would you expect to happen to the degree of randomness involved in the determination of this average?

Chapter 3

Descriptive Statistics for Populations

The simplest form of statistical analysis involves describing the information contained within a population that is fully observed. (Remember the distinction between a population and a sample as discussed in Chapter 1.) Typically, it is difficult to just look at the data contained within a population and gain a good sense of its characteristics; this effort usually involves trying to absorb more information than we can usefully interpret. Through the use of descriptive statistics, however, we can summarize these data in a meaningful way so as to make interpretation and comparison simpler. In this chapter we shall present some basic ways of describing populations so as to make them more readily understandable. (The problem of description is applicable to samples as well; some descriptive statistics for samples are discussed in Chapter 4 to the extent that they are the sources of inferences regarding the larger population.)

SUMMARIZING THE ENTIRE POPULATION

Suppose, for example, that we observed a population of 100 families and the number of children within each of these families. The observations within this population would form a string of 100 numbers, and most individuals looking at this string of numbers would find it difficult to get a firm grasp on the characteristics of the population. However, suppose that we grouped together those families with the same number of children, and then asked how many have no children, how many have one child, etc. For a hypothetical population we find the following:

Number of children	Number of families
0	19
1	27
2	36
3	13
4	4
5	1

This simple table gives us a much clearer picture of the population's characteristics than we would gain by observing a string of 100 numbers. We see, for example, that families with more than three children are unusual and that families with two children are most common. We might also organize the table so that the right-hand column is the fraction of families in the population with that number of children; this is particularly helpful when (as is normally the case) the size of the population is not a nice round number like 100. Doing this, we would get

Number of children (X)	Proportion of families [f(X)]
0	.19
1	.27
2	.36
3	.13
4	.04
5	.01

The number of children in a family is a random variable, which we will call X; the proportion of families with a particular value of X is referred to as the *relative frequency* of that value of X, written $f(X)$. The above table is the *relative-frequency distribution* of the values of X in this population, and it suggests to us how the families are distributed among the various possible outcomes.

With the relative-frequency distribution, we summarize the

population by listing the possible values of X and the proportion (relative frequency) of the population characterized by that value. We have in the example above illustrated this approach in the context of a discrete random variable. What do we do when we have a continuous random variable? The strategy used above (list the possible values of X and the proportion of the population characterized by each of those possible values) is an inappropriate strategy in this case; for a continuous random variable, the recurrence of individual values (which is the basis for the relative-frequency distribution in the discrete case) is unlikely to occur because a continuous random variable can take on any value within some range.[1]

We often deal with this problem by creating categories of values of the random variable; that is, we create a set of ranges of values within which the continuous random variable X can fall and then ask as to the relative frequency of values within these ranges (without concern for where within a range the individual values happen to lie). For example, suppose that we observe a population of 20 adult males and that X equals their annual incomes in dollars per year. The values of X are:

8235, 10461, 12685, 15083, 18213, 22138, 3423,

6417, 11231, 13142, 26841, 9019, 12507, 23411,

24702, 16148, 20405, 14895, 19049, 16917

(Income is best viewed as a continuous random variable; the fact that all values are to an even dollar is simply the result of the degree of precision with which we measure the quantity.) To illustrate the technique suggested above, let us organize the data into *cells* having a width of $5000. In a table we then list the relative frequency of outcomes within the various cells:

X (dollars)	f(X)
0–4999	.05
5000–9999	.15
10,000–14,999	.30
15,000–19,999	.25
20,000–24,999	.20
25,000–29,999	.05

[1] This issue will be developed in a different context in Chapter 6 when we discuss probability distributions.

This table gives us a far more readily understandable picture of this population than the earlier string of 20 numbers did.

The way we set up the cells was in many ways arbitrary, but it was not completely so. The size of the cells is really a compromise between too little and too much detail, and in some sense we decide upon the width of the cells so that where an observation is *within* the cell is of minor importance; that is, we are not focusing on the variation of observations within a cell and are saying that differences within a cell are uninteresting for the problem under investigation. This leads us in the current problem to think of individuals with incomes between $20,000 and $24,999 as more or less the same.

Obviously, the degree of detail needed for a proper investigation of the question under consideration plays an important role in determining what the appropriate cells are. Nevertheless, an appropriate choosing of the cells can lead to a quite informative description of the underlying characteristics of a population.

In Figure 3.1 we draw a geometric representation of the income example, constructing what is often referred to as a relative-frequency *histogram,* which is a bar graph in which the bar for each cell extends over the width of that cell. We have plotted the midpoint of each cell to provide the viewer with a "typical" value within that cell. A

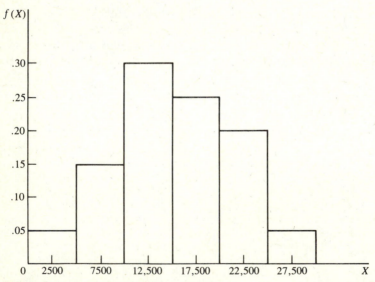

Figure 3.1   A relative-frequency histogram.

picture is often worth a thousand words, as patterns of variation in the random variable are often most clearly seen in a visual representation.[2]

It is not always the case that the optimal way to proceed is with cells of identical width. Consider, for example, the following table, which characterizes the distribution of family income in the United States in 1986:[3]

Income (X)	Fraction of families [f(X)]
0–2499	.018
2500–4999	.028
5000–7499	.037
7500–9999	.041
10,000–12,499	.049
12,500–14,999	.048
15,000–19,999	.097
20,000–24,999	.098
25,000–34,999	.181
35,000–49,999	.196
50,000+	.207

Notice how the cell widths get wider as incomes rise; the reasoning is that a difference between, say, $4000 and $7000 is of more interest than that between $44,000 and $47,000. The table is still a very useful way of characterizing the population, and has some advantages over one in which all cells are of equal width. Obviously, the changes in cell widths must be considered when interpreting the information presented in the table.[4]

---

[2] We could, of course, plot relative frequencies for a discrete random variable as well. Here we might simply have vertical lines rising from the horizontal axis giving the relative frequency of each possible value of X. (We would not use bars because the random variable cannot extend over a range of values but can only take on specific values.)

[3] These data are from U.S. Bureau of the Census, Current Population Reports, *Money Income of Households, Families, and Persons in the United States: 1986*, Series P-60, No. 159 (U.S. Government Printing Office, Washington, DC, 1988), table 11.

[4] Note that graphing and interpreting a histogram would not be as straightforward as in our earlier example in view of the different widths of the cells.

PROBLEMS

3.1 In a population of 20 families, we observe the number of cars owned by each family. The observations are as follows:

0, 2, 2, 1, 3, 1, 1, 2, 2, 2, 3, 4, 1, 1, 2, 2, 0, 0, 2, 1

Tabulate the relative-frequency distribution of this variable.

3.2 In the population of problem 3.1, we also observe the number of different jobs the primary wage earner has held during the past five years. The observations are as follows:

1, 3, 2, 2, 0, 2, 1, 2, 1, 3, 4, 3, 1, 1, 5, 2, 1, 3, 1, 2

Tabulate the relative-frequency distribution of this variable.

3.3 Consider a state in which there are ten colleges and universities. Tuition and fees (in dollars) at the institutions are as follows:

8315, 12124, 4239, 14138, 16050,

9327, 7136, 9051, 13124, 9250

Using cells with midpoints of 6500, 11500, and 16500, graph the relative-frequency histogram for this variable.

3.4 There are 20 flower vendors in a city, with the following profits (in dollars) on a given day:

52.24, 68.01, 73.13, 38.26, 65.06, 58.26, 48.57,

53.42, 71.29, 64.13, 30.05, 48.27, 51.43, 52.67,

49.28, 48.16, 59.04, 63.23, 67.15, 70.21

Using cells with midpoints of 35, 45, 55, 65, and 75, graph the relative-frequency histogram for this variable.

SUMMARY STATISTICS

The methods of description just considered are often very helpful ways of summarizing the information in a population. Often, however, we want to summarize the population to an even greater extent; one common reason for this is so as to facilitate comparisons between different populations. This leads us to the consideration of *summary statistics,* which are individual numbers calculated from the information in a population to give us a "snapshot" of the population's characteristics.

The two most common types of summary statistics are measures of central tendency (i.e., the typical value in the population) and measures of dispersion (the degree of variation in the values). We shall now consider these in turn.

To begin, let us consider a population consisting of seven values: 3, 4, 4, 5, 6, 7, and 9. Consider three measures of central tendency.

1. The *mode* is the value that occurs most frequently, that is, that has the highest relative frequency. In the example under consideration, the mode is 4.
2. The *median* is (essentially) the value such that half of the observations are above it and half are below it. Formally, to calculate the median we rank the values in ascending order (as is already done above). When the number of observations (m) is an odd number,[5] the median is the $\left(\dfrac{m+1}{2}\right)$th value in that order; in the current example, $m = 7$ and the median is then the fourth value in the ascending sequence (i.e., 5). [When the number of observations is an even number, we average the two values in the middle. That is, we average the $\left(\dfrac{m}{2}\right)$th and the $\left(\dfrac{m}{2}+1\right)$th value in the ascending sequence.]
3. The *mean,* generally represented by the Greek letter μ, is the average value in the population. That is, $\mu = \dfrac{1}{m}\sum\limits_{i=1}^{m} X_i$. In the current example, $\mu = \dfrac{38}{7}$ or 5.4.

Notice that the three measures of central tendency yield different values in the current example. Let us pursue this matter further.

Consider a new population having the values 1, 2, 3, 3, 3, 4, and 5. For this population, the mode, median, and mean are all

---

[5] We let m represent the size of a population, reserving n for the size of a sample.

equal to 3. Consider, on the other hand, the population having the values 1, 2, 3, 3, 3, 4, and 54. For this population, the mode and median are both 3, but the mean equals 10. In this case we get very different measures of central tendency depending on the measure we use. Why? The value of 54 is no different from a value of 5 (as in the previous example) as far as the median and mode are concerned; for the median all that matters is that it is a value "above the middle," and for the mode all that matters is that it is a value that occurs only once. For the mean, however, the value of 54 is quite different from the value of 5 because the magnitude of its difference from the middle value matters in the calculation of this measure of central tendency.

The two most commonly used measures are the median and the mean; we shall work primarily with the mean in this text. The use of the mean does not imply that it is the "best" measure of central tendency; which measure is "best" depends on the problem at hand and the interests of the investigator. However, the mean is the most direct foundation for our future work, particularly in regard to regression analysis.

EXAMPLE ■
Figure 3.2 contains an example of a distribution that is not symmetric; that is, the behavior of relative frequencies to the

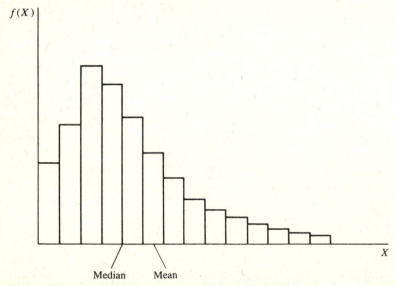

Figure 3.2 An asymmetric relative-frequency histogram.

right of the peak is not the same as to the left of the peak. In particular, the distribution is said to be "skewed" to the right, as the right-hand tail extends further than does the left-hand tail. In such distributions the mean exceeds the median; the values in the right-hand tail "pull" the value of the mean upward and above the median. Whether the median or mean is appropriate depends in part on the problem at hand. Wage incomes in the United States have often been found to have this pattern; for example, in 1979 the median earnings for males 15 years or older was $13,172 as compared with a mean of $15,441.[6] ∎

A measure of central tendency is a single number that gives us an idea about a particular characteristic of a population. It is useful in evaluating that population as well as in comparing the characteristics of different populations (e.g., comparisons of mean family incomes in different countries). But these measures do not tell us everything about a population, and we are often interested in characteristics that are not picked up by the mean (or some other measure of central tendency).

Consider the two (discrete) random variables X and Y, which have the following populations:

$$X: \quad 1, 2, 3, 4, 5, 6, 7$$

$$Y: \quad 3, 3, 3, 4, 5, 5, 5$$

These populations each have a mean of 4, yet they are very different populations. How so? The Y values are much more clustered about the mean of 4 than the X values are; that is, the X values are more *disperse* than those of the Y population. How might we try to quantify this?

Let us first emphasize that by dispersion we are here referring to a specific characteristic—the extent and degree to which individual observations differ from the mean. How do individual observations differ from their mean? For X_i this deviation is $X_i - \mu_X$ (where μ_X is the mean of the X population). Clearly, we want to in some way aggregate the $X_i - \mu_X$ terms in order to develop a measure of dispersion for X. How might we do this?

[6] U.S. Bureau of the Census, *1980 Census of Population, Vol. 1 Characteristics of the Population, Chapter D: Detailed Population Characteristics, part 1* (U.S. Government Printing Office, Washington, DC, 1984), table 295.

One possibility would be to calculate a mean deviation from the mean—that is, to take the individual deviations from the mean and average them. For X the result would be

$$\frac{1}{m} \Sigma(X_i - \mu_X)$$

The problem with this method of aggregation is that for some observations $X_i - \mu_X$ is positive and for some observations it is negative, and these positives and negatives cancel out. That is, when we think of dispersion we think of the two directions from the mean equivalently, but this method treats them as opposites and therefore they, at least partially, offset each other. In fact, it will *always* be the case that this method of aggregation leads to a value of zero, no matter what the pattern within the population! To see this, we simply note that

$$\Sigma(X_i - \mu_X) = \Sigma X_i - m\mu_X = \Sigma X_i - m \left(\frac{1}{m}\right) \Sigma X_i = 0$$

and this finding will be true regardless of the values of the individual X's because it comes directly from the definition of the mean. This result implies that in the above example this aggregation will take on a value of 0 for both the X and Y populations even though the X population is (intuitively) more disperse than the Y population is.

The problem is that the positive (above the mean) and negative (below the mean) deviations perfectly cancel each other when we simply add them. One solution to this problem would be to take the mean *absolute* deviation from the mean (i.e., take the absolute values of the deviations before we add them). Although this method is a fairly good one, it turns out that there is one that is more convenient to work with. This measure, in which we square the deviations (thereby making them all nonnegative) before we add them, is known as the *variance* of the random variable X and is formally defined as

$$\sigma_X^2 = \frac{1}{m} \sum_{i=1}^{m} (X_i - \mu_X)^2$$

Here we use the symbol σ^2 to denote the variance; σ_X^2 is the variance of the random variable X. To calculate the variance it is often useful to set up a type of spreadsheet:

X_i	$X_i - \mu_X$	$(X_i - \mu_X)^2$
1	−3	9
2	−2	4
3	−1	1
4	0	0
5	1	1
6	2	4
7	3	9
		$\Sigma = 28$

and thus $\sigma_X^2 = (1/7)28 = 4$. If were to undertake a similar calculation for Y, we would find $\sigma_Y^2 = (1/7)6 = 6/7$. By comparing σ_X^2 to σ_Y^2 we gain insights as to the relative degree of dispersion in the two populations, and thus we have a method for quantifying the extent to which the X population is more disperse than the Y population is.

Although the variance is useful for the purpose of comparing the degrees of dispersion from the mean in different populations, an intuitive interpretation is difficult to come by. The main reason has to do with the units in which the variance is measured. If, for example, the X variable above were measured in persons (perhaps X is family size) then the units of σ_X^2 would be square persons, a unit of measurement for which there is no obvious intuitive interpretation! In general, the variance is measured in the square of the units of the observations.

We often then speak of the square root of the variance, known as the *standard deviation,* which has units the same as those in which the variable is measured. We generally represent this as $\sigma(=\sqrt{\sigma^2})$. The standard deviation is, intuitively, a type of "typical deviation from the mean." Since the variance is 4 for the X values above, the standard deviation is 2.

The definitions of the mean and variance described above are applicable to any random variable, regardless of whether it is continuous or discrete. In the case of a discrete random variable, however, there are some shortcuts available that are quite important because they set the stage for future ways of viewing means and variances.[7]

CALCULATION OF MEANS AND VARIANCES FROM RELATIVE-FREQUENCY DISTRIBUTIONS

Consider a population from earlier in this chapter, where X represented the number of children. In our population of 100 families, we had the following relative-frequency distribution of X:

Number of children (X)	Proportion of families [$f(X)$]
0	.19
1	.27
2	.36
3	.13
4	.04
5	.01

Suppose that we were interested in calculating the mean and variance of this population. Obviously, given the relative frequencies and the population size, we could enumerate this population: There would be 19 values of 0, 27 values of 1, and so forth. From this enumerated population we could then use the formulas above to calculate the mean and variance of X. These calculations clearly would be somewhat tedious; however, there is a simpler way to proceed. In calculating the mean we use the formula

$$\mu = \frac{1}{m} \sum_i X_i$$

where the Σ is over the m individual observations making up the population (the i's). Think about what this would involve in the problem at hand. In calculating the sum of the X's we would add

[7] An alternative measure of dispersion is the *coefficient of variation,* defined as σ/μ. This measures dispersion in proportional terms as opposed to absolute terms (as with the standard deviation). For certain problems it might be more appropriate; a standard deviation of $1000 when the mean income is $10,000 suggests in most contexts greater inequality in the distribution of income than the same standard deviation with a mean of $25,000. This difference would be apparent in the coefficients of variation: .1 versus .04. (Note that the coefficient of variation is unit-free, whereas the standard deviation depends upon the units of measurement.)

up 19 values of 0, 27 values of 1, 36 values of 2, 13 values of 3, 4 values of 4, and 1 value of 5. That is,

$$\frac{1}{m} \sum_i X_i = \frac{1}{m} \sum_{i*} X_{i*} n(X_{i*})$$

The term on the left is the sum over the individual observations (i's) of the values of X_i. The term on the right is a sum over the possible values of X (the $i*$'s, of which there are six) of that value of $X(X_{i*})$ times the number of times that value of X occurs, or $n(X_{i*})$. The term on the right can be reexpressed as

$$\sum_{i*} X_{i*} \frac{n(X_{i*})}{m}$$

Note, however, that $n(X_{i*})/m$ is simply the fraction of the population having the value X_{i*}; that is, it is the relative frequency of X_{i*}. Thus, the mean of this population can be calculated as

$$\sum_{i*} X_{i*} f(X_{i*})$$

This set of instructions simply tells us, for each possible value of X, to calculate that value of X times its relative frequency and then add up these products. For the problem at hand the mean would be

$$\mu = 0(.19) + 1(.27) + 2(.36) + 3(.13) + 4(.04) + 5(.01)$$

$$= 1.59$$

Note that this procedure suggests that the mean is simply a weighted average of the possible values of X, where the weights are given by the relative frequencies. We might, for simplicity, then say that the mean of X can be calculated as $\Sigma X f(X)$ where the $i*$ subscripts are omitted for simplicity.

Note that the X in this case was a discrete random variable. If X were instead a continuous random variable, we might use this general formula in the context of the cells from a relative-frequency histogram; here we might let the X_{i*}'s be the cell midpoints and the $f(X_{i*})$'s be the relative frequencies of outcomes in these cells. If we then calculated $\Sigma_{i*} X_{i*} f(X_{i*})$ we would only achieve an approximation of μ, however, because each observation in a cell is assumed to equal the midpoint of that cell. We shall therefore only use this formula for discrete random variables; for a continuous random variable we must calculate

$$\mu = \frac{1}{m} \sum_i X_i$$

where the sum is now across the individual observations.

A similar analysis pertains to the calculation of the variance of a discrete random variable. The variance of X is

$$\sigma^2 = \frac{1}{m} \sum_i (X_i - \mu)^2$$

where the sum is over the individual observations. Every time X is 0, we add up $(0 - \mu)^2$; how many times is X equal to 0? $n(0)$. A similar statement holds for the other possible values. Thus,

$$\sigma^2 = \frac{1}{m} \sum_{i*} (X_{i*} - \mu)^2 n(X_{i*})$$

where the $i*$'s as before represent the possible values of X; that is,

$$\sigma^2 = \sum_{i*} (X_{i*} - \mu)^2 \frac{n(X_{i*})}{m}$$

or

$$\sigma^2 = \sum_{i*} (X_{i*} - \mu)^2 f(X_{i*})$$

In the example at hand, the variance would be

$$\sigma^2 = (0 - 1.59)^2(.19) + (1 - 1.59)^2(.27)$$
$$+ (2 - 1.59)^2(.36) + (3 - 1.59)^2(.13)$$
$$+ (4 - 1.59)^2(.04) + (5 - 1.59)^2(.01)$$
$$= 1.2419$$

The standard deviation would then be about 1.11 children.

As with the case of the mean, to achieve an exact answer this formula can be used only with a discrete random variable. These alternative formulas are nevertheless extremely important, not only for the additional insights into means and variances just obtained but also because the discussion of means and variances in the context of probability distributions (Chapter 5) will build upon them quite directly.

PROBLEMS

3.5 For the data from problem 3.1 calculate the mean and variance both from the actual observations and from the relative-frequency distribution. What is the median of this population?

3.6 For the data from problem 3.2 calculate the mean and variance both from the actual observations and from the relative-frequency distribution. What is the median of this population?

3.7 For the data of problem 3.3 calculate the mean and variance.

3.8 For the data of problem 3.4 calculate the mean and standard deviation.

3.9 The relative frequency distribution of X is as follows:

X	f(X)
1	.2
2	.3
3	.4
4	.1

Find the mean and variance of X.

3.10 Suppose that there are 2 million inhabitants of a country in which the mean gross (before tax) income is $20,000 per year. Suppose that the mean individual pays taxes of $3000 per year. What is the total disposable income (gross income minus taxes) per year for the country as a whole?

3.11 Consider a college where tuition and fees are $17,000. The mean student receives a $2000 grant against tuition from the college. If total revenues (tuition net of grants) for the college are $22,500,000 for the year, how many students are enrolled at the college?

3.12 For a discrete X, we have shown that

$$\sigma^2 = \sum_{i*} (X_{i*} - \mu)^2 f(X_{i*})$$

Prove that an alternative formula is

$$\sigma^2 = \sum_{i*} X_{i*}^2 f(X_{i*}) - \mu^2$$

Illustrate this result using the data from problem 3.9.

MEANS AND VARIANCES OF LINEAR FUNCTIONS

In Chapter 2 we discussed the concept of a linear function and suggested its importance. We here briefly extend the concepts of means and variances to the cases of linear functions to show that if $Y_i = a + bX_i$ (i.e., if Y is a linear function of X) and the mean and variance of X are known, then the mean and variance of Y can be directly determined. We shall present these results as a series of properties.

Property 1:   If $Y_i = a + bX_i$, then $\mu_Y = a + b\mu_X$.

To prove this property let us remember that the linearity characteristic of the Σ sign tells us that

$$\sum_{i=1}^{m} Y_i = ma + b \sum_{i=1}^{m} X_i$$

where m is the size of the populations. If we multiply both sides by $1/m$ we get

$$\frac{1}{m} \sum_{i=1}^{m} Y_i = \left(\frac{1}{m}\right) ma + b \left(\frac{1}{m}\right) \sum_{i=1}^{m} X_i$$

which simplifies to

$$\mu_Y = a + b\mu_X$$

proving the property.

Property 2:   If $Y_i = a + bX_i$, then $\sigma_Y^2 = b^2 \sigma_X^2$.

To prove this property let us recall that

$$\sigma_X^2 = (1/m)\Sigma(X_i - \mu_X)^2$$

and similarly for Y. The linear relation in conjunction with property 1 suggests that

$$Y_i - \mu_Y = a + bX_i - (a + b\mu_X)$$

(since $\mu_Y = a + b\mu_X$, we have subtracted the same quantity, μ_Y, from both sides of the linear relationship). This equation simplifies to

$$Y_i - \mu_Y = b(X_i - \mu_X)$$

Squaring both sides gives us

$$(Y_i - \mu_Y)^2 = b^2(X_i - \mu_X)^2$$

Summing both sides and multiplying by $1/m$ gives us

$$\frac{1}{m} \sum_{i=1}^{m} (Y_i - \mu_Y)^2 = b^2\left(\frac{1}{m}\right) \sum_{i=1}^{m} (X_i - \mu_X)^2$$

which simplifies to

$$\sigma_Y^2 = b^2 \sigma_X^2$$

proving the proposition. Notice in particular that the value of the constant term a does not influence the variance of Y because it is the same for each observation and therefore does not contribute to the pattern of variation in the random variable Y.

Property 3:   If $Y_i = a + bX_i$, then $\sigma_Y = |b|\sigma_X$, where $|\ \ |$ is the absolute value sign.

This result comes directly from property 2. If $\sigma_Y^2 = b^2\sigma_X^2$, then

$$\sqrt{\sigma_Y^2} = |b|\sqrt{\sigma_X^2}$$

(A standard deviation cannot be negative so we take the positive square root of b^2—thus the absolute value signs around b.) But since the standard deviation is just the square root of the variance, this equation simplifies to

$$\sigma_Y = |b|\sigma_X$$

proving the proposition.

EXAMPLE ■

Suppose that the renters in a particular city pay their monthly rent (including, for simplicity, utilities) X_i plus an annual "renter's fee" of $50 to the city. Suppose that the mean monthly rent is $500, with a standard deviation of $150. Let us find the mean and standard deviation of annual housing expenditures Y.

Each year a renter pays X_i times 12 plus the one-time $50 fee. Annual expenditures Y are then

$$Y_i = 50 + 12X_i$$

which is a linear function. The mean of Y can be calculated as

$$\mu_Y = 50 + 12\mu_X = 50 + 12(500) = 6050$$

(Can you convince yourself that this makes sense intuitively?) The variance of annual expenditures is

$$\sigma_Y^2 = (12)^2\sigma_X^2 = 144(150)^2 = 3,240,000$$

Or, the standard deviation of Y is

$$\sigma_Y = |12|\sigma_X = 12(150) = 1800$$

(Note that 1800 is the square root of 3,240,000.) Since the $50 fee is the same for everyone, it does not contribute to the variation in Y_i. Thus, annual expenditures have a mean of $6050 and a standard deviation of $1800. ∎

PROBLEMS

3.13 X has a mean of 3 and a variance of 9. If

$$Y_i = 2 + 4X_i$$

find the mean and variance of Y.

3.14 Annual entertainment expenses in a population of 100 families have a mean of $2040 and a standard deviation of $360.

(a) Find the mean and standard deviation of monthly expenses.

(b) What is the total amount of monthly entertainment expenses for this population?

SUMMARY REMARKS

In this chapter we have discussed some relatively simple, yet often highly informative, descriptive statistics for summarizing the information contained within a known population. Of course, we generally do not know the entire population of interest. The problem of inference typically becomes, then, one of using the information in a sample to infer the characteristics of the (unknown) statistics describing the population as a whole. We now turn to a preliminary discussion of sampling and some basic methods of forming such inferences.

REVIEW PROBLEMS FOR CHAPTER 3

3.15 Consider an economy with a flat rate tax system. Each dollar of income over $5000 is taxed at 20%. (Income below $5000 is tax-free.) In general,

$$T = .2(Y - 5000)$$

or

$$T = -1000 + .2Y$$

where T is taxes and Y is income. Suppose that the population mean income is $20,000 and that the population standard deviation of income is $8000. All families have at least $5000 of income.

(a) Find the mean of T.
(b) Find the standard deviation of T.
(c) If the population contains 20 million families, what is the government's total tax revenue?

3.16 Suppose that the population for X has 9 observations:

$$2, 4, 3, 6, 8, 2, 5, 7, 3$$

Suppose that $Y = 2 + X^2$. Find the mean of Y. Then find the median of Y.

*3.17 Suppose that we observe three random variables, W_i, X_i, and Y_i, for the m observations in a population.

(a) If $W_i = X_i + Y_i$ for each i, what is the relationship between μ_W, μ_X, and μ_Y?
(b) If $W_i = aX_i + bY_i + c$ for each i, where a, b, and c are constants, what is the relationship between μ_W, μ_X, and μ_Y? How is part (a) a special case of part (b)?

*3.18 The coefficient of variation for a random variable is defined as σ/μ. This coefficient is an alternative to the standard deviation as a measure of dispersion; the coefficient of variation measures that dispersion in proportional terms. If $Y_i = a + bX_i$, what is the relationship between the coefficients of variation of Y and X?

DISCUSSION QUESTIONS FOR CHAPTER 3

1. As mentioned in the text, income distributions are typically ones in which the mean exceeds the median. For what type of issue might it be more appropriate to consider the median income as the appropriate measure of a typical income? For what type of issue might it be more appropriate to consider the mean income as the appropriate measure of a typical income?

2. Suppose that $Y_i = X_i^2$ (i.e., Y is a nonlinear function of X). Explain why $\mu_Y \neq (\mu_X)^2$ and then provide an intuitive rationale for this lack of equality. (A numerical example may be helpful.)

Chapter *4*

Sampling and Basic Methods of Inference

In Chapter 3 we described some methods for summarizing the information contained within a known population. Unfortunately, we generally do not know the entire population but observe merely a part of it, and our job is then to attempt to use this partial information to infer the characteristics of the larger population. In this chapter we begin to pursue this issue, discuss some basic ways of forming inferences, and set the stage for a future discussion of the enhancements cf these basic methods that would allow us to learn more about the population of interest.

SAMPLES AND RANDOM SAMPLES

Recall from our discussion in Chapter 1 that a population is the entire group of persons or things in which we are ultimately interested. A sample, on the other hand, is a subset of the population, and we use the information in the sample to make inferences regarding the characteristics of the population as a whole.

The fact that some samples are more useful than others for making inferences about the underlying population has implications for the way we go about choosing our sample observations (the so-called process of "sampling"). A *random sample* is a special type of sample in which each element of the population has the same chance of being selected. In general, random samples are much more useful for the process of inference than are other types of samples, as we shall now discuss.

Conceptually, we can visualize the process of taking a random sample as follows: Imagine that we are interested in the incomes of the families residing in a particular city; our population is then all families who live within the city (and this population is of size m). If we were to take each family and put its name on a piece of paper, put these pieces of paper into a giant drum and mix them thoroughly, and then blindly draw n of them, the chosen families would then represent a random sample of size n. Why? Since we blindly choose the pieces of paper, each family has the same chance of being selected.

In practice the process of choosing a random sample is often more complicated. In the problem at hand, we might consider choosing the families from some official roster in the public records. Contrast the following two techniques: (1) We (arbitrarily) choose the families from the city's records of deed holders of residential property; and (2) we (arbitrarily) choose the families from the tax returns filed for the city's income tax (which, for simplicity, all families are required to file regardless of income).

Method 1 yields a sample that clearly is not a random sample. Why? In order to be included, a family must own its residence. This implies that families with relatively high incomes have a greater chance of being selected than do families with relatively low incomes in that families with relatively high incomes are more likely to own their homes. Since income is the characteristic of families which we are interested in, our sample is then likely to contain proportionally more relatively high-income persons (and proportionally fewer relatively low-income persons) than are present in the population as a whole. This sample would then likely overstate the incomes in the population as a whole.[1]

---

[1] It is, of course, possible that our particular sample would not overstate the incomes in the population as a whole; if, by chance, we actually picked a large number of relatively low-income homeowners, we might in fact have a sample with incomes that are typically below the population mean. The point is, however, not whether this situation is possible but rather whether it is likely. If we were to repeat the process of taking a random sample many times, we would just as often overstate as understate the incomes in the population as a whole. With method 1, however, it we were to take many samples, more often than not we would overstate the incomes in the overall population.

Method 2 is substantially better in this regard because both renters and homeowners must pay this income tax, and we are therefore more likely to get a sample that well represents the population as a whole. Yet this method might still not yield a perfect random sample. For example, low-income families might be less likely than other families to file or pay their taxes; homeless families who actually live in the city might not file an income tax due to their lack of an official address within the city. It is then still possible that not all families have the same chance of being included in our sample. Although method 2 yields a sample which is closer to a random sample than that from method 1, it may not be perfectly so. In practice, however, method 2 might be as close to a random sample as is feasibly attainable.

Although it might be simple to visualize the taking of a random sample at the conceptual level, the large size of many interesting populations often makes this difficult in practice. With such a large population, how do you go about listing its component elements so that each family has the same chance of being selected? We are, in fact, often forced to work with samples such as those coming from method 2 above, and we often proceed as if they are indeed random samples. But we need to be aware of our sampling technique and the possible ways in which our sample might not be representative of the population as a whole, since it is clear that a sample that is nonrandom in a significant way can lead to quite erroneous inferences.

EXAMPLE ■

Suppose that you are interested in determining the mean summer income of all students at your college or university, for financial aid analysis purposes. Suppose that you take all students enrolled in an introductory economics course as your sample. Is this a random sample?

The question is whether all students (in terms of their summer income, not in terms of their course of study) have the same chance of being selected. In many colleges and universities, students studying economics are disproportionately male; that is, the percent of students in economics courses who are male exceeds the percent of students at the college or university who are male. The evidence is fairly clear that (for whatever reason) males on average earn more than females do in the U.S. economy as a whole. If this finding is true for all workers, it might very well be true for workers in summer jobs.

Thus, if the introductory economics course is disproportionately male, using it as the basis for an inference regarding the population mean might very well lead to an estimate that tends to be too large. ∎

SAMPLE STATISTICS

Our population consists, say, of m observations on the random variable X: $X_1, X_2, \ldots, X_m$ (or X_i, $i = 1, \ldots, m$). We take a random sample of size n ($n < m$) from this population; this sample consists of X_i, $i = 1, \ldots, n$. (Be careful, this notation does not imply that the first n observations in the population are the ones that show up in the random sample; we have labeled them that way merely for convenience.)

What is X_1? It is the value of X for the first observation in our sample. If the variable of interest were family income, X_1 would be the income of the first family randomly selected from the population. The value of X_1 is then determined in part by an element of chance, since which family is first in our sample is determined by chance. X_1 is then a random variable, and by analogy, so too are the X_2 through X_n that comprise the rest of our sample.

A random sample is then composed of a set of random variables. We are going to use these random variables to infer the characteristics of the population as a whole. What types of characteristics? Things like the mean and the variance of the population, which are referred to as *population parameters.* Note, and this is very important, that these population parameters are not random variables but are rather fixed numbers. Why? There is no element of chance involved in (for example) calculating a population's mean, inasmuch as each observation in the population is going to be counted once in this calculation (i.e., there is no element of chance involved in this calculation because *all* elements of the population are included in the calculation). The same sort of argument can be made for any population characteristic, and we conclude that population parameters are constant (fixed) numbers and not random variables. Nevertheless, they are unknown. Why? Because we do not observe the entire population.

Our job is then to use elements that are random (the observations in our sample) to infer the characteristics of constant numbers that are unknown. This dichotomy between randomness and constancy is a key force in the types of inferences we make, their quality, and a discussion of how their quality can be enhanced. To begin to pursue these issues, let us now consider some more concrete ways of making inferences.

Suppose we wanted to learn about the (unknown) population mean μ. A logical way to proceed would be to say that a good guess of μ is the mean of the observations in our sample (the so-called *sample mean*). That is, to guess the value of μ we might form the sample mean

$$\bar{X} = \frac{1}{n} \sum_{i=1}^{n} X_i$$

Note that

$$\mu = \frac{1}{m} \sum_{i=1}^{m} X_i$$

where m is the size of the entire population and the Σ is over that entire population (i.e., from 1 to m). The sample mean $\bar{X}$ is our typical way of forming an estimate of the population mean μ. (Chapter 10 contains a detailed discussion of estimation and the criteria by which we choose to estimate an unknown population parameter in a particular way.)

Note that whereas μ is a constant (its value is fixed and not determined by an element of chance), $\bar{X}$ is a random variable; the value of $\bar{X}$ depends on which X_i's show up in our sample, and this result is determined by chance. It is thus suggested that $\bar{X}$ *may,* but need not, equal μ, and that if $\bar{X}$ in fact does equal μ it does so by chance. The sample mean is then a guess that is possibly (and usually) wrong as a result of the inherent random nature of sampling! We shall return below to this observation.

What about an estimate of the population variance σ^2? Remember that

$$\sigma^2 = \frac{1}{m} \sum_{i=1}^{m} (X_i - \mu)^2$$

There are two ways in which σ^2 is unknown when we observe only a sample: (1) μ is unknown and thus $(X_i - \mu)$ cannot be calculated for any observation; and (2) not all of the X_i's in the population are known. To estimate σ^2 we might substitute $\bar{X}$ for μ, and then sum over the elements of the sample and divide by the sample size n:

$$\frac{1}{n} \sum_{i=1}^{n} (X_i - \bar{X})^2$$

It turns out that, for a technical reason that we shall consider in Chapter 10, we get a better guess of σ^2 if we divide by $(n-1)$ instead of n. Our usual estimate of σ^2 is then the so-called *sample variance*

$$s^2 = \frac{1}{n-1} \sum_{i=1}^{n} (X_i - \bar{X})^2$$

Following the discussion above regarding $\bar{X}$, s^2 is a random variable that is used to infer the value of the constant σ^2. It is then possible (and generally likely) that our guess is wrong.

Finally, the *sample standard deviation* (s) is calculated as the square root of s^2, and it is our basis for inference regarding σ (the population standard deviation of X).

EXAMPLE ■

The values of X_i for a random sample of five observations are as follows:

X_i	$X_i - \bar{X}$	$(X_i - \bar{X})^2$
5	−3	9
8	0	0
9	1	1
6	−2	4
12	4	16
$\Sigma = 40$		$\Sigma = 30$

Since $\Sigma X_i = 40$, we know that $\bar{X} = 8$. In the middle column we calculate $X_i - \bar{X}$ for each observation, and in the last column we square these quantities. Since $\Sigma(X_i - \bar{X})^2 = 30$, we know that $s^2 = (1/4)(30) = 7.5$.   ■

PROBLEMS

4.1 Suppose that we have a random sample of ten observations, as follows:

$$4, 7, 6, 11, 12, 14, 8, 9, 10, 9$$

Calculate $\bar{X}$ and s^2 for this sample.

4.2 In a random sample of six individuals, we find the following weekly wages (in dollars per week):

$$375, 406, 723, 208, 411, 359$$

Calculate the sample mean and sample standard deviation of weekly wages.

4.3 The population of a city contains 2000 employed males and 1500 employed females. In a random sample of 20 males you find their sample mean weekly wages to be $510, with a sample standard deviation of $200. In a random sample of 15 females you find their sample mean weekly wages to be $380, with a sample standard deviation of $150. Estimate the total wage income (per week) for this city.

4.4 A sample of size 3 consists of the following observations:

$$2, 3, 4$$

(a) Find $\bar{X}$.
(b) Find $\Sigma(X_i - \bar{X})$ for this sample.
(c) Prove that, for any sample,

$$\sum_{i=1}^{n} (X_i - \bar{X}) = 0$$

THE RANDOM NATURE OF THE PROBLEM

So far we have suggested ways of estimating the population parameters; however, we have also suggested that, in view of the random nature of samples, these (educated) guesses are often wrong. How can we respond to this problem?

In a basic or simple inference we make our guess, perhaps acknowledge the problems with the random nature of this guess, and then stop. In a more sophisticated inference, however, we attempt to say more about the pattern of randomness involved in the problem. For example, we might try to determine a "likely margin of error" for our estimate; we might be able to say that our guess $\bar{X}$ of μ is likely to be off by at most k units. This inference is much more sophisticated than an estimate is because it acknowledges (and, in fact, quantifies) the limits of our knowledge. We often see such a construction in results concerning public opinion polls. For example, when asked "Is the president doing a good job?" perhaps 46% of those sampled said yes with a margin of error of plus or minus 3%. What is this 3%? The poll is of a randomly chosen sample of individuals, but of course our real interest is with the population as a whole. The pollster is saying that he or she has determined that it is very likely that the 46% figure for the sample is within 3% of the corresponding figure for the population as a whole; in other words, it is very likely that the percent of the population that thinks the president is doing a good job is between 43% and 49%. This inference is more sophisticated than a simple guess, and it is one that does not give us a false sense of precision.

Now, where does the 3% figure come from? The 3% figure is a measure of the maximum likely extent to which the sample proportion differs from the population proportion. Why does it differ at all? Random variability. The 3% figure is then a measure of a maximum plausible degree of random variability in the sample figure. How did the pollster calculate this figure? He or she must have known something about the pattern of randomness of the sample proportion.

Probability theory is the study of the behavior of random variables, and from this study we are able to determine such things as likely margins of error, which form the basis for more sophisticated inferences. We therefore in Section II of this text turn to the development of some basics from probability theory, and in Section III we begin to use these concepts to make inferences that are more sophisticated than simple estimates or guesses.

SUMMARY REMARKS

In this chapter we have discussed the basics of using the information in samples to infer the characteristics of a population as a whole. We have also suggested the limitations of simple inferences (estimates or guesses) and what we need to develop in order to rise above these limitations. We now begin this development.

REVIEW PROBLEMS FOR CHAPTER 4

4.5 Suppose that we take a random sample of 200 families and find the following numbers of families with various numbers of children:

Number of children	Number of families
0	29
1	58
2	81
3	24
4	8

Find the sample mean and sample variance.

4.6 Suppose that we observe a random sample of ten X's, and find $\bar{X} = 3$. If $Y_i = 5 + 2X_i$, find $\bar{Y}$. If $s_X = 5$, find s_Y.

4.7 We have random samples of 9 "type A" batteries and 12 "type B" batteries. Type A batteries have a sample mean life of 10 hours. The overall sample mean life (for both types) is 15 hours. Find the sample mean life of type B batteries.

*4.8 Following up on problem 4.6, suppose that we observe a random sample of n values of X (i.e., X_i, $i = 1, \ldots, n$) and calculate the sample mean $\bar{X}$ and the sample variance s_X^2. If $Y_i = a + bX_i$ for each X_i, where a and b are constants, prove that

(a) $\bar{Y} = a + b\bar{X}$
(b) $s_Y^2 = b^2 s_X^2$
(c) $s_Y = |b| s_X$

Explain carefully how these properties relate to those determined in Chapter 3 for the population mean, variance, and standard deviation of a linear function.

*4.9 The sample variance is defined as

$$s^2 = \frac{1}{n-1} \sum_{i=1}^{n} (X_i - \bar{X})^2$$

Prove that the sample variance may be equivalently expressed as

$$s^2 = \frac{1}{n-1} \left[\sum_{i=1}^{n} X_i^2 - n(\bar{X})^2 \right]$$

Use this alternative formula to recalculate s^2 for the data in problem 4.1.

DISCUSSION QUESTIONS FOR CHAPTER 4

1. John takes a random sample of five students from a college's population and calculates the sample mean SAT score. Mary takes a random sample of 15 students from this college's population and calculates the sample mean SAT score for her sample. Which sample mean would you expect to be a more reliable guess of the population mean? Be clear what you mean by reliability in this context. Can you generalize this conclusion?

2. Remember our conclusion that a sample statistic such as $\bar{X}$ is a random variable. Suppose that someone challenges that conclusion by arguing that in a particular sample, where $\bar{X}$ equals 1, $\bar{X}$ is a constant since the number "1" is just a fixed number (i.e., "1" is always "1"). How would you respond to this argument?

Section II

PROBABILITY THEORY

Chapter 5

Probability Theory

In this chapter the foundations of probability theory are developed, applied, and interpreted. We have already discussed populations and samples, random variables, and the role played by random variables in the process of using samples to infer the characteristics of populations. Now we proceed to the study of random variables and their behavior so that we may eventually be able to enhance our ability to make inferences about unobserved populations. The study of random variables begins with a discussion of probability theory.

WHAT IS PROBABILITY?

In Chapter 3 we developed the concept of a relative-frequency distribution as a way of characterizing and summarizing a population; the relative frequency of an outcome is the proportion of the population for which that outcome occurs. Suppose that we have a standard die, with the six sides containing patterns of dots representing the numbers from one to six. If we were to take this die and roll it thousands and thousands of times, we would expect that just about one-sixth of the time it would turn up a "six." We then say that the probability of a "six" is one-sixth. The probability of an outcome is then its *long-run relative frequency*—that is, what the relative frequency of the outcome would be after a large number of repetitions of the experiment.

In some contexts it is appropriate (when determining a probability) to think in terms of repeated plays of a particular game or experiment. At other times we have a population that is essentially fixed in size. Suppose that 15% of the families within a particular country have zero children; we would then say that the relative frequency of zero children is .15. If we draw a family at random from this population, what is the probability of picking one with no children? This probability is clearly .15; in this context, the probability of an outcome is just its relative frequency within the population as a whole. We see again that the concept of probability is really another way of thinking about a relative frequency.

This discussion of the definition of a probability is, to be sure, somewhat oversimplified. There are, in fact, differing views as to what exactly a probability is. In the *subjectivist* view, for example, probabilities can be determined only by some judgment made by the observer concerning the degree of certainty with which the outcome in question will occur. This view is in contrast to a pure *frequentist* view, in which probabilities are based upon observed relative frequencies, or to an *axiomatic* view, in which we do not attempt to define probability but merely state the rules that probabilities must follow. For our purposes, however, the above discussion in the general context of the frequentist approach provides the most useful working definition of probability.[1]

Let us now begin to think about some properties of probabilities. The probability of an outcome is the proportion of the population (either actual or hypothetical) for which that outcome occurs. Since a proportion of a population must be between 0 and 1 (inclusive) we know that, for any event A,

$$0 \leq \Pr(A) \leq 1$$

where $\Pr(A)$ is read "the probability that event A occurs." (An "event" may be any subset of the set of all possible outcomes; see the discussion of the "sample space" below.) This is a very important and often forgotten result; it tells us that if we calculate a probability

---

[1] For a more complete discussion of alternative views on probability see, for example, John E. Freund and Ronald E. Walpole, *Mathematical Statistics,* 4th ed. (Englewood Cliffs, NJ: Prentice-Hall, 1987), pp. 25–26.

that is outside the range from 0 to 1, then we *must* have made a mistake!

Let us consider a simple example to illustrate the concept of probability and the links between probabilities and relative frequencies. Suppose that a family plans to have three children and is wondering about the likelihoods with which various numbers of boys and girls will occur. For simplicity, let us suppose that for each child the two sexes are equally likely. What then is the probability that the family will have three boys?

The *sample space* is defined as the set of all possible outcomes. In this example, there are eight possible outcomes: for each of two possibilities for the first child (boy or girl) there are two possibilities for the second and therefore 2×2 possibilities for the first two; for each of 2×2 possibilities for the first two children there are two possibilities for the third and therefore $2 \times 2 \times 2 = 8$ possibilities for the three children. (In general, if there are n possibilities for each of m draws, there are n^m possible outcomes.) These eight outcomes are:

$$BBB$$

$$GBB$$

$$BGB$$

$$BBG$$

$$BGG$$

$$GBG$$

$$GGB$$

$$GGG$$

where, for example, *BBB* represents the outcome a boy on child one, a boy on child two, and a boy on child three. By assumption these eight possibilities are equally likely, and we should therefore presume that if we were to observe thousands and thousands of families with three children then just about one-eighth of them would have the outcome *BBB*. We say therefore that the probability of having three boys is $1/8$.

What is the probability of having two girls and one boy? There are three of eight outcomes for which this occurs (*BGG, GBG,* and

GGB); since all outcomes are equally likely, we conclude that this probability is 3/8.

In this last problem we have actually introduced the notion of "combining" different events; we now turn to this matter in a more general context.

UNIONS AND INTERSECTIONS

Sample spaces can very quickly become very large (for a family having six children, for example, there are $2^6 = 64$ possible outcomes), and a solution based upon a listing of the sample space can therefore become tedious (if not overwhelming) to obtain. We must then begin to think in more depth about the nature of probability and the events in which we are interested. A first step in this direction is to develop further ways of describing and combining various events by considering the subject of unions and intersections.

We are often interested in determining the probability of an event that can be expressed as some combination of two or more other events; the notation to deal with such situations is important because it often suggests ways of determining the probability of interest.

We define *A intersect B* (written $A \cap B$) as the set of all outcomes that are in *both A* and *B*. For example, if *A* represents all individuals who are female, and *B* represents all those who are college graduates, then $A \cap B$ would represent all individuals who are both females and college graduates (or female college graduates). $\Pr(A \cap B)$ is then the probability of choosing a person who has both of these characteristics.

We define *A union B* (written $A \cup B$) as the set of all outcomes that are in *either A* or *B,* or both. For example, if *A* represents persons with a B.A. degree in economics and *B* represents persons with a B.A. in political science, then $A \cup B$ would represent persons with a degree in one or the other field (or both), and $\Pr(A \cup B)$ would represent the probability of choosing a person with a B.A. in one (or both) of these fields.

Suppose that in this last example we were interested in calculating $\Pr(A \cup B)$ from our knowledge concerning $\Pr(A)$ and $\Pr(B)$. Could we simply say

$$\Pr(A \cup B) = \Pr(A) + \Pr(B)?$$

Suppose that 3% of individuals have a B.A. in economics and 2% have a B.A. in political science. The percentage of the population with a B.A. in either economics or political science would be 5% (3% + 2%) only if there were no individuals with B.A.'s in both fields (double majors). If there were double majors, however, we would be double-counting them if we used the above technique. To avoid this problem we must subtract the percent of individuals who are double majors. Suppose that 1% of individuals are double majors; the percent of the population with a B.A. in either economics or political science or both would then be

$$Pr(A \cup B) = .03 + .02 - .01 = .04$$

(where we've expressed probability as a decimal rather than as a percent).[2] To generalize, we write

$$Pr(A \cup B) = Pr(A) + Pr(B) - Pr(A \cap B)$$

Notice that in the case where A and B have no outcomes in common the last probability is 0 (and there is no problem with double counting). In such a case, A and B are said to be *mutually exclusive* ($A \cap B$ is an empty set); the probability of the union of mutually exclusive events is the sum of the individual probabilities with which they occur.

Let us use this example to present the visual technique known as a *Venn diagram*. Remember that we are choosing a person from the population as a whole; A represents those individuals with a B.A. in economics (3% of the population), and B represents those with a B.A. in political science (2% of the population). A Venn diagram approach to characterizing the relationship between these two events is contained in Figure 5.1.

The rectangle S represents the sample space (the set of all possible outcomes), which in this case is all individuals within the population. Area A represents those individuals with a B.A. in economics, area B represents those individuals with a B.A. in political science, and A and B are drawn so that the area they have in common (shaded in the figure) represents $A \cap B$ (those with B.A.'s in both disciplines).

---

[2] In general, in the process of computing probabilities we shall express them as decimals even if they are originally given as percents. Of course, since percent means hundredths, percents and decimals are equivalent ways of expressing the same quantities.

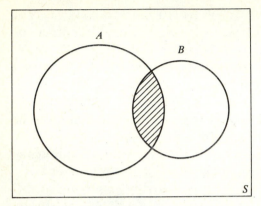

Figure 5.1   Venn diagram.

In such a visual representation we may, through the use of a conceptual experiment, view probability as represented by area. Imagine that S is a square dart board with an area of one square unit. We blindly throw a dart onto the board in a manner such that (1) the dart will *always* land on the board and (2) each point on the board is equally likely. In such a situation, what is $Pr(A)$? It is simply the area of A divided by the area of S, and this value is just the area of A since the area of S is one square unit. Similarly, $Pr(B)$, $Pr(A \cup B)$, and $Pr(A \cap B)$ are all represented by the area of the appropriate region. We have in essence created a visual representation of the concept of probability as area. We now ask: What is the area of $A \cup B$? If we were to add the area of A to the area of B, we would double count $A \cap B$. Thus

$$\text{area}(A \cup B) = \text{area}(A) + \text{area}(B) - \text{area}(A \cap B)$$

Since area represents probability, we have confirmed the formula for $Pr(A \cup B)$ given earlier. This visual representation, and in particular the conceptual experiment of viewing probability as represented by area, is often a very convenient way of bringing out the intuition behind an issue or problem.

Another useful relationship in determining probabilities is based upon the concept of a *complement*. The complement of any event A (call this new event A^c) consists of all outcomes in the sample space that are not a part of A. In the previous example, A represented individuals with a B.A. in economics; A^c would then represent all individuals who do not have a B.A. in economics. Since 3% of the population has a B.A. in economics (i.e., $Pr(A) = .03$), it must be the case that 97% of the population does not have this degree (i.e., $\Pr(A^c) = .97$). This result suggests the general rule that

$$\Pr(A) = 1 - \Pr(A^c)$$

for any event A—the so-called *complement rule*. This rule is very useful because there are many problems for which $\Pr(A^c)$ is much easier to determine than $\Pr(A)$ is. The complement rule then essentially gives us a choice as to which way to go about determining the probability of interest.

The calculation of probabilities of intersections is often difficult when they are not directly given or implied in the problem. To enhance our ability to calculate probabilities in general and probabilities of intersections in particular we turn now to a consideration of the concept of conditional probability. This concept is useful not only for broadening the range of probabilities we can determine but also for bringing forward an important issue and some important distinctions related to it.

PROBLEMS

5.1 You roll a die thousands and thousands of times, and let X equal the number showing on the die.

(a) What would you expect the relative frequency distribution of X to be?

(b) What does this imply about the probabilities associated with the various values of X?

5.2 You toss a coin four times, and observe whether the result is "heads" or "tails" on each toss.

(a) What is the sample space?

(b) What is the probability of getting zero heads? one head? two heads? three heads? four heads?

(c) What assumption have you implicitly made concerning the relative probabilities of the various outcomes? Would this be applicable if the coin were unbalanced such that a "heads" was more likely than a "tails"?

5.3 You toss a coin ten times and observe the sequence of heads and tails that results.

(a) How many possible outcomes are there?

(b) What's the probability that you'll get at least one "heads"?

5.4 Suppose that we consider the individuals in a class who live with both their father and their mother. Of these students, 82% have a father who works and 58% have a mother who works; 47% of students in this class have both parents working. What percent of students in this class (who live with both their father and their mother) have at least one parent working?

5.5 Belgium is a bilingual country, in which the two languages are French and Dutch. Suppose that in a small town you discover that 51% of individuals can speak French, and 83% of individuals can speak Dutch. If 3% of individuals in this town speak neither French nor Dutch (they are "foreigners"), then what percent of individuals speak both French and Dutch?

5.6 Consider three events A, B, and C that are all a part of the same sample space. Develop a general formula for determining $\Pr(A \cup B \cup C)$.

5.7 Consider a company that has a retirement plan for its employees in which the money contributed for each employee can be divided (at the employee's choice) between stocks, bonds, and certificates of deposit (CDs). An employee may choose to put all of his or her money into one investment (e.g., stocks) or may divide it between the three (say, one-third into each, or half each into stocks and bonds with nothing into CDs). You have the following information regarding employees' decisions:

> 60% have at least some money in CDs
> 12% have all their money in CDs
> 10% have all their money divided between stocks and CDs
> 16% have all their money divided between bonds and CDs
> 20% have all their money in stocks
> 66% have at least some money in stocks

(a) What percent of employees have money in all three?

(b) What percent of employees have all their money divided between stocks and bonds?

(c) What percent of employees have all their money in bonds?

(d) What percent of employees have at least some money in bonds?

5.8 Consider a high school in which 48% of students are boys and 52% of students are girls. One-half of the boys are on either the track or swimming teams (or both); similarly, one-half of the girls are on either the track or swimming teams (or both). One-fourth of the boys are on the track team, and one-twelfth of the boys are on both teams. One-fifth of the girls are on the swimming team but not the track team, and one-fourth of the girls play only one of these two sports.

(a) What percent of students are boy swimmers?

(b) What percent of girls are swimmers?

(c) What percent of all students are swimmers?

CONDITIONAL PROBABILITY

When we try to infer something about an unknown characteristic of a person or thing, a fundamental principle is to use all of the information available. In trying to determine whether a job candidate will do an effective job if hired (something we do not know), we consider information about that person's characteristics (e.g., is he or she a college graduate?). Now, imagine that we ask: What is the probability that this person will perform the job effectively? It is likely that the answer we give to that question will be different if we know that he or she is a college graduate than if we do not have that piece of information. Clearly, the fact that we do not know the candidate's educational status becomes an additional source of randomness; however, if we do have that information, a source of uncertainty has been eliminated. For many jobs, the probability that the candidate will perform effectively is higher if we know that he or she is a college graduate than if we are unaware of his or her level of educational attainment.

This last distinction is between a *conditional* and an unconditional probability; when we know the candidate's educational status, we evaluate the probability that he or she will perform the job effectively *conditional* on that piece of information.

Let us consider Figure 5.2, using the dart board metaphor from earlier in this chapter; that is, *A* occurs if the dart lands in ring *A*, etc. These rings, of course, could represent any of a wide variety of

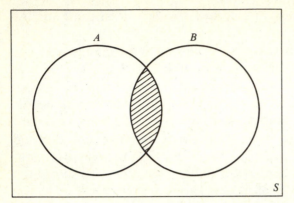

Figure 5.2 Illustrating a Venn diagram based on the dart board metaphor.

phenomena (a specific example would be where A represents the event that the job candidate will do the job effectively and B the event that he or she is a college graduate), and so we must imagine that the areas of the respective regions perfectly represent the probabilities of the real-world phenomena with which they are associated.

To view this issue in a general context imagine that you throw a dart into the board and are interested in whether or not it lands in A. You would initially evaluate the probability of this occurring as the area of A (since area represents probability). Suppose, however, that you are now told that the dart has landed somewhere in B. You do not know whether it is in A, and of course that is what you are interested in. How would you now evaluate the probability that the dart has landed in A? Given that the dart is in B, the only way that it can also be in A is if it is somewhere within $A \cap B$. This implies that of the (now-limited) possible locations (represented by the area of B) a particular portion (represented by the area of $A \cap B$) is associated with event A occurring. The probability that A occurs given that B has occurred is then

$$\frac{\text{area}(A \cap B)}{\text{area}(B)} = \frac{\Pr(A \cap B)}{\Pr(B)}$$

We say that this is the probability that the dart is in A, *given* that it is in B. Notice that this value is generally different from $\Pr(A)$ (which is an unconditional probability). To return to our earlier example, it will often be the case that the probability that a job candidate will perform a job effectively given that he or she is a college graduate is higher than the corresponding probability if we do not know that this condition has occurred.

This discussion suggests a definition of conditional probability:

Definition of Conditional Probability The conditional probability of A given B, which we shall write as $\Pr(A \mid B)$, is

$$\Pr(A \mid B) = \frac{\Pr(A \cap B)}{\Pr(B)}$$

Let us use this definition in the context of an example in which we choose a person at random from the U.S. population (for 1986) and are interested in the race and poverty status of this person.[3] Suppose that A represents the event that the person we pick is "poor" (i.e., lives in a family with an income below the official poverty level), and B represents the event that he or she is white. If we pick a person at random the probability that he or she is poor is $\Pr(\text{poor})$. If we pick a person at random whom we then identify as white the probability that he or she is poor(given white) is $\Pr(\text{poor} \mid \text{white})$. In 1986, 13.6% of all individuals in the population were poor, and 9.3% were poor and white. Since 84.8% of individuals were white, the probability that our chosen person is poor given that he or she is white is

$$\Pr(\text{poor} \mid \text{white}) = \frac{\Pr(\text{poor} \cap \text{white})}{\Pr(\text{white})} = \frac{.093}{.848} = .110$$

Notice that this result is different from the unconditional probability of picking a poor person (.136). Why? Whites were less likely than nonwhites to be poor, and therefore the condition that the person you picked is white lowers the probability that he or she is poor. (To convince yourself that whites are less likely than nonwhites to be poor, calculate the probability of picking someone who is poor given that he or she is nonwhite. You will get a probability of .283.)

APPLICATIONS OF CONDITIONAL PROBABILITY

An important application of conditional probability is in the area of regression analysis; we shall discuss this topic in Section IV. But it is also the case that conditional probability gives us a "trick" that

[3] The data in this example are from U.S. Bureau of the Census, Current Population Reports, *Poverty in the United States: 1986,* Series P-60, No. 160 (Washington, DC: U.S. Government Printing Office, 1988), table 1.

can be used fairly easily to evaluate some probabilities that would otherwise be quite complicated; it therefore broadens for us the general applicability of the concept of probability. Let us consider an example.

Assume that we have a normal 52-card deck of playing cards. Suppose that we draw two cards from this deck *without replacement.* By "without replacement" we mean that the first card is not put back into the deck before the second one is chosen; obviously, this is a different game than if the first card were returned to the deck and the deck reshuffled before choosing the second card (sampling *with replacement*). What is the probability of getting two aces? Remember the definition of conditional probability:

$$\Pr(A \mid B) = \frac{\Pr(A \cap B)}{\Pr(B)}$$

The situation at hand is one of a fairly large class of problems for which we know $\Pr(A \mid B)$ and $\Pr(B)$; we can therefore solve indirectly for $\Pr(A \cap B)$ by noting that

$$\Pr(A \cap B) = \Pr(A \mid B) \Pr(B) \qquad (5.1)$$

In the problem under consideration, let A_2 be the event an ace on the second card and A_1 the event an ace on the first. Clearly, $\Pr(A_1) = 4/52$. $\Pr(A_2 \mid A_1)$, which is the probability of an ace on the second card *given* that the first card is an ace, must be 3/51. Why? If the first card is an ace, then out of 51 cards remaining there are three aces. If we treat A_2 as "A" and A_1 as "B" we then get

$$\Pr(A_2 \cap A_1) = \Pr(A_2 \mid A_1) \Pr(A_1) = \frac{3}{51} \times \frac{4}{52} = \frac{1}{221}$$

What is the intuition here? Let us think about how many possibilities there are for this "hand" of two cards. Clearly, there are 52 possibilities for the first card. Now, for each of these 52 possibilities for card 1 there are 51 possibilities for card 2; thus, there are 52 × 51 possibilities for the two-card hand (where order matters—that is, ace of hearts followed by ace of spades is a different hand than ace of spades followed by ace of hearts). These possibilities are equally likely since the cards are randomly drawn from the deck. How many of the possibilities contain two aces? For each of four ways of getting an ace on the first card there are three ways of getting an ace on the second; there are then 4 × 3 ways of getting two aces. Since there are 4×3 ways of getting two aces, and 52×51 ways of choosing two cards, the long-run relative frequency (i.e., the probability) of two aces would be

$$\frac{4 \times 3}{52 \times 51}$$

which is identical to what we earlier determined.

Suppose now that we choose three cards from a deck, without replacement. What is the probability of getting three aces? That is, what is the probability of $(A_3 \cap A_2 \cap A_1)$ where A_3 is the event an ace on the third card? The key is to apply the notion of conditional probability, but how do we do this given that we now have three events instead of two?

Whether we have three events or two depends on how we view the problem. We can rephrase our problem as

$$\Pr[A_3 \cap (A_2 \cap A_1)]$$

and think of $(A_2 \cap A_1)$ as a *single* event. Thus

$$\Pr[A_3 \cap (A_2 \cap A_1)] = \Pr[A_3 \mid (A_2 \cap A_1)] \Pr(A_2 \cap A_1)$$

We have already determined that

$$\Pr(A_2 \cap A_1) = \frac{3}{51} \times \frac{4}{52}$$

In addition,

$$\Pr[A_3 \mid (A_2 \cap A_1)] = \frac{2}{50}$$

since, given that the first two cards are aces, there are two aces out of 50 cards remaining. We then conclude that

$$\Pr[A_3 \cap (A_2 \cap A_1)] = \frac{2}{50} \times \frac{3}{51} \times \frac{4}{52} = \frac{1}{5525}$$

To generalize this result to a set of k events, note that, in the above example ($k = 3$),

$$\Pr(A_3 \cap A_2 \cap A_1) = \Pr[A_3 \cap (A_2 \cap A_1)]$$
$$= \Pr[A_3 \mid (A_2 \cap A_1)] \Pr(A_2 \cap A_1)$$
$$= \Pr[A_3 \mid (A_2 \cap A_1)] \Pr(A_2 \mid A_1) \Pr(A_1)$$

[since $\Pr(A_2 \cap A_1) = \Pr(A_2 \mid A_1) \Pr(A_1)$]. For a set of events B_1, $\ldots$, B_k, therefore,

$$\Pr(B_1 \cap \cdots \cap B_k) = \Pr(B_k \cap \cdots \cap B_1)$$

$$= \Pr[B_k \mid (B_{k-1} \cap \cdots \cap B_1)]$$

$$\times \Pr[B_{k-1} \mid (B_{k-2} \cap \cdots \cap B_1)]$$

$$\times \cdots \times \Pr(B_2 \mid B_1) \times \Pr(B_1) \quad (5.2)$$

which generalizes Eq. (5.1). To illustrate the use of Eq. (5.2), suppose that we have a bowl containing six red and four blue marbles. If we pick four marbles without replacement, the probability that all four are blue is, by Eq. (5.2),

$$\frac{1}{7} \times \frac{2}{8} \times \frac{3}{9} \times \frac{4}{10}$$

Notice that there is a type of "chain rule" effect here; notice also that it is easier to determine the right-side probabilities from right to left when the events in question are the result of a series of draws from a population, because this way we move from earlier to later events in time.

PROBLEMS

5.9 Suppose that $\Pr(A) = .2$, $\Pr(B) = .3$, and $\Pr(A \cap B) = .1$.
   (a) Find $\Pr(A \mid B)$.
   (b) Find $\Pr(A \mid B^c)$, where B^c is the complement of B.

5.10 Consider an economy in which the overall unemployment rate is 18%. Moreover, 20% of individuals are college graduates, and 2% of all individuals are unemployed college graduates.

   (a) If I pick a college graduate at random, what is the probability that he or she is unemployed?
   (b) If I pick a person at random from those who are not college graduates, what is the probability that he or she is unemployed?
   (c) If I pick an unemployed person at random, what is the probability that he or she is a college graduate?

5.11 You pick four cards (without replacement) from a regular deck of playing cards. Find the probabilities associated with the following outcomes:

(a) four aces
(b) aces on the first two cards, and jacks on the next two
(c) aces on the first three cards, and a nonace on the fourth
(d) exactly three aces
(e) at least three aces

5.12 A firm receives a shipment of 100 personal computers for use by its staff. Of these 100 computers, five are defective and not usable without repair. In a particular office, there are five employees.

(a) What is the probability that each of these five employees gets a computer that is in working order?
(b) What is the probability that at least one of these five employees gets a computer that is in working order?

5.13 Jake the Flash, a star basketball player, has a two-shot foul with no time remaining and the score tied. Jake normally makes 60% of his foul shots. What is the probability that he will then win the game for his team? What assumption are you implicitly making in determining your answer?

5.14 In a large hat you have three blue, four red, and five purple marbles. You pick three marbles from this hat, without replacement.

(a) What is the probability of getting two reds and one blue, in that order?
(b) What is the probability of getting two reds and one blue in any order?
(c) What is the probability of getting three reds, given that the first marble is red?
(d) What is the probability of getting three primary colors (blue or red) given that the first marble is blue?
(e) What is the probability of getting at least one red marble?

5.15 Suppose that 15% of the boys at a particular high school play basketball, and these boys represent 7% of the entire student body (boys and girls). What percent of the student body is girls?

5.16 In the United States in 1986, 48.7% of persons age 25-plus were males. Of these males, 23.8% were college graduates. In addition, 20.5% of all persons (males and females) were college graduates. [Source: U.S. Bureau of the Census, Current Population Reports, *Money Incomes of Households, Families, and Persons in the United States: 1986,* Series P-60, No. 159 (Washington, DC: U.S. Government Printing Office, 1988), table 27.]

(a) What proportion of persons 25-plus were female college graduates?

(b) What proportion of females 25-plus were college graduates?

INTUITION AND CONDITIONAL PROBABILITY

Conditional probability often leads us to results that on the surface appear anti-intuitive and therefore hinder our understanding of the material. For this reason we will now present two examples to show that seemingly anti-intuitive results are often in fact quite intuitive *if* they are viewed in the proper context; in so doing, we will suggest some "tricks" for uncovering hidden intuition. The point is that probability theory is usually quite intuitive if only looked at in the right way. These examples will also apply the concepts we have developed in fairly sophisticated ways, and they will therefore contribute to our ability to determine complicated types of probabilities.

EXAMPLE ■

Suppose that we have three cups: *A, B,* and *C. A* contains two red marbles, *B* contains two blue marbles, and *C* contains one red and one blue marble. Our experiment is to pick a cup, then pick a marble from this cup (noticing its color), and then pick the second marble from this cup.

Suppose that we have picked a cup and the first marble. This marble is red. We now ask: Given that this marble is red, what is the probability that the second marble we pick from the cup will also be red?

The most tempting answer to this question is 1/2. If the first marble is red, then we have either *A* or *C.* In one of these (*A*) the second marble picked will be red. Thus the probability that the second marble is red must be 1/2. Is this valid logic?

Most individuals are surprised to learn that in fact the correct answer to this question is 2/3; they therefore presume that this is a case where probability theory leads to anti-intuitive results. Before providing the reasoning as to why 2/3 is in fact an intuitively pleasing answer, let us provide the theoretical argument as to why it is correct.

Formally, we are determining the probability that the second marble is red (R_2) *given* that the first one is red (R_1). This is a conditional probability, which we shall write as

$$\Pr(R_2 \mid R_1) = \frac{\Pr(R_2 \cap R_1)}{\Pr(R_1)} \qquad (5.3)$$

The numerator on the right-hand side is the easy part; the only way to get reds on both the first and the second marbles is if we have picked cup A. This is an unconditional probability, and since the three cups are equally likely at the beginning of the experiment this probability equals 1/3. The denominator $\Pr(R_1)$, however, is more involved.

There are two conceptually separate ways of getting R_1: a red on the first draw from cup A [which we can refer to as $(R_1 \cap A)$] and a red on the first draw from cup $C(R_1 \cap C)$. We can then write the event R_1 as

$$R_1 = (R_1 \cap A) \cup (R_1 \cap C)$$

The events $(R_1 \cap A)$ and $(R_1 \cap C)$ are the two separate ways of achieving R_1; notice that these two events have no intersection (they are mutually exclusive). Remember that when two events have no intersection, the probability of their union is simply the sum of the individual probabilities, and so

$$\Pr(R_1) = \Pr(R_1 \cap A) + \Pr(R_1 \cap C)$$

The probabilities on the right-hand side can be broken down into conditional probabilities as we saw in our earlier examples. Doing this, we get

$$\Pr(R_1) = \Pr(R_1 \mid A)\Pr(A) + \Pr(R_1 \mid C)\Pr(C)$$

We know the four probabilities on the right side: $\Pr(R_1 \mid A) = 1$; $\Pr(A) = 1/3$; $\Pr(R_1 \mid C) = 1/2$; $\Pr(C) = 1/3$. Thus,

$$\Pr(R_1) = \left(1 \times \frac{1}{3}\right) + \left(\frac{1}{2} \times \frac{1}{3}\right) = \frac{1}{2}$$

Going back to Eq. (5.3), we can then say that

$$\Pr(R_2 \mid R_1) = \frac{\Pr(R_2 \cap R_1)}{\Pr(R_1)} = \frac{1/3}{1/2} = \frac{2}{3}$$

which is the correct solution to this problem.   ■

To see why this result should not really be viewed as anti-intuitive, that is, to see why the first intuition (leading to an answer of 1/2) is not correct, let us consider a slight variation on this problem. This slightly varied version will show the fallacy of our initial intuition and suggest the intuition behind the correct solution of 2/3.

Suppose that we have three new cups: A', B', and C'. A' contains two red marbles, B' contains two blue marbles, and C' contains one red and 999,999 blue marbles. The experiment is, as before, to pick a cup and then pick, consecutively, two marbles from it.

We pick the cup and then the first marble. It is red. What is the probability that the second marble we pick will also be red? Intuitively, we are almost sure that we have cup A' because a red marble from C' is, though possible, very unlikely. Since we are almost sure that we have A' we should be almost sure that the second marble we pick will be red. If we were to determine the probability using the formal method just outlined we would find that the probability that the second is red given that the first is red is 1,000,000/1,000,001 (can you verify this?), and this is consistent with our intuition.

The key to this intuition is the recognition that, given that the first marble is a red one, it is very unlikely that we have cup C'. How does this extreme example shed light on the earlier one involving cups A, B, and C? Recall our first insight as to the answer to that problem:

> If the first marble is red, then we have either A or C. In one of these (A) the second marble picked will be red. Thus the probability that the second marble is red must be 1/2.

This logic is clearly not applicable to the A', B', C' problem; it is clearly not the case that, given that the first marble is red, A' and C' are equally likely (and the final step in the logic is based on this assumption). Although A' and C' are equally likely at the start of the problem (unconditionally), they are *not* equally likely *given* that the first marble is red (since a red marble from C' is so unlikely). But, and this is the key, if A' and C' in the second problem are not equally likely (given that the first marble is red) then it must also be the case that A and C (in the first problem) are *not* equally likely (given that the first marble is red). Since A contains two reds but C contains one red and one blue, cup A is more likely to have occurred than cup C is (given that the first marble is red). Our original analysis was then invalid.

Since cup A is *conditionally* more likely than cup C, we know that the probability that the second marble is also red must be greater than $1/2$. In fact, there is some sense to saying that cup A is conditionally twice as likely as cup C (since it's all red as opposed to C, which is half red) and therefore not being surprised that the probability that the second marble is red (given that the first is red) is $2/3$.

What is interesting about this problem is that, although the intuition is hidden, it is still there. Often our intuition can be brought out by considering a more extreme version of the problem (e.g., one with different and more extreme numbers). The nature of the original problem essentially fooled us into thinking that an equality of likelihoods was present when in fact it was not.

We have therefore uncovered a convenient trick for bringing out intuition: see what happens in a more extreme version of the problem. The original problem had an intuitive answer; we just needed to search a bit for that intuition. We will now present a second example with a seemingly anti-intuitive result and argue for the fact that it is intuitive once we properly view the problem.

EXAMPLE ■

Suppose that there is a disease that is known to affect 1% of the population. There is a test for this disease that, though relatively accurate, is not perfect: of those with the disease, 99% will test positive; of those without the disease, 95% will test negative. Clearly, false positives (as well as false negatives) are possible; 5% of those without the disease will test positive (a false positive), and 1% of those with the disease will test negative (a false negative).

Suppose that you take the test, and test positive. What is the probability that you in fact don't have the disease?

Your first inclination might be to say that since only 5% of those without the disease test positive it must be the case that there is only a 5% chance that you do not have the disease (and thus a 95% chance that you do). Is this right?

The 5% figure is Pr (test positive | do not have disease);

that is, it is a conditional probability where the condition is based on whether or not one has the disease. But this is not the probability being asked for in the question!

The question of interest may be rephrased as: Given that you have tested positive, what is the probability that you do not have the disease? Algebraically, this probability is Pr(do not have disease | test positive). But this probability is different from the one mentioned in the preceding paragraph, from which our first answer came. What is going on here?

The original probabilities concerning the accuracy of the test are conditioned upon whether one has or does not have the disease. But an individual patient does not know whether he or she has the disease (that is the whole reason for taking the test). The patient in fact wants to know the probability of having or not having the disease given the result on the test (positive or negative).

It turns out that the probability that you do not have the disease given a test result that is positive is 5/6 (about 83%). Even with a positive test it is relatively unlikely that you have the disease! How could our initial guess have been so wrong? Because we answered the wrong question! Let us begin by deriving the correct solution using set algebra.

Let Y represent the event that you have the disease, and let N represent the event that you do not have the disease. Let "pos" represent the event a positive test, and let "neg" represent the event a negative test. Our problem is then to determine

$$\Pr(N \mid \text{pos})$$

From the definition of conditional probability, we know that

$$\Pr(N \mid \text{pos}) = \frac{\Pr(N \cap \text{pos})}{\Pr(\text{pos})} \qquad (5.4)$$

Let us determine the numerator and denominator on the right side separately. For the numerator, we know that

$$\Pr(N \cap \text{pos}) = \Pr(\text{pos} \cap N)$$

$$= \Pr(\text{pos} \mid N)\,\Pr(N)$$

$$= (.05)(.99)$$

For the denominator of Eq. (5.4), let us begin by noting that

$$\text{pos} = (\text{pos} \cap Y) \cup (\text{pos} \cap N)$$

Here we have simply decomposed the event "pos" into (1) a positive test when you have the disease and (2) a positive test when you do not have the disease. The right-hand side of this equation is the union of mutually exclusive events [(pos ∩ Y) and (pos ∩ N)], and we know from earlier in this chapter that the probability of the union of exclusive events is the sum of the individual probabilities. Thus

$$\Pr(\text{pos}) = \Pr(\text{pos} \cap Y) + \Pr(\text{pos} \cap N)$$

If we break down the right-hand side into conditional probabilities, we get

$$\Pr(\text{pos}) = \Pr(\text{pos} \mid Y)\Pr(Y) + \Pr(\text{pos} \mid N)\Pr(N)$$
$$= (.99)(.01) + (.05)(.99)$$

If we substitute the results for $\Pr(N \cap \text{pos})$ and $\Pr(\text{pos})$ into Eq. (5.4), we get

$$\Pr(N \mid \text{pos}) = \frac{(.05)(.99)}{(.99)(.01) + (.05)(.99)} = \frac{5}{6}$$

which is the answer we earlier asserted.   ∎

Although this example in a technical (and complicated) manner gives us the solution to the problem, we are left with little intuition as to what it is that drives this result. (This result, in fact, is an application of Bayes' theorem, which is discussed more fully in the appendix to this chapter.) Fortunately, there is an alternative way of viewing the problem that is much more successful in that regard.

Let us imagine that we have a population of 1,000,000 people, which perfectly represents the probabilities given earlier: 1% have the disease, etc. Consider Figure 5.3, in which we construct a diagram to analyze this hypothetical population.

At the top of the diagram are the 1,000,000 persons who make up the population. Since 1% have the disease, we know that 10,000 individuals have the disease and 990,000 do not, as shown in the second line of the diagram. On the third line we begin to consider the effectiveness of the test.

Of those with the disease 99% will test positive. Thus, of the 10,000 persons with the disease we know that 9900 will test positive and 100 will test negative (these are the false negatives). We also

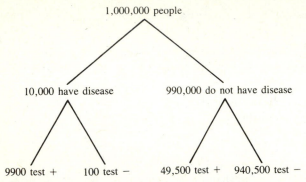

Figure 5.3 An illustration of Bayes' theorem.

know that 5% of those without the disease test positive; thus, of the 990,000 without the disease 49,500 will test positive (these are the false positives) and 940,500 will test negative. The third line of the diagram then divides the population into four subgroups, manifesting the four possible combinations of the two characteristics under question.

What is the probability of not having the disease given that you have tested positive? Looking at the third line of the figure, we know that 9900 + 49,500 = 59,400 persons will test positive. Of them, 49,500 do not have the disease. The probability of not having the disease given that you have tested positive is then 49,500/59,400 = 5/6. Even if you test positive, chances are that you do not have the disease!

What drives this result? A look at the figure provides the answer. There are many more people who test positive who do not have the disease (false positives) than people who test positive and do have the disease (true positives) simply because the vast majority of people do not have the disease! Although the test is reasonably accurate, the false positives greatly outweigh the true positives because of the rareness of the disease! Notice, however, that the probability of having the disease given a positive test result (1/6) is substantially higher than the probability of having the disease before the test result is known (.01). The test is then still indicative of a substantially increased probability of having the disease.

This type of hypothetical population can often be constructed to get further insights into the problem, evaluate a probability, and bring out the underlying intuition. This construction is then another potentially helpful trick for enhancing the intuition behind probabilities.

PROBLEMS

5.17 In a particular state, 40% of law school graduates pass the bar exam on their first try. Of those taking the bar exam for the second time, 70% pass. If you've just graduated from law school, what is your chance of (eventually) passing the bar exam in this state? (Suppose that a maximum of two tries are allowed.)

5.18 The Smith and Jones families live in adjacent houses that share a common yard. The Smith family has two children, a boy and a girl. The Jones family has six children, all of whom are boys.

(a) Suppose that all the children are in the yard playing. If you pick a child at random, what is the probability that you will choose a boy?

(b) All the children are inside their appropriate houses. You pick a house at random, and then pick a child from within that house. What is the probability that this child is a boy?

5.19 Consider a virus that is spread by transfusions of blood contaminated by the virus; a person receiving contaminated blood has a 20% chance of becoming infected with the virus. Suppose that 1000 persons each receive one pint of blood and that five of them wind up becoming infected by the virus. What fraction of the pints of blood used do you expect were contaminated? Derive your answer both intuitively and using the algebra of probabilities.

5.20 Suppose you have three cups: *A, B,* and *C. A* contains three red marbles, *B* contains two red marbles and one blue marble, and *C* contains one red and two blue marbles.

(a) You pick a cup at random, and then a marble from that cup. It is red. What is the probability that if you picked a second marble from this cup that it would also be red?

(b) Suppose now that you add a fourth cup *D,* which contains three red marbles. You pick a cup at random, and then a marble from that cup. It is red. What is the probability that if you picked a second marble from this cup that it would also be red?

(c) Suppose instead that cup *D* contained four red marbles. How would your answer to part (b) change? Carefully explain.

5.21 In an example presented earlier in this chapter, we noted that, in the United States in 1986, 13.6% of all individuals were poor, 9.3% of all individuals were poor whites (thus 4.3% were poor nonwhites), and 84.8% of all individuals were white (thus 15.2% were nonwhite). From these data we determined that 11.0% of whites were poor and 28.3% of nonwhites were poor. Now, suppose that we pick a poor person at random from this population. What is the probability that he or she is nonwhite?

5.22 Consider a hypothetical city in which 80% of individuals are native English speakers, 15% are native Spanish speakers, and 5% are native speakers of a language other than English or Spanish. Suppose that 80% of native English speakers, 90% of native Spanish speakers, and 60% of native speakers of some other language are high school graduates.

(a) If I pick a person at random from this population, what is the probability that he or she is a high school graduate?

(b) If I pick a person at random who turns out to be a high school graduate, what is the probability that he or she is a native English speaker?

5.23 A company's management is worried about employee drug use and is instituting a policy of mandatory drug tests. If an employee is a drug user, there is an 85% chance that he or she will test positive. If the employee is not a drug user, there is a 95% chance that he or she will test negative. Individuals who test positive are fired. We do not know the fraction of employees who are drug users. We do know, however, that 10% of all employees tested positive.

(a) What fraction of employees are drug users?

(b) Of those who test positive, what fraction are in fact not drug users?

INDEPENDENCE

Conditional probability involves evaluating the probability of one event occurring based upon knowledge as to whether or not some other event has occurred. There are special cases, however, when knowing whether or not event B has occurred in no way influences the probability that event A will occur. In such situations, events A and B are said to be *independent*.

A simple example illustrates this concept. Suppose that I toss a coin two times and am interested in the possibility that the second toss turns up heads. Does knowing whether the first toss was heads or tails in any way influence my evaluation of the probability that the second toss will be heads? Of course not. The two tosses are then said to be independent.

Definition of Independence   Two events A and B are independent if and only if $Pr(A \mid B) = Pr(A)$—that is, if the conditional and unconditional probabilities are equivalent.

To illustrate, consider Figure 5.4 (which we shall interpret in the context of the dart board metaphor used earlier in this chapter). Four areas have been labeled with small letters: a represents the area of that part of A that is not in B; b represents the area of that part of B that is not in A; c represents the area of the intersection of A and B; and d represents the area of that part of S that is in neither A nor B. Note that $a + b + c + d = 1$.

If I blindly throw the dart then $Pr(A)$ is simply $a + c$. If I throw the dart and know that it has landed in B, then the probability that it is also in A [which is $Pr(A \mid B)$] is $c/(b + c)$. Why? The dart is somewhere in B, which has an area $b + c$; of that area, only the area c is in A.

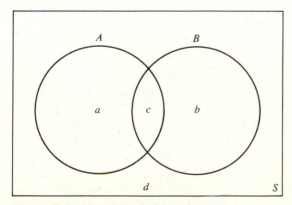

Figure 5.4   Visualizing independence.

If $a + c = c/(b + c)$—that is, if $\Pr(A) = \Pr(A \mid B)$—then knowing that B has occurred does not make it more (or less) likely that A has occurred. The events would then be independent.

For a numerical example, suppose that $a = .3$, $b = .15$, and $c = .1$. The (unconditional) probability that A occurs [$\Pr(A)$] is $a + c = .4$. Given that B occurs, the (conditional) probability that A occurs [$\Pr(A \mid B)$] is $c/(b + c) = .1/(.1 + .15) = .4$. These probabilities are the same! In this case, knowing that B occurs does not change our evaluation of the likelihood that A will occur.

Notice that the probability that A occurs given that B *does not* occur [$\Pr(A \mid B^c)$, where B^c is the complement of B] is $a/(a + d)$. In this example, $d = .45$ and $a/(a + d) = .3/(.3 + .45) = .4$, which is the same as the unconditional probability of A! Clearly, knowing the status of B is unhelpful in determining the probability that A will occur. In all cases, A occurs with probability .4.

In the case where A and B are independent, knowing whether one event has occurred is no help in determining whether or not the other will occur. A gambler betting on A would not first pay money to learn whether B had occurred, for this knowledge is of no help. Notice that if $\Pr(A \mid B) \neq \Pr(A)$, then knowing whether B has occurred does influence the likelihood that A will occur (our gambler would pay some amount of money for this information); in this case, A and B are not independent. [Since $\Pr(A \mid B) \neq \Pr(A)$ implies that A and B are not independent, we say that A and B are independent *if and only if* $\Pr(A \mid B) = \Pr(A)$.]

At a basic level, applied statistical analysis is often concerned with trying to determine whether or not there is a relationship between two variables (and at a more sophisticated level with trying to determine some of the specifics of that relationship). At this point we view the concept of independence, and the presence or lack of this characteristic, as a way of learning whether such relationships are present at all. Let us consider an example.

In the United States in 1986, the following numbers (all expressed in thousands) characterized the situation of persons in regard to the poverty level of income.[4] Of 238,534 persons, 32,370 had incomes below the poverty level (BPL) and therefore were considered to be "poor." Of 202,282 whites, there were 22,183 with incomes BPL, and of 36,252 nonwhites, there were 10,187 with incomes BPL. Is poverty independent of race?

---

[4] U.S. Bureau of the Census, *Poverty in the United States: 1986,* table 1 (see note 3).

Notice that these numbers suggest that 11.0% of whites but 28.1% of nonwhites were poor, suggesting that race and poverty are related (and not independent). To show this formally, let's ask the question: Is

$$Pr(\text{poor} \mid \text{white}) = Pr(\text{poor})?$$

The left side is .110. The right side can be calculated from the data on the population as a whole, and is .136. Therefore, it is clear that race and poverty are not independent.[5]

INDEPENDENCE AND PROBABILITY

The definition of independence

$$Pr(A \mid B) = Pr(A)$$

is equivalent to

$$\frac{Pr(A \cap B)}{Pr(B)} = Pr(A)$$

or

$$Pr(A \cap B) = Pr(A)\,Pr(B)$$

This very important result is applicable to a broad variety of probability problems; it implies that if two events are independent, then the probability of their intersection is the product of the individual probabilities. Here we have yet another possible way of evaluating probabilities of intersections.

Suppose that we draw two cards from a normal deck *with replacement* (we return the first card to the deck and reshuffle before picking the second). If we are sampling with replacement, successive

---

[5] Notice that although the poverty rate among nonwhites exceeds the poverty rate among whites, there are still more poor whites than poor nonwhites (22,183 versus 10,187); the reason, of course, is that there are so many more white persons than nonwhite persons in the population as a whole. It is important to realize that a comparison of the numbers of poor whites versus poor nonwhites captures a different concept than a comparison of the poverty rates among whites versus nonwhites does; any confusion here is in essence a confusion between unconditional and conditional probabilities. The poverty rates by race are conditional probabilities, $Pr(\text{poor} \mid \text{white})$ and $Pr(\text{poor} \mid \text{nonwhite})$. The probability of choosing a person from the population as a whole who is a poor white is $Pr(\text{poor} \cap \text{white})$, and the probability of choosing a person who is a poor nonwhite is $Pr(\text{poor} \cap \text{nonwhite})$; these are unconditional probabilities that reflect the relative numbers of poor whites and poor nonwhites in the population as a whole.

draws from the deck are independent; that is, the probability of an outcome for the second card is unaffected by knowledge concerning the first card. In general, repeated sampling with replacement from a population leads to a set of observations that are independent, because the process of restoring the population to its original form (by replacement) implies that the result of an earlier draw has no effect on the result of a later draw. On the other hand, if we were engaged in repeated sampling without replacement then successive draws would not be independent (since the result of an earlier draw would have an effect on the possibilities for later draws).

For applied purposes, however, there is an exception to the rule just stated for sampling without replacement—an exception that is very often relevant in economics. If we are sampling without replacement a relatively small number from a relatively large population then there is no noticeable effect on later draws of the results of earlier draws. In such a situation, successive draws may be said to be approximately independent for application purposes. In economics, our samples are very often samples drawn without replacement from a large population, for example, a sample of 50,000 members of the labor force drawn so as to make an estimate of the unemployment rate for the economy as a whole. Therefore, often the types of samples we have can be characterized as being composed of independent events, an important property in our discussion of sampling theory in Chapter 9.

To begin to apply the concept of independence, let us consider an example. Suppose that we draw two cards from a normal deck with replacement, and ask: What is the probability that both cards we pick are aces? That is, what is $\Pr(A_1 \cap A_2)$ where A_1 is an ace on card one and A_2 is an ace on card two? Since A_1 and A_2 are independent,

$$\Pr(A_1 \cap A_2) = \Pr(A_1) \Pr(A_2) = \frac{4}{52} \times \frac{4}{52} = \frac{1}{169}$$

The concept of independence can be extended to more than two events. For example, three events A, B, and C are mutually independent if information concerning the other two does not influence the probability of the event in question occurring. The mutual independence of A, B, and C would occur if and only if

$$\Pr(A \mid B) = \Pr(A \mid C) = \Pr(A \mid (B \cap C)) = \Pr(A)$$

$$\Pr(B \mid A) = \Pr(B \mid C) = \Pr(B \mid (A \cap C)) = \Pr(B)$$

$$\Pr(C \mid A) = \Pr(C \mid B) = \Pr(C \mid (A \cap B)) = \Pr(C)$$

When A, B, and C are mutually independent events,

$$\Pr(A \cap B \cap C) = \Pr(A)\,\Pr(B)\,\Pr(C)$$

In general, if $A_1, \ldots, A_k$ are a set of mutually independent events, then

$$\Pr(A_1 \cap A_2 \cap \cdots \cap A_k) = \Pr(A_1)\,\Pr(A_2) \cdots \Pr(A_k)$$

and this result is the key to solving a wide variety of probability problems.[6]

Suppose that we draw four cards from a normal deck, with replacement. From our earlier discussion, we know that sampling with replacement (or, sampling without replacement a relatively small number from a relatively large population) leads to a set of events that are mutually independent. The probability of choosing four aces would then be

$$\Pr(A_1 \cap A_2 \cap A_3 \cap A_4) = \Pr(A_1)\,\Pr(A_2)\,\Pr(A_3)\,\Pr(A_4)$$

$$= \frac{4}{52} \times \frac{4}{52} \times \frac{4}{52} \times \frac{4}{52} = \frac{1}{28,561}$$

We will build on the concept of independence in future chapters as we explore in more detail the possible ways that events (and variables) can be related to each other. But for now, we may view it as the most basic way of addressing whether such relationships exist.[7]

PROBLEMS

5.24 Suppose we have ten balls in a large drum; these balls are numbered 1 through 10. We pick one ball from the drum and define the following events:

---

[6] For a formal definition of mutual independence in the case of more than two events and of the complications which arise in attempting to make such a definition, see Freund and Walpole (note 1), pp. 60–61. It should be noted that if A, B, and C are each pairwise independent with the other two (i.e., A and B are independent, B and C are independent, and A and C are independent), it is still possible that A, B, and C are not mutually independent.

[7] Does independence suggest a different way of addressing probability problems than conditional probability does? Consider the game of picking two cards with replacement from a deck. If we used the conditional probability model, we would say that $\Pr(A_2 \cap A_1) = \Pr(A_2|A_1)\,\Pr(A_1)$. But since A_2 and A_1 are independent, we know that $\Pr(A_2|A_1) = \Pr(A_2)$. We may then simplify to

$$\Pr(A_2 \cap A_1) = \Pr(A_2)\,\Pr(A_1)$$

which is the result suggested in the above discussion of independent events.

A: the ball has an even number on it
B: the ball has a number greater than 7 on it
C: the ball has a number less than 5 on it

(a) Are *A* and *B* independent? Rigorously justify your answer.

(b) Are *A* and *C* independent? Rigorously justify your answer.

(c) Are *B* and *C* independent? Rigorously justify your answer.

(d) Note that *B* and *C* are exclusive. Comment on this in conjunction with your answer to part (c), and generalize the result.

5.25 Suppose you roll two dice. Let *A* represent the event that the first die is a four, let *B* represent the event that the total showing on the two dice is six, and let *C* represent the event that this total is seven. Two of these events are independent. Which two? Explain.

5.26 You have an unfair coin, which turns up heads two-thirds of the time. You toss this coin three times.

(a) What is the probability of getting zero heads? one head? two heads? three heads? Explain why these probabilities are different than in the case where we have a normal coin (which turns up heads half the time).

(b) If *A* represents the event three heads, and *B* represents the event heads on the first toss, are *A* and *B* independent? Rigorously justify your answer.

(c) If *C* represents the event a heads on the second toss, are *B* and *C* independent? Rigorously justify your answer.

5.27 You draw three cards from a regular deck, with replacement. Find the probabilities of getting:

(a) any ace, any ace, jack of hearts (in that order).
(b) any ace, jack of hearts, any ace (in that order).
(c) any ace, any ace, jack of clubs (in that order).
(d) any two aces and the jack of hearts (in any order).
(e) any two aces and any jack (in any order).

5.28 Suppose that for each of five antibiotics there is a 20% chance that it will be an effective treatment for a particular infection. What is the probability that a person with this infection can then be effectively treated? What assumption did you implicitly make in reaching your answer?

5.29 You pick one card from a regular deck of playing cards.

   (a) Is the event "a king" independent of the event "a heart"?
   (b) Is the event "a king" independent of the event "a royal card" (i.e., jack, queen, king, or ace)?
   (c) Is the event "a heart" independent of the event "a royal card"?

5.30 In the United States in 1986, the median income of males who worked year-round and full-time was $25,894. For similarly defined females, this median was $16,843. Explain why this finding suggests that income (for such workers) was not independent of sex. (*Hint:* What does the median tell us about the distribution of income in a population?) [*Source:* U.S. Bureau of the Census, Current Population Reports, *Money Incomes of Households, Families, and Persons in the United States: 1986,* Series P-60, No. 159 (Washington, DC: U.S. Government Printing Office, 1988), table 27.]

5.31 You toss a fair coin ten times, and it turns up heads each time. What is the probability that if you toss it an eleventh time you will finally get a tails? Explain.

SUMMARY

In this chapter we have presented a discussion of the basics of probability theory. We have paid particular attention to the intuitive foundations of the approach and have tried to discuss the main concepts without belaboring the technical complications.

   Probability theory is ultimately concerned with characterizing patterns of randomness. As we have discussed, randomness is the key element that we must understand further in order to increase the sophistication with which we infer population characteristics from sample information. In the next chapter we shall focus on probability as applied to random variables and probability distributions as a method for characterizing their behavior.

REVIEW PROBLEMS FOR CHAPTER 5

5.32 In a college in which 55% of students are male, 10% of males are on either the basketball team or the tennis team (or both) and 12% of females are on either the basketball team or the tennis team (or both). Moreover, 7% of females are on the basketball team, 1% of females are on both teams, 2% of males are on both teams, and 4% of the males play tennis only.

(a) What is the probability of picking a female, given that we have a tennis player?

(b) What is the probability of picking a student who is on both teams, given that we have a tennis player?

(c) What is the probability of picking a male, given that we have a student who is on neither the basketball nor the tennis team?

(d) Is being on the tennis team independent of sex? Explain.

5.33 Consider a state lottery in which there are two games. In each, six numbers from 0 to 50 are randomly chosen (without replacement, so there can be no duplicates), and you win $10,000,000 if you have a ticket on which your six numbers match the six numbers that were chosen. There are two types of tickets you can purchase. With the first, you pay $1 for a set of six numbers that are randomly generated by computer; with the second, you pay $2 for a ticket on which you choose the six numbers. Which game makes more sense to play?

5.34 After lightning destroyed his barn, farmer Brown rebuilt the barn with confidence that it could not happen again. After all, the chance of lightning hitting in the same place twice is almost zero. Comment on farmer Brown's decision.

5.35 A poker player is dealt a five-card hand.

(a) What is the probability that he or she has four of a kind (i.e., four cards of the same denomination) in this hand?

(b) What is the probability that he or she has a flush (all five cards in the same suit)? [For simplicity, include the possibility of a straight flush (see problem 5.38).]

(c) Suppose that the first three cards are of the same denomination (e.g., three aces, three kings, etc.). What is the probability that this player will wind up with either four of a kind or a full house (three cards of one denomination *and* two of another—e.g., three aces and two kings)?

5.36 Suppose that one-third of new businesses fail in one year. Of those remaining after one year, one-third fail in their second year of operation.

(a) What is the probability that a new business will survive for two years?

(b) What is the probability that a new business will fail in its second year?

(c) Out of a group of five new businesses, what is the probability that at least one will survive two years? Upon what assumption is your answer based?

5.37 A new drug has been developed for the treatment of a particular condition, and it is very effective as a treatment for this condition. There is concern that a possible side effect of this drug is the development of blurry vision.

To investigate the frequency of blurry vision as a side effect, a study was conducted in which a large number of individuals with this disease participated. Half were randomly selected to receive the drug; the remaining half received a placebo. (Patients did not know which they received.) It was found that 3% of individuals receiving the drug complained of vision problems, and 2% of individuals receiving the placebo complained of vision problems.

You are a patient receiving this drug. You experience blurry vision while taking the drug, and question your doctor as to whether this is a likely side effect of the drug or unrelated to the drug. Your doctor cites the above study and concludes that since there was only a 1% increase in the incidence of blurry vision in the above study, it is very unlikely that your blurry vision was caused by the drug.

Comment on your doctor's conclusion.

*5.38 A "straight flush" is a poker hand in which you have five cards in sequence of the same suit (e.g., 3, 4, 5, 6, 7 of hearts). The ace may be either the low or high card, but not both; that is, ace, 2, 3, 4, 5 and ace, king, queen, jack, 10 of the same suit both form straight flushes, but king, ace, 2, 3, 4 does not. What is the probability of receiving a straight flush in a five-card poker hand?

*5.39 A "straight" is a poker hand that is like a straight flush, except that the five cards (though in sequence) need not be of the same suit. What is the probability of getting a straight (other than a straight flush) in a five-card poker hand?

*5.40 Consider a class in which there are 40 students enrolled.

(a) John is a student in this class. What is the probability that at least one other student in the class has John's birthday (day and month, but not necessarily the year)?

(b) What is the probability that, among the 40 students, there are at least two students with the same birthday? [*Hints for both parts:* (1) Assume that each birthday is equally likely; and (2) ignore February 29 and the resulting complications.]

*5.41 At a medical laboratory, the tags labeling the blood vials drawn from three persons (*A, B,* and *C*) have fallen off. The technician, to save time, arbitrarily places the tags back on the vials.

(a) What is the probability that none of the vials is relabeled correctly? One? Two? Three?

(b) Suppose that this test detects the presence of a particular antibody, and for the sake of argument, suppose that *C* has this antibody but *A* and *B* do not. (The patients and lab, of course, do not know this.) What is the probability that no one will get the correct test *result* (even if it is associated with someone else's blood)? One person will? Two persons? Three?

*5.42 A country is about to have a presidential election, but its system for determining the winner is somewhat involved. The country has five provinces (*A, B, C, D,* and *E*) and the candidate who wins a province gets all the "electoral votes" in that province. The winner of the presidential election is then the candidate receiving a majority of the electoral votes. Province *A* has 100 electoral votes, and provinces *B* through *E* have 30 electoral votes each.

Mary has a 60% chance of winning province *A,* a 40% chance of winning *B,* a 30% chance of winning *C,* a 20% chance of winning *D,* and a 10% chance of winning *E.* What is the probability that she will win the election? Upon what assumption does this answer depend?

*5.43 There are four adjacent parking spaces, and two cars will park in (two of) these spaces. Drivers choose their spaces randomly from those that are available (all four spaces are initially available). What is the probability that the two cars will be in adjacent spaces?

Repeat this for five parking spaces. Do you see a pattern? Generalize your answer to the probability that the cars are in adjacent spaces when there are n parking spaces.

APPENDIX TO CHAPTER 5

In this appendix we briefly discuss a result known as Bayes' theorem. We have implicitly used this result in this chapter, and we here formalize the logic developed earlier. [For a further discussion see Freund and Walpole (note 1), pp. 62–66 and Thomas H. Wonnacott and Ronald J. Wonnacott, *Introductory Statistics for Business and Economics,* 4th ed. (New York: John Wiley & Sons, 1990), pp. 93–97 and 583–584.]

Suppose that we have two exclusive events, B_1 and B_2, and one of them must occur. We know the probabilities with which these events occur [$\Pr(B_1)$ and $\Pr(B_2)$] and also the probabilities with which a third event A occurs given $B_1[\Pr(A \mid B_1)]$ or $B_2[\Pr(A \mid B_2)]$.

Bayes' Theorem In the above situation,

$$\Pr(B_1 \mid A) = \frac{\Pr(A \mid B_1)\Pr(B_1)}{\Pr(A \mid B_1)\Pr(B_1) + \Pr(A \mid B_2)\Pr(B_2)} \quad \text{(A5.1)}$$

and

$$\Pr(B_2 \mid A) = \frac{\Pr(A \mid B_2)\Pr(B_2)}{\Pr(A \mid B_1)\Pr(B_1) + \Pr(A \mid B_2)\Pr(B_2)} \quad \text{(A5.2)}$$

To prove Eq. (A5.1), note that

$$\Pr(B_1 \mid A) = \frac{\Pr(B_1 \cap A)}{\Pr(A)}$$

Now

$$\Pr(B_1 \cap A) = \Pr(A \cap B_1) = \Pr(A \mid B_1)\Pr(B_1)$$

which is the numerator of Eq. (A5.1). Note also that

$$A = (A \cap B_1) \cup (A \cap B_2)$$

since either B_1 or B_2 occurs. Since B_1 and B_2 are exclusive, the events $(A \cap B_1)$ and $(A \cap B_2)$ are exclusive. Thus

$$Pr(A) = Pr(A \cap B_1) + Pr(A \cap B_2)$$

$$= Pr(A \mid B_1) Pr(B_1) + Pr(A \mid B_2) Pr(B_2)$$

which is the denominator of Eq. (A5.1). The proof for Eq. (A5.2) follows similarly.

Let us illustrate this theorem in the context of an example from earlier in this chapter. There is a disease which is known to affect 1% of the population. There is a test for the disease; of those with the disease 99% test positive, and of those without the disease 95% test negative. In the text we argued that the probability that an individual with a positive test result does not in fact have the disease is 5/6. Here we confirm this result using Bayes' theorem.

Let B_1 be the event "patient does not have the disease" and B_2 be the event "patient does have the disease." Let A be the event "a positive test result." We know $Pr(B_1) = .99$, $Pr(B_2) = .01$, $Pr(A \mid B_1) = .05$, and $Pr(A \mid B_2) = .99$. If the patient receives a positive test result, the probability that he or she does not have the disease is $Pr(B_1 \mid A)$. By Eq. (A5.1)

$$Pr(B_1 \mid A) = \frac{Pr(A \mid B_1) Pr(B_1)}{Pr(A \mid B_1) Pr(B_1) + Pr(A \mid B_2) Pr(B_2)}$$

$$= \frac{.05 \times .99}{(.05 \times .99) + (.99 \times .01)} = \frac{5}{6}$$

This solution is the same one we arrived at earlier.

Bayes' theorem can be extended to a set of events $B_1, \ldots, B_k$, but we here only present the $k = 2$ case. The intuition behind Bayes' theorem can best be achieved by reviewing the intuitive argument for the above solution that was presented earlier in this chapter.

DISCUSSION QUESTIONS FOR CHAPTER 5

1. Suppose that you toss an allegedly fair coin ten times, and each time it turns up heads. Do you believe that the probability of a heads on the eleventh toss is 1/2? Construct an argument for why it would be, and then construct an argument for why it

might not be. [*Hints:* (1) In either case, successive tosses are independent; and (2) note the use of the word "allegedly."]

2. Suppose we determine that two characteristics of the individuals in a population (e.g., race and poverty) are not independent. What kinds of things might we investigate in order to establish more completely the nature of the relationship between these two variables?

Chapter 6

Probability Distributions

In this chapter, we will build upon the concepts presented in Chapter 5 by introducing and developing the concept of a probability distribution. Probability distributions are interesting and helpful because they provide (1) new methods for calculating certain types of probabilities and (2) often useful models for characterizing the behavior of sample statistics. This second characteristic will be of particular importance as we proceed to develop more sophisticated ways of making inferences in Section III of this book.

DISCRETE PROBABILITY DISTRIBUTIONS

We begin by considering the case of a discrete random variable X. (The notion of a probability distribution in the discrete case is a bit different than in the continuous case, and we must therefore be careful to distinguish between discrete and continuous random variables.) The probability distribution of X is a way of fully characterizing the behavior of this random variable:

Definition of a Discrete Probability Distribution The probability distribution of the discrete random variable X, written $p(X)$, is a listing of the possible values of X and the probabilities with which they occur.

As an example, suppose that we toss a coin once and let X equal the number of heads. The probability distribution of X would then be

X	p(X)
0	.5
1	.5

Notice that in the left column we have listed the possible values of X and in the right column the probabilities with which these individual values of X occur. Formally, for a discrete random variable X, $p(a) = \Pr(X = a)$ where a is a possible value of X.

Notice that if we were to sum $p(X)$ over all the possible values of X we would find that $\Sigma_X p(X) = 1$ (or, just $\Sigma p(x) = 1$).[1] This result means that one of the possible values of X must occur; in the example just considered, we will get either heads or tails in our toss of the coin.

If we were to take this coin and toss it thousands and thousands of times, what would we expect the relative-frequency distribution of X (the number of heads) to look like after all these tosses? Presumably, we would expect the following:

X	f(X)
0	.5
1	.5

Notice that the relative-frequency distribution is the same as the probability distribution. This result should not surprise us once we recall that the notion of probability we are working with is that of a long-run relative frequency.

Remember that in the case of a relative-frequency distribution of a discrete random variable the mean and variance can be calculated by the following formulas:

---

[1] Notice that we write Σ_X, which indicates that the sum is over the possible values of X. The i subscript on the X's is typically omitted here; note that the sum is over the possible values of X and not over individual observations on X (the context in which the i subscript was earlier used).

$$\mu = \sum_X X f(X)$$

$$\sigma^2 = \sum_X (X - \mu)^2 f(X)$$

Given our notion of probability as a long-run relative frequency, we may then define the mean and variance of a discrete random variable (calculated from its probability distribution) as

$$\mu = \sum_X X p(X)$$

$$\sigma^2 = \sum_X (X - \mu)^2 p(X)$$

For the probability distribution considered above (X is the number of heads when we toss a coin once) the mean and variance are

$$\mu = 0(.5) + 1(.5) = .5$$

$$\sigma^2 = (0 - .5)^2(.5) + (1 - .5)^2(.5) = .25$$

The mean and variance are, as before, measures of central tendency and dispersion.

An alternative formula for the variance of a discrete random variable, which is often easier for calculation purposes, is

$$\sigma^2 = \left[\sum_X X^2 p(X) \right] - \mu^2$$

(You are asked in one of the problems following this section to prove that the variance may be rewritten this way.) In the above example,

$$\sigma^2 = 0^2(.5) + 1^2(.5) - (.5)^2 = .25$$

which agrees with our earlier result.

We now turn to an example of a relatively sophisticated discrete probability distribution. This case is helpful for many probability problems, and it will also serve to increase our understanding of the concept of a probability distribution.

PROBLEMS

6.1 The random variable X has the following probability distribution:

X	p(X)
0	.4
1	.3
2	.2
3	.1

Find the mean and variance of X.

6.2 Suppose that I roll a fair die and let X represent the number that shows up. What is $p(X)$? μ? σ^2?

6.3 Consider a family that plans to have three children. Assuming that each child is equally likely to be a boy or girl, find the probability distribution of the number of girls among the three children. Then find the mean and standard deviation of the number of girls.

6.4 Consider an unfair coin that turns up heads one-fourth of the time. I toss the coin two times and let X equal the number of heads. Find the probability distribution of X, and then find its mean and variance. Repeat for the situation of tossing the coin three times.

*6.5 The variance of a discrete random variable is defined as

$$\sigma^2 = \sum_X (X - \mu)^2 \, p(X)$$

Prove that an alternative formula for this variance is

$$\sigma^2 = \left[\sum_X X^2 \, p(X) \right] - \mu^2$$

and illustrate this result using the X from problem 6.1.

BINOMIAL DISTRIBUTION

Suppose that we take a fair die and roll it six times. We can easily determine the probability that all six rolls turn up a "six" by using the technique suggested in our discussion of independence in Chapter 5:

$$\Pr(\text{all sixes}) = \Pr(S_1 \cap S_2 \cap S_3 \cap S_4 \cap S_5 \cap S_6)$$
$$= \Pr(S_1)\Pr(S_2)\Pr(S_3)\Pr(S_4)\Pr(S_5)\Pr(S_6)$$
$$= (\tfrac{1}{6})^6$$

where S_1 is the event a six on the first roll, and so on. Suppose that we now asked: What is the probability of getting exactly two sixes (and therefore four nonsixes) in our six rolls of the die?

This problem is more complicated than the previous one because there is more than one possible arrangement of two sixes (and four nonsixes); in the earlier problem there was only one possible arrangement of six sixes. One possible arrangement (of two sixes and four nonsixes) occurs when the first two rolls are the sixes:

$$\Pr(S_1 \cap S_2 \cap E_3 \cap E_4 \cap E_5 \cap E_6) = (\tfrac{1}{6})^2(\tfrac{5}{6})^4$$

where E_3 represents the event "something else" (i.e., a nonsix) on the third roll, and so on. Another possible arrangement occurs when the last two rolls are the sixes:

$$\Pr(E_1 \cap E_2 \cap E_3 \cap E_4 \cap S_5 \cap S_6) = (\tfrac{5}{6})^4(\tfrac{1}{6})^2$$

Notice that this probability is the same as the probability when the first two rolls gave us the sixes!

There are many different arrangements of two sixes and four nonsixes; each one of them, however, occurs with exactly the same probability given above. To find the probability of getting two sixes and four nonsixes in *any* order, we then simply take this common probability and multiply it by the number of possible arrangements of two sixes and four nonsixes. [We are in essence determining the probability of the union of all the possible arrangements; since the individual arrangements are mutually exclusive, our desired result is achieved by adding their individual probabilities. Alternatively, we multiply the (common) probability of each individual arrangement by the number of possible arrangements.]

But how do we determine the number of possible arrangements? The key to this is a counting rule; this rule is the main innovation (for our purposes) in this section.

In the problem at hand, we have six rolls of the die, and there are two possibilities for each roll: a six and a nonsix. The number of ways of arranging two sixes and four nonsixes in six rolls is

$$\frac{6 \times 5 \times 4 \times 3 \times 2 \times 1}{(2 \times 1) \times (4 \times 3 \times 2 \times 1)} = 15$$

It is difficult to provide intuition for this formula, but there is a story that is helpful. We are trying to determine the number of ways of arranging six objects, two of which are S's and four of which are E's. The total number of ways of arranging six unique objects is

$6 \times 5 \times 4 \times 3 \times 2 \times 1 = 720$: for each of six possibilities for the first there are five for the second; for each of 6×5 for the first two there are four for the third, and so on. In the problem at hand, however, the two S's and the four E's are indistinguishable. That is, if the "first S" is the first roll and the "second S" is the second, this is an outcome that is indistinguishable from the possibility that the "second S" is on the first roll and the "first S" is on the second roll. A similar fact holds for the E's, and this conclusion implies that $6 \times 5 \times 4 \times 3 \times 2 \times 1 = 720$ overstates the number of distinguishable arrangements.

There are 720 possible arrangements if the two S's are distinguishable from each other and the four E's are distinguishable from each other. Since there are 2×1 ways of arranging two unique S's, each distinguishable arrangement (e.g., S's on the first and third is distinguishable from S's on the first two) involving two S's is represented 2×1 times in the 720 figure; we must then deflate 720 by 2×1 in order to arrive at the number of distinguishable arrangements. Similarly, since there are $4 \times 3 \times 2 \times 1$ ways of arranging four unique E's, each distinguishable arrangement involving four E's is represented $4 \times 3 \times 2 \times 1$ times in the 720 figure; we must then also deflate 720 by $4 \times 3 \times 2 \times 1$ in order to arrive at the number of distinguishable arrangements. Thus, given the indistinguishability of S's and E's, the number of arrangements of two S's and four E's is

$$\frac{6 \times 5 \times 4 \times 3 \times 2 \times 1}{(2 \times 1) \times (4 \times 3 \times 2 \times 1)} = 15$$

which is what we had earlier argued to be the case. Thus, the probability of getting two sixes and four nonsixes in any order is

$$15(\tfrac{1}{6})^2(\tfrac{5}{6})^4 = .2009$$

To simplify notation, we typically define the concept of a *factorial*:

$n! = n \times (n-1) \times (n-2) \times \cdots \times 1$ for any nonnegative integer n.

For example,

$$3! = 3 \times 2 \times 1$$

$$2! = 2 \times 1$$

$$1! = 1$$

$(0! = 1$ by convention$)$.

Thus, the number of ways of arranging two S's and four E's in six rolls of the die may be expressed as

$$\frac{6!}{2!4!}$$

To generalize, in a set of n trials each having two possible outcomes ("success" and "failure"), the number of ways of arranging s successes [and $(n - s)$ failures] is

$$\frac{n!}{s!(n - s)!}$$

As an additional illustration, we may ask: How many ways of arranging five successes in six trials are there? There must be six; the one failure can show up on any of the six trials, and the five successes are indistinguishable. Using the formula above would confirm this intuition.

We have in essence already developed the binomial distribution; let us now formalize it.

Definition of the Binomial Distribution   If we have a set of n independent trials, each of which has two possible outcomes (success and failure), the probability distribution of the number of successes s, $p(s)$, is

$$\frac{n!}{s!(n - s)!} \pi^s(1 - \pi)^{n-s}$$

where π is the probability of a success on any individual trial.

Let us justify this. The first term

$$\frac{n!}{s!(n - s)!}$$

is the counting rule described above; this gives us the number of ways of arranging s successes [and $(n - s)$ failures] in n trials. The remainder of the formula,

$$\pi^s(1 - \pi)^{n-s}$$

is the probability of getting s successes and $(n - s)$ failures in a particular arrangement; the s successes occur with probability π each, and the $(n - s)$ failures occur with probability $(1 - \pi)$ each.

This finding is based on the condition that the successive trials are independent.[2]

As an example, suppose that we pick five persons from a population. We know that 15% of the population is "poor." If we let s represent the number of poor persons in our sample, the probability distribution of s is

s	$p(s)$
0	$\dfrac{5!}{0!5!}(.15)^0(.85)^5 = .4437$
1	$\dfrac{5!}{1!4!}(.15)^1(.85)^4 = .3915$
2	$\dfrac{5!}{2!3!}(.15)^2(.85)^3 = .1382$
3	$\dfrac{5!}{3!2!}(.15)^3(.85)^2 = .0244$
4	$\dfrac{5!}{4!1!}(.15)^4(.85)^1 = .0022$
5	$\dfrac{5!}{5!0!}(.15)^5(.85)^0 = .0001$

where the probabilities are rounded to four decimal points. (Note that whether we are sampling with or without replacement is insignificant in this case since the population is very large.)

Since the binomial distribution has many useful applications, values of the probabilities have been tabulated for us so as to save calculation time. Table I (page 382) contains an abridged collection of useful binomial probabilities. Recall that a binomial probability is determined once we know three pieces of information: the number of independent draws n, the probability π of a success on an individual draw, and the number of successes s for which we are determining the probability. In Table I there is a block of rows for values of n from 1 to 10, and one column for each value of π in .05 increments. The probability of a particular number of successes can then be read directly off the table; for example, if n is 4 and π is .3, then $p(2) = .2646$.

---

[2] Remember that the probability of the intersection of independent events is the product of the individual probabilities. If the successive draws are not independent, then the binomial distribution will not apply; the individual draws would then not be characterized as having the same probability of a success.

We present the binomial distribution for several reasons. It gives us a method of evaluating some relatively complicated probabilities that are potentially applicable to economic problems. It also shows how a model of a probability distribution can be developed in a general context and then applied to a wide variety of situations. And by exploring the content of this rather sophisticated model we gain a more complete understanding of what probability and probability distributions are all about. This knowledge is particularly helpful as preparation for our discussion of the more complicated concept of the probability distribution of a continuous random variable.

The vast majority of random variables we encounter in economics, however, are not discrete. Variables such as price, income, wealth, and the money supply are continuous; how do we proceed in such cases?

PROBLEMS

6.6 A hat contains ten marbles, three of which are red. If I pick five marbles from the hat with replacement, what is the probability distribution of the number of red marbles? Find the mean and variance of this probability distribution.

6.7 Suppose that one-third of families in a given city have two cars. If I pick three families at random from this population, what is the probability that I will pick exactly one family that has two cars?

6.8 Suppose that 10% of U.S. families have a vacation home. If I pick ten families at random from the U.S. population, what is the probability that at least one of those picked has a vacation home?

6.9 Suppose that 25% of students at a particular college have SAT scores over 1200. If I pick five students at random from this student body, what is the probability that at least two of those picked have SAT scores over 1200?

6.10 In the United States in 1987, the civilian labor force participation rate for white females age 16 and older was approximately 56%. [Source: *Economic Report of the President: 1989* (Washington, DC: U.S. Government Printing Office, 1989), table B-37.] If I pick four females at random from this population, what is the probability that at least one of those picked is a member of the civilian labor force? At least two?

6.11 In the United States in 1986, 20.47% of individuals age 25 and over had four years or more of college education. [Source: U.S. Bureau of the Census, Current Population Reports, *Money Income of Households, Families, and Persons in the United States: 1986,* Series P-60, No. 159 (Washington, DC: U.S. Government Printing Office, 1988), table 27.] If I pick seven individuals at random from this population, what is the probability that a majority of those picked are college graduates? (For simplicity, round the 20.47% figure to 20%.)

6.12 Suppose that there are two hats. Hat A contains one red and four blue marbles. Hat B contains two red and three blue marbles. If I pick a hat at random, and then four marbles (with replacement) from that hat, what is the probability that I will have two red and two blue marbles?

6.13 Consider a population in which 80% of males and 60% of females are employed. In this population, 55% of individuals are females. If I pick five persons at random from this population, what is the probability that no more than one of those chosen is not employed?

6.14 Consider the binomial distribution in which $n = 4$ and $\pi = .25$.

(a) Tabulate this distribution and calculate its mean μ.

(b) Notice that in this case $\mu = n\pi$. Provide an intuitive explanation for why, in general, the mean of the binomial distribution is $n\pi$.

*6.15 Look at Table I. Notice that for $n = 5$, $p(3)$ when $\pi = .8$ equals $p(2)$ when $\pi = .2$.

(a) Explain verbally why this is the case.

(b) In general, with a set of n independent trials, $p(s_0)$ when $\pi = \pi_0$ always equals $p(n - s_0)$ when $\pi = 1 - \pi_0$, for any values $s = s_0$ and $\pi = \pi_0$. Prove this assertion.

PROBABILITY DISTRIBUTIONS FOR CONTINUOUS RANDOM VARIABLES

In the case of a discrete random variable, our strategy in compiling the probability distribution is to list the possible values of the random

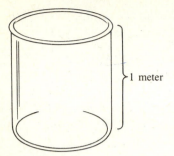

Figure 6.1 A cylindrical water container.

variable and the probabilities with which they occur. Can this strategy be applied to the case of a continuous random variable?

Remember that a continuous random variable is one that can take on any value within some range of values; as an example, the continuous random variable X might take on any of the values between 0 and 1. The first problem that we would encounter if we were to use the same strategy as in the discrete case (list the possible values and the probabilities with which they occur) is that there are an infinite number of possible values of X; how can we then go and list them all?[3]

The second problem, which is more fundamental, is that the strategy used in the discrete case does not allow us to distinguish between different continuous random variables. To illustrate this problem, we must explore more fully the concept of probability as applied to continuous random variables.

Suppose that we have a cylindrical container, into which water is flowing from a tap (see Figure 6.1). The water flows from the tap at a constant rate, and it takes 1 minute for the container to be filled. The cylinder is 1 meter high. Imagine that we stop the water at some time during a minute in which the water is flowing into the container; we do this in a way that is purely arbitrary, and thus each time during the minute is equally likely. X is a random variable that represents how high from the base of the cylinder the water is when we turn off the tap.

Clearly, X can take on any number between 0 and 1 (presuming that we can measure height with infinite accuracy). X is then a

---

[3] Actually, this problem is not insurmountable; it is possible, for example, to consider a discrete random variable that has an infinite number of possible values but where the probabilities can be specified through a particular formula. An example of this would be if $p(X) = (1/2)^X$, for $X = 1, 2, 3, \ldots$. This X is discrete even though there are an infinite number of possible values of it. The "fatal" problem with using the strategy from the discrete case when attempting to describe the behavior of a continuous random variable will be brought out below.

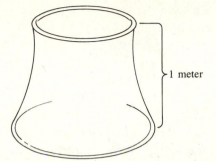

1 meter

Figure 6.2 A noncylindrical water container.

continuous random variable. What is the probability that X equals .1? That is, what is the probability that the water is exactly .1 meter off the ground when we turn the tap off? Presuming that we can measure with infinite accuracy and presuming that there is no tendency to stop the water at one time as opposed to another, this probability must be zero as .1 is but one of an infinite number of equally likely possible outcomes.

What is the probability that X equals .2? This must also be zero, by the same logic. The fact is that for any specific value of X, call it a, $\Pr(X = a) = 0$. So if we were to list the possible values of X and the probabilities with which they occur, all we would have is a long string of zeros!

Note that $\Pr(X = .1) = 0$ does not imply that this event is impossible; it is possible that the water will be exactly .1 meter off the ground when we turn off the tap. The problem is that no matter how many times we play this game, we would never expect this to occur; since .1 is but one of an infinite number of equally likely values of X, its relative frequency is expected to be zero.

Let us be careful about what we are saying here. For a continuous random variable X, it will in general be the case that the probability associated with any specific value of X will be zero. This does not, however, imply that this value is impossible; after all, we know that one of the possible values of X will occur. In the case of a continuous random variable, a probability of zero is not the same as impossibility. For the X from our cylinder example, $X = 2$ is *impossible*, whereas $X = .1$, though possible, has a probability of zero.[4]

While a general characteristic of continuous random variables is that the probability of each specific value is 0, this does not imply

---

[4] Of course, in this example we would say $\Pr(X = 2) = 0$; that is, impossibility does imply a zero probability (it is the converse of this statement that does not hold).

that all continuous random variables have the same pattern of behavior.[5] In Figure 6.2 we consider another water receptacle that has a height of 1 meter; this receptacle, however, is not a perfect cylinder because it is wider at the bottom than at the top. It is still the case that if the water were to run out of the tap for 1 minute, the container would be full. The experiment is still to stop the water flow at some arbitrary time during the minute where one time is as likely as any other. If Y equals the height of the water off the bottom of the container (note *not* the fraction of the container by volume which is full) then Y is a continuous random variable. But Y is a different continuous random variable than the X considered above (in the case of the cylindrical container). To see this, let us compare $\Pr(X < .5)$ to $\Pr(Y < .5)$. Given that the first receptacle is a perfect cylinder, $\Pr(X < .5) = .5$, but given that the second is wider at the bottom that at the top, $\Pr(Y < .5) > .5$. But if we were to list the possible values of the random variables and the values with which they occur, these different random variables would appear identical (all probabilities would be zero)! The strategy from the discrete case is then unsatisfactory for the continuous case, since X and Y behave differently.

Our argument as to why X and Y are different suggests a solution to this problem. The difference between X and Y is that values "near zero" are for Y disproportionately likely, whereas for X this is not the case. We see the difference between X and Y by thinking in terms of *ranges of values* as opposed to specific values of the random variables. Looking at ranges of values, as opposed to specific values, is then the way to approach the characterization of the pattern of behavior of a continuous random variable.

How does this observation help us to characterize the behavior of X? Note that we can make statements such as $\Pr(0 \le X \le .1) =$

---

[5] There are random variables that are neither purely continuous nor purely discrete, but are in essence a combination of the two. As an example, consider a set of cars, 80% of which are running and 20% of which are not. Let X be the revolutions per minute of the engine of a car randomly chosen. This random variable equals 0 if the car chosen is one of those which is not running, but is (say) at least 500 if the car is running. X then equals 0 with probability .2; this part of the behavior of the random variable is of a discrete nature. But the range of possibilities $X \ge 500$ is of a continuous nature; X is then a random variable that is in part discrete and in part continuous. An economic example of such a variable might be wage incomes of individuals randomly chosen from the population as a whole; those not working all have a wage income of $0. The wage incomes of those with employment, however, would be a continuous random variable.

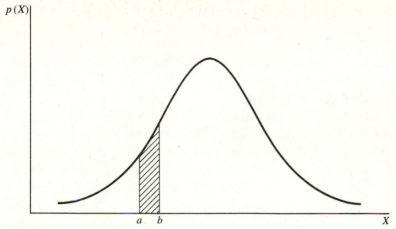

Figure 6.3 A probability density function. Area under the curve represents probability.

.1 or $\Pr(0 \leq X \leq .01) = .01$. In fact, for any range of values of X, we can determine the probability that X takes on a value within that range as the height of the range as a fraction of the total height of the cylinder (or just the height of the range since the height of the cylinder is one).[6] And as long as we are referring to a range of values (as opposed to a specific value), the problem of zero probability discussed above does not present itself. The lost detail due to the necessity of referring to ranges of values is not of practical significance, however, for we can make the range as small as we want (so that variation within the range is not interesting for practical purposes) and still have a meaningful way of characterizing the distribution of probability.

Our general way of representing the probabilistic behavior of a random variable is to construct a curve known as the *probability density function;* Figure 6.3 presents an example of one such curve. On the horizontal axis we represent the values of X (this is a different X than that associated with the water in the receptacle example which we were just considering), and on the vertical axis we measure $p(X)$—the probability density function—which we define as follows:

Definition of the Probability Density Function   The probability density function $p(X)$ is a function that is constructed such that,

---

[6] Calculating probabilities for Y would be more complicated, but it could still be done by using the geometric characteristics of the receptacle to determine how the height of the water changes as we progress through a minute.

for any interval (a, b), the probability that X takes on a value in that interval is the total area under the curve between a and b.

In Figure 6.3, $p(X)$ is drawn such that the shaded area represents $\Pr(a \leq X \leq b)$; $p(X)$ is drawn so that this would be the case regardless of the placement of a and b.

Note that $p(X)$ has a different meaning in the case of a continuous random variable than in the case of a discrete one. In particular, $p(a)$ does *not* (in the continuous case) represent $\Pr(X = a)$; we know that such a probability is zero for any continuous random variable. It is important then to affirm that, with a probability density function, probability is represented not by the height of the curve but rather by the area under the curve. Although $p(X)$ represents an inherently different thing in the continuous case than in the discrete case, in both cases it is our method for describing the behavior of a random variable. Do not let the fact that we use the notation $p(X)$ in both the discrete and continuous cases fool you into thinking that the meanings are the same!

The probability density function and its interpretation affirm for us several things regarding continuous random variables. The probability that X takes on a value between a and a, that is, the probability that X equals a, is the area under $p(X)$ between a and a. This is the area of a line segment, and since a line segment has no width, this area (and thus the probability) is zero. Notice that $\Pr(a < X < b)$ is the same as $\Pr(a \leq X \leq b)$ since the probability that X is exactly a or b is zero. (Thus, with a continuous random variable there is really no need to be careful about weak or strict inequality signs for the purpose of determining a probability.) It is always the case that $p(X) \geq 0$; if $p(X)$ were ever negative (i.e., if the probability density function were ever below the horizontal axis) then a negative probability would be possible (some area "under" the curve is negative in this case). Notice finally that the total area under $p(X)$, over the range of possible X's, must be equal to one; this indicates that one of the possible values of X must occur and that a probability cannot exceed one.

Probability density functions come in many shapes, some rise to a peak and then decline, as in Figure 6.3. Others are flat throughout, such as the one drawn in Figure 6.4. Some extend over the entire number line from $-\infty$ to ∞ (e.g., Figure 6.3), whereas others extend over a limited range (e.g., Figure 6.4). But in all cases, probability is represented by the area under the curve.

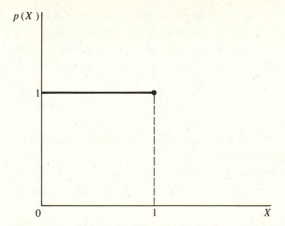

Figure 6.4 The probability density in the case of the cylindrical water container.

EXERCISE ■

Remember the X and Y variables used earlier in this chapter having to do with the models of water running into a cylinder? X was the height of the water when the receptacle was as in Figure 6.1 and Y was the height of the water when the receptacle was as in Figure 6.2. In Figure 6.4 we have in fact drawn the probability density function of this X. Can you see why? Given that in Figure 6.4 we see the probability density function of X, how would you then draw the probability density function for Y? ■

Some students may find it helpful to think about the probability density function in terms of some concepts from calculus. This is really nothing more than a way of reexpressing what we have already communicated in words. (If you are unfamiliar with calculus, just view the following discussion as a restatement of what we developed earlier.) For the probability density function $p(X)$ it is the case that

$$\Pr(a \le X \le b) = \int_a^b p(X) \, dX$$

The symbol $\int$ is the integral sign; this is a mathematical symbol that instructs us to "operate" on what follows in a particular way. Having done this operation, the result is the area under $p(X)$ between (the limits of integration) a and b. Integration is then a mathematical technique for calculating the area under a curve between two points, and this is exactly the problem at hand when determining a probability for a continuous random variable from its probability density function.

We shall not actually calculate any integrals. The reason is that the $p(X)$ functions most useful for applied purposes have integrals that are extremely difficult to determine. In any case, they have been precalculated by others in a way that makes them generally applicable and available. (We shall consider this in more detail below in our discussion of the normal distribution.) The concept of the integral can nevertheless be useful in order to help in understanding a probability density function without getting into the mechanics of how integrals are actually calculated.

We can refer to the mean and variance of a probability density function. Remember that in the case of a *discrete* random variable X, we said

$$\mu = \sum_X X p(X)$$

$$\sigma^2 = \sum_X (X - \mu)^2 p(X)$$

In the case of a continuous random variable Y, we calculate the mean and variance as follows:

$$\mu_Y = \int_{-\infty}^{\infty} Y p(Y) \, dY$$

$$\sigma_Y^2 = \int_{-\infty}^{\infty} (Y - \mu_Y)^2 p(Y) \, dY$$

Notice that these formulas are similar to those for the discrete case, except that the $\int$ sign has replaced the Σ sign and the term dY has been added at the end. The dY term simply indicates that we are allowing Y to vary over the limits of integration ($-\infty$ to ∞). The conclusion then is that the $\int$ sign is much like a Σ sign. This analogy is in fact correct; the process of integration may be loosely thought of as the process of summation over a continuous region, point by point.

We shall not consider problems for the calculation of the mean and variance of a probability density function. It is important to note, however, that the mean and variance have the same interpretation in the continuous case as in the discrete case.

We shall now consider two of the many possible general shapes of probability density functions and suggest some real-world phenomena with which they are likely to be associated.

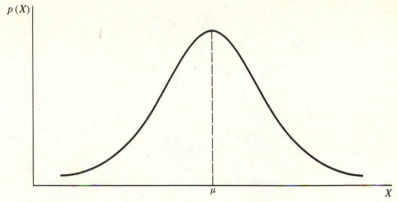

Figure 6.5 A symmetric probability density function.

EXAMPLE ■

In Figure 6.5 we present a very important general shape. This $p(X)$ suggests a random variable whose pattern of behavior is symmetric; that is, the decline in $p(X)$ as we move away from the peak is the same regardless of in which direction we move (right or left). Ranges of values of X that are close to μ occur more frequently than those of equal width that are further away from μ. The distribution tails off gradually in each direction toward the horizontal axis; it may or may not reach it, although it is typically modeled as approaching this axis asymptotically.[7]   ■

It turns out that this general shape is very important, as we shall discuss below; IQ scores (where $\mu = 100$) might be modeled with this shape.[8]

EXAMPLE ■

Figure 6.6 contains the second general type of shape, which is very important in economics. The curve emanates from the origin; $p(X) = 0$ for all $X < 0$. The curve tails off asymptotically toward the horizontal axis as X gets bigger and bigger. A classic

---

[7] See our discussion of the normal distribution below.

[8] IQ scores, of course, do not range from $-\infty$ to ∞ but rather from 0 to some maximum number (perhaps 200 or so). The shape in Figure 6.5 is then a model of—an approximation of—the distribution of IQ scores, and the key characteristic of this shape is its symmetry as opposed to its range.

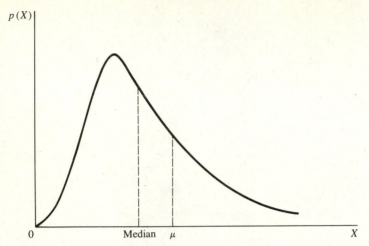

Figure 6.6   An asymmetric probability density function.

example of this is wage incomes of employed individuals. One's wages, of course, cannot be negative, but they can be very high; the extent to which they can exceed the middle is far greater than the extent to which they can be below the middle. This situation immediately suggests an asymmetry, which is a key distinguishing characteristic of its general shape. Because of this asymmetry, the median and the mean are different, as illustrated.[9]   ■

PROBLEMS

6.16 X is a continuous random variable which has the following probability density function:

$$p(X) = 0 \qquad \text{for } X < 0$$

$$p(X) = 1/6 \qquad \text{for } 0 \le X \le 1$$

$$p(X) = 4/6 \qquad \text{for } 1 < X \le 2$$

$$p(X) = 1/6 \qquad \text{for } 2 < X \le 3$$

$$p(X) = 0 \qquad \text{for } X > 3$$

[9] One of the discussion questions at the end of this chapter asks you to explain why.

(a) Draw a graph of this probability density function.
(b) Verify that the total area under the probability density function is one.
(c) Find $\Pr(0 < X < 3/4)$.
(d) Find $\Pr(X > 2.5)$.
(e) Find $\Pr(0 < X < 1.5)$.
(f) Find $\Pr(.5 < X < 2.5)$.

*6.17 Suppose that $Y = 2X$, where X is the random variable described in problem 6.16.

(a) What is $p(Y)$?
(b) Verify that the total area under $p(Y)$ equals one.

MODELS OF PROBABILITY DENSITY FUNCTIONS

There are many different models of probability density functions that are quite useful in applied analysis. In economics, most continuous random variables can be characterized by either the symmetric shape suggested in Figure 6.5 or the asymmetric shape suggested by Figure 6.6. Typically, we take our knowledge or intuition regarding the general shape and then use a specific model of a probability density function to characterize the behavior of the random variable.[10]

In the case of the symmetric shape of Figure 6.5, we most commonly use a model known as the *normal* distribution. This probability density function is fundamentally important in applied statistical analysis, for reasons that will be discussed in Chapter 9. We now move to a discussion of this distribution.[11]

THE NORMAL DISTRIBUTION

The normal distribution is a probability density function of the general shape given by Figure 6.5. The distribution reaches a peak at

[10] This approach does not imply that we believe that the probability density function being used perfectly characterizes the random variable (although, of course, we hope it is at least a reasonable approximation); we often use the probability density function to model the behavior of the random variable so that we may further understand its characteristics and behavior.

[11] In the case of the general shape of Figure 6.6, several different probability density functions are potentially useful; the *chi-square* distribution, discussed in the appendix to this chapter, is one such distribution.

the mean and tails off asymptotically toward the horizontal axis in both directions. A key characteristic of this distribution is that it is *symmetric about the mean;* and therefore the right-hand side is the mirror image of the left-hand side. The formula for the normal probability density function is

$$p(X) = \frac{1}{\sigma\sqrt{2\pi}}\, e^{-.5[(X-\mu)/\sigma]^2}$$

where e is the base of the natural logarithm scale (about 2.72) and π is the ratio of a circle's circumference to its diameter (about 3.14). If we were to plot this function, we would get the general shape given in Figure 6.5; we omit this exercise since it is done most easily with calculus.

Given this formula for $p(X)$, we can determine the probability associated with any range of values of X by calculating the appropriate area under the curve. As discussed above, this process would involve taking the integral of this function and evaluating this integral over the appropriate range. Given the commonness of this distribution (and the difficulty of the integration) this work has effectively been done for us. That is, by the appropriate use of specific probabilities under a *special* normal distribution we can indirectly determine probabilities regarding *any* normal distribution in which we might be interested.

Notice that there is a different normal distribution for each value of μ and for each value of σ^2; any of these distributions, of course, would have the general symmetric shape we have suggested. As μ increases, the curve is "moved" further to the right; as σ^2 increases, its distribution becomes more disperse.

Regardless of the mean and variance, however, all normal distributions are directly related to each other. Consider the following special case of the normal distribution.

Definition of the Standard Normal Distribution   A random variable Z is said to be *standard normal* if it is normally distributed with a mean of 0 and a variance of 1.

In general, if X is normal with mean μ and variance σ^2 we abbreviate this result as

$$X \sim N(\mu, \sigma^2)$$

where the first number in the parentheses represents the mean and the second number represents the variance. Thus, $Z \sim N(0, 1)$, and Z is the usual notation for a standard normal random variable. Its shape is given in Figure 6.7; as we can see, it is symmetric about its mean of 0.

Table II (page 388) contains probabilities associated with the standard normal distribution. The numbers in the body of this table are, for some value a, $Pr(Z \geq a)$. The shaded area in Figure 6.7 is a geometric representation of such a probability. The value of a to one decimal place is given along the left-hand margin of Table II; by moving to the appropriate column of the table we add a second decimal place to the value of a. For example, $Pr(Z \geq 1.35) = .0885$.

Notice that once the value of a is above about 3 the probabilities are very small. This observation suggests that the right-hand tail beyond 3 contains a very small mass of probability. But what about the other side of the distribution? And in general, how do we calculate probabilities that are of a form other than $Pr(Z \geq a)$ for some positive number a?

It turns out that the information contained in Table II is all that we need in order to be able to calculate whatever probability regarding a random variable Z that is of interest to us. The keys to this are remembering general characteristics of probability density functions (e.g., the total area under the curve is one) and specific properties of the standard normal distribution (specifically, that it is symmetric about its mean of zero). To this end, let us consider a set of examples to illustrate some general procedures.

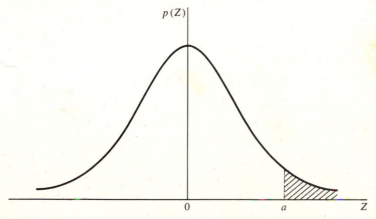

Figure 6.7 The standard normal distribution.

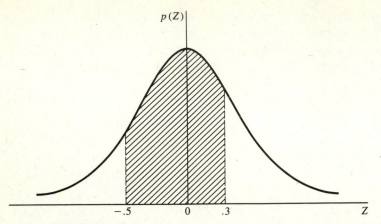

Figure 6.8   Calculating probabilities with the standard normal distribution.

EXERCISES ■

1. Determine $\Pr(Z < -1)$. Since the standard normal distribution is symmetric about 0, we know $\Pr(Z < -1) = \Pr(Z > 1)$.[12] Thus, $\Pr(Z < -1) = .1587$.

2. Determine $\Pr(Z < 1)$. The complement rule tells us that $\Pr(Z < 1) = 1 - \Pr(Z \geq 1)$. But, we know that $\Pr(Z \geq 1) = .1587$. Thus, $\Pr(Z < 1) = 1 - .1587 = .8413$.

3. Determine $\Pr(-.5 < Z < .3)$. Consider Figure 6.8. The probability of interest has been shaded in this figure. Since the total area under the curve (from $-\infty$ to ∞) is 1, then

$$\Pr(Z \leq -.5) + \Pr(-.5 < Z < .3) + \Pr(Z \geq .3) = 1$$

   or

$$\Pr(-.5 < Z < .3) = 1 - \Pr(Z \leq -.5) - \Pr(Z \geq .3)$$

   or

$$\Pr(-.5 < Z < .3) = 1 - \Pr(Z \geq .5) - \Pr(Z \geq .3)$$
$$= 1 - .3085 - .3821$$
$$= .3094$$

4. Determine $\Pr(1 < Z < 2)$. The key here is to notice that

$$\Pr(1 < Z < 2) = \Pr(Z > 1) - \Pr(Z \geq 2)$$

---

[12] Remember that, since Z is continuous, $\Pr(Z > 1) = \Pr(Z \geq 1)$.

(Can you draw a figure like Figure 6.8 to illustrate this?) Therefore,

$$\Pr(1 < Z < 2) = .1587 - .0228 = .1359 \qquad \blacksquare$$

Of course, few if any real-world variables have the standard normal distribution. Even if a variable is normally distributed, chances are that it has a mean different from 0 or a variance different from 1. Of what use then is the standard normal table (Table II)? The key is that any random variable that has a normal distribution can be transformed into a standard normal random variable, and the probability of interest regarding the original variable can then be determined by reference to the newly created standard normal.

Suppose $X \sim N(\mu_X, \sigma_X^2)$, where we have added the subscripts to distinguish X's mean and variance from those of other random variables, and either $\mu_X \neq 0$ or $\sigma_X^2 \neq 1$ (or both). There is some probability regarding X that we want to determine. Our strategy is to transform X into a Z and then find the probability regarding X by determining the equivalent probability regarding Z.

Consider a new random variable Y, which is defined as follows:

$$Y = \frac{-\mu_X}{\sigma_X} + \left(\frac{1}{\sigma_X} \right) X$$

Let us note several characteristics of Y.

First, this is an example of a one-to-one function, that is, for each value of one variable there is one and only one value of the other. The implication is that for a range of values of X there is some unique range of values of Y that occurs only when the original range of values of X occurs; we can then determine the probability that X falls within that range by figuring out the probability that Y falls within the corresponding range.

Second, notice that this is an example of a linear function. As we saw in Chapter 2, if the function is of the form

$$Y = a + bX$$

then Y is a linear function of X. In this case, $a = -\mu_X/\sigma_X$ and $b = 1/\sigma_X$. This result is important because it suggests the relationship between the means and variances of X and Y.

Third, if X is normally distributed, then Y too will be normally distributed. Although the formal proof of this statement is beyond the scope of this book, we can provide some intuition into why this result is not surprising. The intercept term $-\mu_X/\sigma_X$ simply "picks up" the distribution of X and moves it to the left or right; clearly, this operation does not change its general shape. The coefficient $1/\sigma_X$ changes the degree of dispersion in the distribution (i.e., makes it skinnier or fatter) but it still does not change its general shape. So we expect that if X is normal, Y will also be normal, *but will have a different specific normal distribution.*

The reason we have made Y the particular function of X given above is that this transformation *always* creates a Y that has a standard normal distribution. Since Y is a one-to-one function of X, for a range of values of X (for which we are trying to determine the probability) there is some corresponding range of values of Y (which then occurs with equal probability). Since Y is a linear function of X, we can determine the mean and variance of Y from our information concerning the mean and variance of X. And since Y is normal, if the transformation is such that $\mu_Y = 0$ and $\sigma_Y^2 = 1$, we can then evaluate the probability of interest based upon the standard normal distribution (since Y will in fact then be a Z). We must show now that, with this transformation, Y will always have such a mean and variance.

Remember that if $Y = a + bX$ then $\mu_Y = a + b\mu_X$ and $\sigma_Y^2 = b^2\sigma_X^2$. Thus

$$\mu_Y = \frac{-\mu_X}{\sigma_X} + \left(\frac{1}{\sigma_X}\right)\mu_X = 0$$

and

$$\sigma_Y^2 = \left(\frac{1}{\sigma_X}\right)^2 \sigma_X^2 = 1$$

Since Y is normal, we then know that $Y \sim N(0, 1)$; that is, Y is a Z. If we transform X in this way, we can then determine the probability that X falls within the range of interest by determining the probability that $Y(Z)$ falls within the corresponding range (and the standard normal table will help us here). The *standardization transformation*

$$Z = \frac{-\mu_X}{\sigma_X} + \left(\frac{1}{\sigma_X}\right)X$$

can be rewritten as

$$Z = \frac{X - \mu_X}{\sigma_X}$$

which is the form in which it is typically presented (often the subscripts are omitted for simplicity).

In general, if we take a normal and subtract its mean and divide by its standard deviation we will change the normal into a standard normal. Let us consider an example.

EXAMPLE ■

Suppose that scores from a standardized test are normally distributed with a mean of 50 and a standard deviation of 10. What proportion of the population will have a score over 65?

If X represents an individual's score on the test, the question is: What is $Pr(X > 65)$? (Remember that probability represents relative frequency, so although the question is posed as a relative frequency, it can be implicitly rephrased as a probability.) We know that

$$Pr(X > 65) = Pr\left(\frac{X - \mu}{\sigma} > \frac{65 - 50}{10}\right)$$

[All we've done is to reexpress the inequality by first subtracting the same thing ($\mu = 50$) from both sides and then dividing both sides by the same thing ($\sigma = 10$).] But since $(X - \mu)/\sigma$ is standard normal we may then say

$$Pr\left(\frac{X - \mu}{\sigma} > \frac{65 - 50}{10}\right) = Pr(Z > 1.5)$$

which the standard normal table tells us is .0668.   ■

In this example we evaluated the probability regarding X by calculating an equivalent probability for Z. The reason why we only need a standard normal table (as opposed to one table for every normal distribution) is that all normals can be reexpressed as standard normals by the standardization transformation.

Problems

6.18 Suppose that IQ scores within a population are normally distributed with a mean of 100 and a standard deviation of

10. If we pick a person at random from this population, what is the probability that he or she has an IQ score

(a) above 130?
(b) below 110?
(c) between 90 and 120?
(d) between 85 and 95?
(e) between 115 and 125?
(f) less than 90?
(g) above 85?
(h) above 110 or below 90?

6.19 X is a normally distributed random variable with mean 10 and variance 24.

(a) Find $\Pr(X > 14)$.
(b) Find $\Pr(8 < X < 20)$.
(c) Find the probability that X takes on a value that is at least 6 away from its mean.
(d) Suppose that X^* is defined such that $\Pr(X > X^*) = .10$. What is X^*?
(e) If $Y = 4 + 6X$, find $\Pr(Y > 16)$. [*Hint:* What is $p(Y)$?]

6.20 Suppose that the amount of reserves of a commercial bank at the end of the business day is a normally distributed random variable with mean $10 million and standard deviation $2 million. Required reserves are $7 million. On what fraction of days does the bank not meet its reserve requirements? To what value would the standard deviation need to be reduced in order to reduce this probability to .020?

6.21 Incomes in a population are approximately normally distributed with mean $15,000 and standard deviation $5000. The poverty level of income is $8000.

(a) What is the poverty rate?
(b) What income is the 90th percentile?

6.22 The interest rate in an economy is a random variable which has a normal distribution with mean 8% and standard deviation 2%. Suppose that the level of investment is given by the formula

$$I = 100,000,000 - 5,000,000r$$

where r is the interest rate (as a percent) and I is the level of investment (in dollars). Find the probability that the level of investment exceeds $50,000,000.

6.23 For the standard normal random variable Z, if $\Pr(|Z| > a) = .05$, what is the value of a?

SUMMARY

In this chapter we have discussed the concept of a probability distribution in a wide variety of contexts, as an extension and further development of the concept of probability. The next step is to consider the concept of probability as it relates to more than one random variable, which becomes important as we try to determine the relationships that may or may not exist between random variables. This is the topic of Chapter 8. But first, in Chapter 7 we undertake a digression to consider some technical issues that will facilitate our future analyses.

REVIEW PROBLEMS FOR CHAPTER 6

6.24 Consider a very large population in which 70% of individuals are "green" and 30% of individuals are "purple." We know that 40% of greens are college graduates, and 60% of purples are college graduates.

(a) If I pick five persons at random, what is the probability that I will get three greens and two purples?

(b) If I pick five greens at random, what is the probability that I will get at least two college graduates?

(c) If I pick five persons at random, what is the probability that I will get exactly four college graduates?

6.25 Suppose that the rate of return on stocks is normally distributed with a mean of 9% and a standard deviation of 3%. If I pick five stocks at random, what is the probability that at least two of them will have a return of more than 12%?

6.26 Consider two carpenters, Dick and Jane, both of whom are employed making bookcases. The amount of time needed by Dick to make a bookcase is a normally distributed random variable with mean 6 hours and standard deviation 2 hours.

The amount of time needed by Jane to make a bookcase is a normally distributed random variable with mean 5 hours and standard deviation 4 hours.

(a) For each carpenter, find the probability that a given bookcase will be finished in 8 hours or less.

(b) Explain why the above probability for Dick is greater than that for Jane, even though Jane is, on average, faster.

6.27 Consider a country in which there are six provinces: *A, B, C, D, E,* and *F.* Province *A* has 50 electoral votes, whereas the other provinces have 20 electoral votes each. A presidential candidate wins the election if he or she receives a majority of electoral votes. All of a province's electoral votes go to the candidate receiving the greatest number of votes therein. Candidate John has a 40% chance of winning province *A,* and a 30% chance of winning in each of the other five provinces. What is the probability that John will be the overall winner?

6.28 Let us reconsider problem 6.27. Suppose that everything is the same except that the electoral votes of the individual provinces have changed. They are now as follows: 60 for *A,* 50 for *B,* and 30 for each of *C, D, E,* and *F.* What now is the probability that John will be the overall winner?

6.29 I pick a five-card hand from a normal deck of playing cards, without replacement. What is the probability that I will have "three of a kind"? That is, what is the probability that I will have exactly three cards of the same denomination among my five cards? (For simplicity, the remaining two cards may be, but need not be, of the same denomination.)

*6.30 Consider a state lottery in which three numbers are drawn randomly (without replacement) from the integers from 1 to 100. When purchasing my lottery ticket I choose five numbers (without replacement) from the integers from 1 to 100, and I win $10,000 if among my five numbers I have the three numbers chosen in the lottery.

(a) If I hold a single ticket, what is the probability that I will win the lottery?

(b) What can you say about the likely "profit" for the state, assuming that tickets cost $1 each and that the only way to win a prize is as described above?

APPENDIX TO CHAPTER 6

In Figures 6.5 and 6.6 we suggested two general shapes of probability density functions that are often encountered in economics. The shape of Figure 6.5 is the basis for the normal distribution. The general shape given in Figure 6.6 is similar to that modeled by the chi-square distribution.[13] This distribution has many important uses at a conceptual level for the material we shall be developing, and the purpose of this appendix is to present a way of thinking about this distribution that will be helpful in our future work.

We begin with an assertion:

Assertion Suppose that $Z_1, Z_2, \ldots Z_k$ are mutually independent random variables all of which have a standard normal distribution. If

$$C = Z_1^2 + Z_2^2 + \cdots + Z_k^2$$

then C has a chi-square distribution with k "degrees of freedom;" that is, $C \sim \chi^2_{(k)}$.[14]

The chi-square distribution has a parameter (k) conventionally known as the "degrees of freedom"; this parameter can be equal to any positive integer. Thus, in order to specify which chi-square distribution is relevant, the degrees of freedom must be specified. The above assertion suggests how the degrees of freedom is in essence the number of independent sources of random variability, and thus the term "degrees of freedom."

To determine the reason a random variable C as defined above would have the general shape of Figure 6.6, let us note first that C cannot be negative since it is a sum of squares (each of which must be nonnegative). Thus $p(C) = 0$ for all $C < 0$. The fact that C is a sum of squares also causes its distribution to be skewed to the right.

Because we are adding nonnegative numbers to form C, the typical value of C increases as we add more nonnegative numbers. That is, as the degrees of freedom increases, the chi-square distribution moves to the right (see Figure A6.1).

Tables giving various important values of the chi-square dis-

---

[13] There are other models of distributions that also have this general shape.
[14] See, for example, John E. Freund and Ronald E. Walpole, *Mathematical Statistics,* 4th ed. (Englewood Cliffs, NJ: Prentice-Hall, 1987), p. 285.

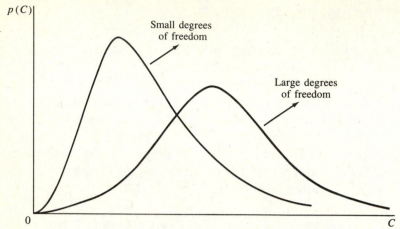

$p(C)$

Small degrees
of freedom

Large degrees
of freedom

0

C

Figure A6.1 Degrees of freedom and the chi-square distribution.

tribution are available; see, for example Table III (page 389). This table provides, for various degrees of freedom (rows) and probabilities α (columns), that value of the chi-square variable C, call it C^*, such that $\Pr(C \geq C^*) = \alpha$. For example, if $C \sim \chi^2_{(2)}$ then $\Pr(C \geq .0100) = .995$.

CHAPTER 6 DISCUSSION QUESTIONS

1. Explain why the mean exceeds the median in Figure 6.6. Use this discussion to explain why the mean and the median are equal in Figure 6.5.

2. Provide an intuitive explanation for why we adopt the convention that $0! = 1$ and *not* $0! = 0$. (*Hint:* Consider the number of ways of arranging n successes in n trials.)

3. Discuss the relationship between the relative-frequency histogram and the probability density function as ways of describing the behavior of a continuous random variable.

Chapter 7

Technical Background on Mathematical Expectation

In this chapter we define the concept of mathematical expectation and suggest some of the important uses to which it can be put. Mathematical expectation gives us a new way of viewing a mean, and this new perspective will be very useful in future chapters.

DEFINITION OF MATHEMATICAL EXPECTATION

We begin with the definition of the concept of an expected value:

Definition of Expected Value   The expected value of a random variable X, written $E(X)$, is the mean of its probability distribution.

We have discussed at length the concept of a mean. With this definition we are not so much presenting a new concept as we are thinking about an old one (the mean) in a slightly different way. First, we here make it clear that when we pick an observation at random from a population, in some sense we "expect" it to be about equal to the mean of that population. If the average height of an American male is 5 feet 9 inches, and if I were to pick a male at random from this population, then I would expect his height to be about 5 feet 9 inches. In addition, the E symbol gives us a mathematical operator that is particularly useful as we begin to consider in depth problems regarding means of functions of a random variable (or random variables).

To pursue this second matter further, suppose that we have a discrete random variable X and some function of it $g(X)$. The function $g(X)$ describes a new random variable [we might say $Y = g(X)$], and it turns out that

$$E[g(X)] = \sum_X g(X)\,p(X) \qquad (7.1)$$

where the sum is over the possible values of X and where $p(X)$ represents X's probability distribution. Notice that in the special case $g(X) = X$, it is implied that

$$E[g(X)] = E(X) = \sum_X X\,p(X)$$

which confirms our formula for the mean of a probability distribution [since $E(X) = \mu$].

To motivate Eq. (7.1), suppose that X is a discrete random variable, and that

$$g(X) = X^2 + 5$$

The distribution of X is as follows:

X	p(X)	g(X)
−1	.2	6
0	.3	5
1	.1	6
2	.4	9

where we have included, for completeness, the values of $g(X)$. What is $E[g(X)]$? Clearly, there are three possible values of $g(X)$: 5, 6, and 9. The probabilities for these three values are, respectively, .3 [$g(X) = 5$ if and only if $X = 0$], .3 [$g(X) = 6$ if X is either −1 or 1], and .4[$g(X) = 9$ if and only if $X = 2$]. That is, the probability distribution of the new random variable $g(X)$ is

g(X)	p[g(X)]
5	.3
6	.3
9	.4

The mean of $g(X)$ is

$$E[g(X)] = \sum_{g(X)} g(X)\, p[g(X)] = 5(.3) + 6(.3) + 9(.4) = 6.9$$

(where the sum is over the possible values of $g(X)$, and where $p[g(X)]$ is the probability distribution of this random variable). If, on the other hand, we used Eq. (7.1), we would calculate

$$E[g(X)] = \sum_{X} g(X)\, p(X)$$

$$= 6(.2) + 5(.3) + 6(.1) + 9(.4) = 6.9$$

where the sum is over the possible values of X. Looking at these two different calculations, it is clear that the only difference is that in the former the two probabilities associated with $g(X) = 6$ are combined, whereas in the latter they are not.[1]

An expression similar to Eq. (7.1) holds for the case of a continuous random variable; this expression, however, is based upon the probability density function and integration. In essence, the values of $g(X)$ are "weighted" by the probabilities associated with the original variable X.[2]

We are often, however, interested in a special type of function— a linear one. Things are easier in this case, as we shall now discuss.

EXPECTED VALUES OF LINEAR FUNCTIONS

Consider the case where $g(X)$ is a linear function of a discrete random variable X; that is,

$$Y = g(X) = a + bX$$

where a and b are constants, and we have given the new random

---

[1] We have purposely chosen an example in which $g(X)$ is not a one-to-one function to illustrate the general applicability of Eq. (7.1).

[2] For the continuous case,

$$E[g(X)] = \int_{-\infty}^{\infty} g(X)\, p(X)\, dX$$

where $p(X)$ is the probability density function of X.

variable the name Y. What can we say about $E(Y)$? Clearly, $E(Y) = E[g(X)]$, and therefore

$$E(Y) = E[g(X)] = \sum_X (a + bX) p(X)$$

where the sum is over the possible values of X. Note, however, that

$$\sum_X (a + bX) p(X) = \sum_X [a\,p(X) + bX\,p(X)]$$

$$= a \sum_X p(X) + b \sum_X X\,p(X)$$

We know that $\Sigma_X p(X) = 1$. In addition, we know that $\Sigma_X X\,p(X) = E(X)$. Thus we conclude that

$$E(Y) = E[g(X)] = a + b\,E(X)$$

This relation is our first indication of the *linearity* property of E; here it has been illustrated in the context of a function of one discrete random variable (in Chapter 8 this will be extended to the context of two or more random variables). Note that this relation is merely another way of representing a result from Chapter 3 concerning the mean of a linear function. The implication is that the expectations operator may be pushed through a linear function of one random variable until it hits the random variable.

Though proven above for a discrete random variable, this result applies for a continuous X as well. This should not be surprising, for when we derived the results regarding means of linear functions in Chapter 3 we did not make the distinction between discrete and continuous random variables.

EXAMPLE ■

Suppose that we consider an economy where the consumption function is given by

$$C = 100 + .9Y$$

where C represents aggregate consumption and Y represents disposable income. Suppose that the determination of disposable income is at least in part due to some sort of random process, but that we know $E(Y) = 2000$. Then

$$E(C) = 100 + .9\,E(Y) = 100 + .9(2000) = 1900 \qquad ■$$

PROBLEMS

7.1 Suppose that we toss a fair coin two times, and let X equal the number of heads. If $Y = 2X^2 + 3$, find $E(Y)$ using Eq. (7.1).

7.2 Suppose that I roll a fair die, and let X equal the number that shows. If $Y = (1/X) + 2$, find $E(Y)$.

7.3 Suppose that the demand function for a good is given by the equation

$$Q^d = 50 - .2P$$

where Q^d is quantity demanded and P is price. If the mean price is \$100, find the mean quantity demanded for this good.

7.4 Suppose that I roll a fair die, and let X equal the number that shows. If $Y = 5 + 4X$, find $E(Y)$ by

(a) using Eq. (7.1).

(b) determining $p(Y)$ and then using this distribution to determine $E(Y)$.

(c) exploiting the linearity property of the expectations operator.

EXPECTED VALUES AND VARIANCES

For a discrete random variable X, we define its variance as

$$\sigma^2 = \sum_X (X - \mu)^2 \, p(X)$$

(see our discussion in Chapter 6). Note that the right-hand side of this equation is nothing more than $E[g(X)]$ when $g(X) = (X - \mu)^2$. This result suggests that we may rewrite the variance as

$$\sigma^2 = E[(X - \mu)^2]$$

or, alternatively, as

$$\sigma^2 = E\{[X - E(X)]^2\} \tag{7.2}$$

[since $E(X) = \mu$]. The variance, then, may be thought of as the expected value of a function of X. This finding is true for the continuous case as well. Note that if we were calculating the variance based upon observation of a complete population, we would form (for both the discrete and continuous cases)

$$\sigma^2 = \frac{1}{m} \sum_{i=1}^{m} (X_i - \mu)^2$$

where the sum is over the individual observations in the population. This result illustrates that the variance is just the mean of a function of X_i, where that function is $(X_i - \mu)^2$. Given the equivalence of μ and $E(X)$, we are provided with an alternative perspective for the origin of Eq. (7.2).

It is important to note that in Eq. (7.2) we *first* square $[X - E(X)]$ and *then* take the expected value, and not the other way around. Although E may be "pushed through" a linear function $(a + bX)$ it cannot be pushed through a nonlinear function—for example, $[X - E(X)]^2$. Thus

$$E\{[X - E(X)]^2\} \neq \{E[X - E(X)]\}^2$$

See the first discussion question at the end of this chapter for a further investigation of this issue.

Equation (7.2) may be rewritten as

$$\sigma^2 = E(X^2) - [E(X)]^2 \tag{7.3}$$

that is, the variance is the expected square of X minus the square of its expected value. This expression suggests, among other things, how if $Y = X^2$, the mean of Y can be calculated directly from information regarding the mean and variance of X. (Can you see how?) In a problem at the end of this chapter you will be asked to derive Eq. (7.3).

SUMMARY

In this chapter we have illustrated the concept of mathematical expectation and the usefulness of the operator E. This operator contains a set of instructions that might be thought of as "take the mean of what follows," and it will facilitate future work in which we must determine the means of functions of one or more random variables.

Our discussions of probability and random variables have been limited so far to the case of one random variable. We now turn to scenarios involving two (or more) random variables and a discussion of probability theory in that context—an important step toward our ultimate development of sophisticated inferences in Section III of this book.

REVIEW PROBLEMS FOR CHAPTER 7

7.5 Suppose that we roll a fair die and let X equal the number that shows. Calculate $E\{[X - E(X)]^2\}$ using Eq. (7.1) and then show that this result equals the variance of X calculated in the typical way (see Chapter 6).

7.6 Using the X from problem 7.5, illustrate that

$$E(X^2) \neq [E(X)]^2$$

7.7 Using the X from problem 7.5, illustrate that

$$E(1/X) \neq 1/E(X)$$

*7.8 Prove that the variance of a random variable X,

$$\sigma^2 = E\{[X - E(X)]^2\}$$

can be rewritten as

$$\sigma^2 = E(X^2) - [E(X)]^2$$

and illustrate this equivalency using the data from problem 7.5.

*7.9 Suppose that X is a random variable with a mean of 5 and a variance of 9. If $Y = 2X^2 + 4$, find the mean of Y. [*Hint:* Rewrite $E(X^2)$ as a function of the mean and variance of X.]

DISCUSSION QUESTIONS FOR CHAPTER 7

1. In this chapter we asserted that

$$E\{[X - E(X)]^2\} \neq \{E[X - E(X)]\}^2$$

Provide (1) an argument for why the right-hand side above is always 0 and (2) an argument that (1) is sufficient to prove the inequality.

2. Provide an intuitive as well as an algebraic justification for why $E(c) = c$ for any constant c.

Chapter 8

Distributions of More Than One Random Variable

In Chapter 6 we discussed probability distributions as a way of describing the behavior of a random variable. The next step is to use the concepts of probability and probability distributions as a way of characterizing the mutual behavior of two or more random variables. Such a characterization allows us to investigate whether the random variables are related, and if they are, to begin to quantify that relationship.

We begin this chapter by developing a theoretical framework for characterizing the behavior of two random variables. Then we consider some methods for quantifying the relationship between random variables. We end with some comments regarding functions of two or more random variables.

JOINT PROBABILITY DISTRIBUTIONS

Remember that in the case of a discrete random variable X, the probability distribution $p(X)$ is a listing of the possible values of X and the probabilities with which they occur.[1] A joint probability

---

[1] We shall for the moment focus on the discrete case; the extension to the continuous case will be suggested later.

distribution (for the discrete case) takes this notion and extends it to the situation of two random variables; the joint distribution characterizes not only the behaviors of the two variables individually but also the ways in which their behaviors are interrelated. Let us develop the formal notion of a joint probability distribution within the context of a simple example.

Consider a family planning to have three children. Suppose that the probability of a boy for each child is .5. Let X equal the number of boys among the three children, and let Y be a second random variable that is equal to 1 if the first child is a boy and 0 if this child is a girl.

There are eight possible outcomes for the three children, and for each of these outcomes we list the values of X and Y:

Outcome	X	Y
BBB	3	1
BBG	2	1
BGB	2	1
GBB	2	0
BGG	1	1
GBG	1	0
GGB	1	0
GGG	0	0

where, for example, *BBB* represents the event boy on child 1, boy on child 2, and boy on child 3. From the assumptions above, we know that these eight outcomes each occur with a probability of 1/8.

Definition of a Joint Probability Distribution   The joint probability distribution of X and Y (for discrete random variables) is a listing of the possible pairs of values of X and Y and the probabilities with which they occur.

In the problem above we would have

(X, Y)	p(X, Y)
(3, 1)	1/8
(2, 1)	2/8
(2, 0)	1/8
(1, 1)	1/8
(1, 0)	2/8
(0, 0)	1/8

where $(3, 1)$ represents the event $[(X = 3) \cap (Y = 1)]$, and $p(3, 1)$ represents the probability with which that event occurs (i.e., $\Pr[(X = 3) \cap (Y = 1)]$). In general,

$$p(a, b) = \Pr[(X = a) \cap (Y = b)]$$

where a and b are possible values of the corresponding random variables.

The joint probability distribution is generally easier to interpret when it is represented in tabular form. In the above example, we can represent the joint distribution as follows:

X \ Y	0	1
0	1/8	0
1	2/8	1/8
2	1/8	2/8
3	0	1/8

The numbers along the top margin represent the possible values of Y, and the numbers along the left margin represent the possible values of X. The number in each cell of the table represents the probability associated with the corresponding pair of values of X and Y. The 1/8 in the top left corner of the body of the table is then $p(0, 0)$; that is, $\Pr[(X = 0) \cap (Y = 0)]$. This table is a convenient and informative way of representing the joint probability distribution because from it we gain a visual description of how probability is distributed across the values of X and Y.

Notice that the sum of the joint probabilities is one. (This is not surprising. Can you explain why?) This fact is often written as

$$\sum_X \sum_Y p(X, Y) = 1$$

We use two Σ signs as we have two random variables; the Σ_X indicates that we sum allowing X to vary over all its possible values, and the Σ_Y indicates that we sum allowing Y to vary over all its possible values. This complicated expression simply means that we sum up what follows $[p(X, Y)]$ allowing (X, Y) to vary over all the possible pairs of values of the random variables. In the context of the tabular representation of $p(X, Y)$, we are summing over the cells in the body of the table.

A joint probability distribution gives us complete information about the behavior of the individual random variables as well as the interaction between them. Let us begin to illustrate this concept using the X and Y variables discussed above.

Suppose that we are interested in $\Pr(X = 0)$ regardless of the value of Y. How can we determine this from the information at hand? Intuitively, $X = 0$ occurs if we are somewhere in the first row of the joint probability distribution table—that is, $(X, Y) = (0, 0)$ or $(0, 1)$. So the probability that $X = 0$ should just be $1/8 + 0$— that is, $p(0, 0) + p(0, 1)$. In general, to determine the probability that X equals a particular value, we add up the probabilities in that row of the table.

Formally, the event $(X = 0)$ can be written as follows:

$$(X = 0) = [(X = 0) \cap (Y = 0)] \cup [(X = 0) \cap (Y = 1)]$$

This equation is simply a formal way of indicating that the event $(X = 0)$ can be decomposed into two possibilities: X equals 0 when Y equals 0, and X equals 0 when Y equals 1. The advantage of this expression is to suggest that the event $(X = 0)$ may be rewritten as the union of two mutually exclusive events $([(X = 0) \cap (Y = 0)]$ and $[(X = 0) \cap (Y = 1)])$. Since the probability of the union of exclusive events is the sum of their individual probabilities, we know that

$$\Pr(X = 0) = \Pr[(X = 0) \cap (Y = 0)] + \Pr[(X = 0) \cap (Y = 1)]$$

or

$$\Pr(X = 0) = p(0, 0) + p(0, 1)$$

Similarly, we can say that

$$\Pr(X = 1) = p(1, 0) + p(1, 1)$$

and, in general,

$$\Pr(X = a) = p(a, 0) + p(a, 1)$$

or

$$\Pr(X = a) = \sum_{Y} p(a, Y)$$

If we refer to the univariate distribution of X as $p(X)$, we can say that

$$p(X) = \sum_{Y} p(X, Y)$$

[For a given X we add $p(X, Y)$, allowing Y to vary over its possible values.]

Similarly, to determine the probability that Y equals a particular value, we add up the probabilities in that column of the table. Algebraically,

$$p(Y) = \sum_{X} p(X, Y)$$

[For a given Y we add $p(X, Y)$, allowing X to vary over its possible values.]

These distributions are often represented in the margin of the joint distribution table as follows:

X \ Y	0	1	$p(X)$
0	1/8	0	1/8
1	2/8	1/8	3/8
2	1/8	2/8	3/8
3	0	1/8	1/8
$p(Y)$	4/8	4/8	

When we calculate $p(X)$ and $p(Y)$ in this manner, we refer to them as the *marginal distributions* of X and Y. But it is important to recognize that they are nothing more than the simple univariate probability distributions that we discussed in Chapter 6. [If we were to calculate the univariate distribution of the number of boys (X) without even thinking about Y, we would get $p(X)$, as shown in the table. A similar statement holds for Y.] What is different about this context is that the individual random variables have been originally represented in conjunction with each other. The advantage of this joint representation, of course, is that we can exploit it to begin to consider what, if any, relationships exist between the random variables.

The first step for us in addressing this issue is to consider an extension of the concept of conditional probability (developed earlier in the context of events) to the situation of two random variables. In the X and Y example considered above, suppose that we knew that $Y = 0$. What could we then say about the possible values of X, and the probabilities with which they occur, *given* that $Y = 0$?

We know, given that $Y = 0$, that the probabilities associated with the four values of X must sum to 1; after all, we know that one of these values will occur. We know that $\Pr(X = 3 | Y = 0) = 0$

since the value $(0, 0)$ has a probability of 0 ($Y = 0$ "rules out" $X = 3$). We then know, given $Y = 0$, that there are three possible values of X (0, 1, and 2). We also know (given $Y = 0$) that $X = 0$ occurs with the same likelihood as $X = 2$, and $X = 1$ occurs with a likelihood twice that. Since the conditional probabilities must sum to one, this implies

X	$p(X \mid Y = 0)$
0	1/4
1	2/4
2	1/4
3	0

(we include $X = 3$ for completeness), where $p(X \mid Y = 0)$ is our way of representing the *conditional probability distribution* of X given that $Y = 0$.

EXERCISE ■

Determine $p(Y \mid X = 1)$, using the logic described above.   ■

Let us generalize this intuition. Formally, the definition of conditional probability tells us that

$$\Pr[(X = a) | (Y = b)] = \frac{\Pr[(X = a) \cap (Y = b)]}{\Pr(Y = b)}$$

where a and b are possible values of the random variables. This is a restatement of the definition of conditional probability (Chapter 5) in the context of the two events $(X = a)$ and $(Y = b)$. Since this statement must be true for each possible value of X and Y (that is, for all relevant a's and b's) we can then say that

$$p(X | Y) = \frac{p(X, Y)}{p(Y)}$$

for all possible pairs of values of X and Y.

Let us use this result to determine the conditional probability distribution of X given that Y equals 0 (which we earlier calculated using our intuition):

X	$p(X \mid Y = 0)$
0	$\dfrac{p(0, 0)}{p_Y(0)} = \dfrac{1/8}{4/8} = 1/4$

1        $$\frac{p(1, 0)}{p_Y(0)} = \frac{2/8}{4/8} = 2/4$$

2        $$\frac{p(2, 0)}{p_Y(0)} = \frac{1/8}{4/8} = 1/4$$

3        $$\frac{p(3, 0)}{p_Y(0)} = \frac{0}{4/8} = 0$$

(where in $p_Y(0)$ we have added the Y subscript for clarity). This result is, of course, consistent with our earlier one.

The conditional probability distribution provides us with some information concerning the relationship between X and Y. Notice, for example, that $p(X \mid Y = 0)$ is different from the unconditional $p(X)$; this observation suggests that there is some sort of relationship between the values of X and Y. Before discussing further the process of learning about the relationship between two random variables from their joint probability distribution, let us consider a real-world example utilizing the notion of conditional probability.

Suppose that we pick a person at random from the U.S. population. Let X represent the individual's race; X may equal "white," "black," or "other."[2] Let Y represent the individual's poverty status; Y may equal "poor" or "not poor." In the United States in 1986, the joint distribution of X and Y was[3]

X \ Y	Poor	Not poor	p(X)
White	.093	.755	.848
Black	.038	.083	.121
Other	.005	.026	.031
p(Y)	.136	.864	

The conditional distributions of Y would be as follows:

[2] X and Y are examples of categorical random variables; the values they take on are not uniquely linked to particular numerical values. We could, of course, arbitrarily assign numerical values to coincide with the possible values of the random variables as long as we remember that this assignment is arbitrary.

[3] These data are derived from U.S. Bureau of the Census, Current Population Reports, *Poverty in the United States: 1986,* Series P-60, No. 160 (Washington, DC: U.S. Government Printing Office, 1988), table 1.

Y	$p(Y \mid X = \text{white}) = \dfrac{p(\text{white}, Y)}{p_X(\text{white})}$
Poor	$\dfrac{.093}{.848} = .110$
Not poor	$\dfrac{.755}{.848} = .890$

Y	$p(Y \mid X = \text{black}) = \dfrac{p(\text{black}, Y)}{p_X(\text{black})}$
Poor	$\dfrac{.038}{.121} = .314$
Not poor	$\dfrac{.083}{.121} = .686$

Y	$p(Y \mid X = \text{other}) = \dfrac{p(\text{other}, Y)}{p_X(\text{other})}$
Poor	$\dfrac{.005}{.031} = .161$
Not poor	$\dfrac{.026}{.031} = .839$

These distributions give us the relative likelihoods of "poor" versus "not poor" for each of the three groups. The conditional probabilities of "poor" are just the poverty rates for the different racial groups. We see from these distributions that the incidence of poverty is lowest among whites and highest among blacks; this finding then gives us important information about the relationship between race and poverty in the United States.

For further investigation of issues regarding the relationship between two random variables, let us now reconsider the concept of independence.

INDEPENDENCE

Recall that independent events (as discussed in Chapter 5) are ones for which knowledge that one has occurred does not in any way influence our evaluation of the probability that the other will occur. To extend this notion to random variables, we say that independence occurs if this lack of influence extends over all the possible values of X and Y.

Consider the following joint probability distribution of X and Y:

X \\ Y	0	1	$p(X)$
0	1/18	2/18	3/18
1	2/18	4/18	6/18
2	2/18	4/18	6/18
3	1/18	2/18	3/18
$p(Y)$	6/18	12/18	

If we did not know X we would say that Y takes on a value of 0 with probability 6/18 and a value of 1 with probability 12/18. Now, suppose that we know that X is 0. What would we then say about the probability distribution of Y (that is, what is the conditional distribution of Y given that $X = 0$)? If $X = 0$, the conditional probabilities of Y are 1/3 for 0 and 2/3 for 1. These are equivalent to the unconditional probabilities for Y! In fact, no matter what the value of X is, Y equals 0 with probability 1/3 and 1 with probability 2/3. (Verify this.) Clearly, knowing the value of X does not in any way influence our evaluation of the probabilities associated with Y. We should then expect that Y is independent of X.

Independence, of course, is a mutual property; that is, if Y is independent of X then X should be independent of Y. Can we verify this? The unconditional probabilities for X are 3/18 (for $X = 0$), 6/18 (for $X = 1$), 6/18 (for $X = 2$), and 3/18 (for $X = 3$). Suppose we know that $Y = 0$. The conditional probabilities for X (given $Y = 0$) would be 1/6 for 0, 2/6 for 1, 2/6 for 2, and 1/6 for 3; this distribution is the same as the unconditional one just mentioned. If you were to calculate the conditional distribution of X given that $Y = 1$, you would find the same thing. (Verify this statement.) Clearly the distribution of probabilities for X is invariant to the information we have concerning Y, implying that X is independent of Y. But since (as we have just illustrated) independence is a mutual property, we would typically say simply that X and Y are independent.

The above argument suggests that X and Y are independent if the conditional and unconditional distributions of X are the same, regardless of the value of Y upon which the probabilities are conditioned. That is,

$$p(X|Y) = p(X)$$

for each possible pair of values of X and Y. Since

$$p(X|Y) = p(X, Y)/p(Y)$$

the condition for independence may be rewritten as

$$p(X, Y) = p(X)p(Y)$$

for each possible pair of values of X and Y. This result suggests the following definition:[4]

Definition of Independent Random Variables   Two discrete random variables X and Y are independent if and only if $p(X, Y) = p(X)p(Y)$ for each possible pair of values of X and Y.

For random variables to be independent, then, the value of the joint distribution must be equal to the product of the marginals for each cell in the table. For example, for $(0, 0)$ we ask:

Does $p(X, Y) = p(X)p(Y)$?

That is, does $p(0, 0) = p_X(0) p_Y(0)$?

Does $1/18 = (3/18) \times (6/18)$? Yes!

In order to show that random variables are independent, we must show that such an equality holds for each cell in the table (that is, for each possible pair of values of X and Y); only then are the random variables independent. If there is one (or more) cell(s) for which the equality does not hold, then the random variables are not independent; in this case, knowing the value of one random variable at least sometimes influences our evaluation of a probability regarding the other. (Note that showing that random variables are independent requires demonstration of an equality for each cell in the joint probability distribution table.)

EXERCISE ■

In the problem just discussed, show that X and Y are independent.   ■

---

[4] A similar definition exists for continuous random variables, where $p(X, Y)$, $p(X)$, and $p(Y)$ represent probability density functions.

EXAMPLE ■

Consider the following joint distribution:

X \ Y	1	2	p(X)
1	.2	.3	.5
2	.3	.2	.5
p(Y)	.5	.5	

Are X and Y independent?

Consider $(1, 1)$. Is $p(1, 1) = p_X(1)\, p_Y(1)$? That is, is $.2 = .5 \times .5$? No! We have therefore found a cell for which $p(X, Y) \neq p(X)\, p(Y)$. This result is sufficient to show that it is not the case that $p(X, Y) = p(X)\, p(Y)$ for *each* cell. Thus X and Y are not independent. Note that to show a lack of independence, this lack of equality need only be shown for one cell; demonstrating a lack of independence is thus generally less work than demonstrating independence [where $p(X, Y) = p(X)\, p(Y)$ must be demonstrated for *each* cell].■

Notice that all we have really done here is to extend the concept of independence from individual events (as in Chapter 5) to random variables (which represent collections of events). Investigating whether or not random variables are independent is a first way of determining whether they are related. But if they are related, *how* are they related? We now consider how we might answer this question.

The above discussion has focused on joint distributions, marginal distributions, conditional distributions, and independence for the case of discrete random variables. In the case of continuous random variables, we must alter the definition of $p(X, Y)$; certainly, $\Pr[(X = a) \cap (Y = b)]$ in the continuous case will equal zero for any a and b. This problem is addressed by basing $p(X, Y)$ in the continuous case on ranges of values of X and Y as opposed to individual values, and by thinking of probabilities as areas under a joint probability curve. Inasmuch as this discussion requires calculus, we do not pursue it further but merely note the intuitive similarity to the discrete case.

PROBLEMS

8.1 Suppose that I toss a fair coin three times. X is a random variable equal to the number of heads among these three tosses, and Y is a random variable equal to 1 if the first toss is heads and 0 otherwise.

(a) Find $p(X, Y)$.

(b) Are X and Y independent? Fully justify your answer.

(c) Provide a verbal explanation of your answer to part (b).

8.2 Consider three universities: Harvard, Columbia, and NYU. Harvard and Columbia are in the Ivy League, but NYU is not. Columbia and NYU are in New York City, but Harvard is not. You have a large population of students, of whom one-third attend each school. You pick two students from this population at random. X is a random variable that equals the number of students chosen who go to school in New York City. Y is a random variable that equals the number of students chosen who attend an Ivy League institution.

(a) Find $p(X, Y)$.

(b) Are X and Y independent? Fully justify your answer.

(c) Provide a verbal explanation of your answer to part (b).

8.3 Redo problem 8.2 presuming that in your population of students 50% attend NYU, 20% attend Harvard, and 30% attend Columbia.

8.4 Consider the following joint distribution of X and Y:

X \ Y	1	2	3
1	.1	.2	0
2	.1	0	.2
3	0	.1	.3

(a) Find the marginal distributions of X and Y.

(b) What is the conditional distribution of X given that Y equals 2?

(c) What is the conditional distribution of Y given that X equals 3?

(d) Are X and Y independent? Fully justify your answer.

8.5 Consider the following joint distribution of X and Y:

X \ Y	0	1
0	.2	.6
1	.05	.15

(a) Find the marginal distributions of X and Y.

(b) What is the conditional distribution of X given that Y equals 0?

(c) What is the conditional distribution of Y given that X equals 1?

(d) Are X and Y independent? Fully justify your answer.

8.6 I toss a fair coin twice. X is a random variable that equals 1 if the first toss is a heads, and 0 otherwise. Y is a random variable that equals 2 if the second toss is a heads, and 1 otherwise. Find the joint distribution $p(X, Y)$.

8.7 Repeat problem 8.6 for an unfair coin that has a probability of heads of one-third.

8.8 The distribution of Y is

Y	$p(Y)$
0	.3
1	.4
2	.3

and the distributions of X conditional on Y are as follows:

X	$p(X \mid Y = 0)$	X	$p(X \mid Y = 1)$	X	$p(X \mid Y = 2)$
0	.4	0	.5	0	.3
1	.4	1	.4	1	.4
2	.2	2	.1	2	.3

Find $p(X, Y)$ and $p(X)$.

8.9 Suppose that $p(X, Y)$ is

X \ Y	1	2	3
1	.1	.2	.1
2	0	.3	.1
3	0	0	.2

Find the mean and variance of X.

8.10 Suppose that $p(X, Y)$ is

X \ Y	0	1
1	.2	.3
2	.1	.1
3	.2	.1

(a) Find $E(X \mid Y = 0)$, the expected value of X given that Y equals 0.

(b) Find $E(X \mid Y = 1)$.

(c) What do your answers to parts (a) and (b) suggest regarding whether or not X and Y are related?

(d) Complement your answer to part (c) by formally determining whether or not X and Y are independent.

COVARIANCE

We illustrate the concept of covariance with an example. Suppose there are two random variables X and Y, and we observe the entire population of each. These variables occur as pairs; for example, X_i might be individual i's education and Y_i his or her income. We plot the pairs of values of the two random variables and arrive at Figure 8.1. [Ignore, for the moment, the dashed lines emanating from $E(X)$ and $E(Y)$ and the markings 1, 2, 3, 4.] If we look at this graphical representation of the two random variables, it becomes clear that some sort of relationship is present; big values of X tend to be associated with big values of Y. The relationship is certainly not a perfect one (as we can see by the lack of a stylized functional relationship); however, some sort of a relationship is clearly present.

In order to develop the concept of the covariance, it is useful to reinterpret the above observation regarding these variables. We plot (in Figure 8.1) $E(X)$ along the horizontal axis and draw a vertical line emanating from that point; any observation to the right of this line is one for which $X > E(X)$. Similarly, we plot $E(Y)$ along the vertical axis and draw a horizontal line emanating from that point; any observation above this line is one for which $Y > E(Y)$. We have essentially divided the graph into four regions or quadrants. In quadrant 1, $X > E(X)$ and $Y > E(Y)$; in quadrant

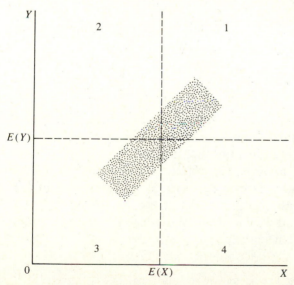

Figure 8.1 Observed values of positively correlated random variables.

2, $X < E(X)$ and $Y > E(Y)$; in quadrant 3, $X < E(X)$ and $Y < E(Y)$; and in quadrant 4, $X > E(X)$ and $Y < E(Y)$.

In this case there is a tendency for our observed points to be in quadrant 1 or quadrant 3; points in quadrants 2 or 4 occur much less frequently. That is, there is a tendency for $X > E(X)$ to be associated with $Y > E(Y)$ (quadrant 1) and for $X < E(X)$ to be associated with $Y < E(Y)$ (quadrant 3). *There is a tendency for above-average values of X to be associated with above-average values of Y and for below-average values of X to be associated with below-average values of Y.* This last statement is a way of restating, in a more specific context, our earlier observation that big values of X tend to be associated with big values of Y.

Clearly, if $X > E(X)$ then $X - E(X) > 0$, and if $Y > E(Y)$ then $Y - E(Y) > 0$. Similarly, if $X < E(X)$ then $X - E(X) < 0$, and if $Y < E(Y)$ then $Y - E(Y) < 0$. Since most of the observations are in quadrants one and three, the most likely combinations are

$$X - E(X) > 0 \quad \text{and} \quad Y - E(Y) > 0$$

$$X - E(X) < 0 \quad \text{and} \quad Y - E(Y) < 0$$

Notice that in both quadrant 1 and quadrant 3,

$$[X - E(X)][Y - E(Y)] > 0$$

Of course, in quadrants 2 or 4 we find

$$[(X - E(X)][Y - E(Y)] < 0$$

Now, suppose that we form the sum

$$\sum_{i=1}^{m} [X_i - E(X)][Y_i - E(Y)] \frac{1}{m} \tag{8.1}$$

(where m is the number of observations and we are summing over all observations).[5] In the problem at hand, this sum is clearly positive, as the observations where $[X_i - E(X)][Y_i - E(Y)] > 0$ clearly outweigh those for which this product is negative.

Imagine now that we had a situation as in Figure 8.2. Here, the observations suggest a negative relationship between X and Y. Notice that in this case the preponderance of observations are in quadrants 2 and 4, and we should then expect that the summation in Eq. (8.1) would be negative.

[5] The purpose of the $1/m$ factor will soon be apparent.

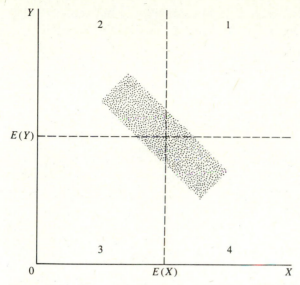

Figure 8.2   Observed values of negatively correlated random variables.

Clearly, not all examples will be as clear-cut as these are. Nevertheless, by calculating the result of Eq. (8.1) and noting its sign we can determine *quantitatively* whether there is a tendency for a positive or negative linear relationship (or no relationship at all) between the random variables.

The above analysis has been in the context of a situation in which we observe the entire populations of two random variables. Let us now elaborate on this within the context of a joint probability distribution of two discrete random variables.

Equation (8.1) is in the context of observations; if we were to put it into the context of a joint probability distribution of two discrete random variables, the comparable formula would be

$$\sum_X \sum_Y [X - E(X)][Y - E(Y)] \, p(X, Y) \qquad (8.2)$$

where the summation is now over the possible pairs of values of X and Y—not the individual observations as in Eq. (8.1). The key to seeing the connection between Eqs. (8.1) and (8.2) is to remember that probability is a long-run relative frequency, and the relative frequency of each observation is $1/m$. Equation (8.2) is a way of expressing the *covariance* between two discrete random variables.

Definition of the Covariance   The covariance between two discrete random variables X and Y, written cov (X, Y), is

$$\sum_X \sum_Y [X - E(X)][Y - E(Y)]\, p(X, Y)$$

EXAMPLE ■
Consider the following joint probability distribution:

X \\ Y	0	1
0	.1	.2
1	.1	.3
2	.2	.1

Note that $E(X) = 1$ and $E(Y) = .6$. The covariance would be

$$\text{cov } (X, Y) = (0 - 1)(0 - .6)(.1) + (0 - 1)(1 - .6)(.2)$$
$$+ (1 - 1)(0 - .6)(.1)$$
$$+ (1 - 1)(1 - .6)(.3)$$
$$+ (2 - 1)(0 - .6)(.2)$$
$$+ (2 - 1)(1 - .6)(.1)$$
$$= -.10$$

This result suggests that there is a negative relationship between X and Y. [Can you provide an intuitive explanation for this by looking at the distribution of $p(X, Y)$ in the above table?]
■

The covariance has been described above in the context of a joint distribution of two discrete random variables. It can also be discussed (for both the discrete as well as the continuous cases) as an expected value, and at times this approach is the most useful way of viewing the situation. To this end, we know that in the case of a single discrete random variable X,

$$E[g(X)] = \sum_X g(X)\, p(X)$$

In the case of a function of two discrete random variables X and Y and a function $g(X, Y)$, it turns out that

$$E[g(X, Y)] = \sum_X \sum_Y g(X, Y)\, p(X, Y)$$

which is most easily viewed as a generalization of the one-variable case. It is therefore suggested that

$$\text{cov}(X, Y) = \sum_X \sum_Y [X - E(X)][Y - E(Y)]\, p(X, Y)$$

may be rewritten as

$$\text{cov}(X, Y) = E\{[X - E(X)][Y - E(Y)]\} \qquad (8.3)$$

That is, the covariance is the expected value of a particular function of X and Y. Equation (8.3) is also valid in the case of continuous random variables, although $E[g(X, Y)]$ in that case depends on integration and a joint probability density function.

Equation (8.3) may be rewritten as

$$\text{cov}(X, Y) = E(XY) - E(X)\, E(Y) \qquad (8.4)$$

that is, the covariance is the expected product of X and Y minus the product of their expected values. [Note that this result implies that $E(XY)$ is generally different from $E(X)E(Y)$, reminding us of the fact that the expected value sign E is not a nonlinear operator.] You will be asked to prove this result in a problem at the end of this section.

Equation (8.4) suggests that, for discrete X and Y, the covariance may be calculated from the expression

$$\text{cov}(X, Y) = \left[\sum_X \sum_Y XY\, p(X, Y) \right] - E(X)\, E(Y) \qquad (8.5)$$

which is easier for calculation purposes than Eq. (8.2) because we no longer need to calculate deviations from means.

EXERCISE ■

In the numerical example above, using Eq. (8.5) to calculate the covariance yields

$$\text{cov}(X, Y) = 0(0)(.1) + 0(1)(.2) + 1(0)(.1) + 1(1)(.3)$$
$$+ 2(0)(.2) + 2(1)(.1) - 1(.6)$$
$$= -.1$$

confirming our earlier result.   ■

When $\text{cov}(X, Y) > 0$, we say that X and Y are *positively correlated;* when $\text{cov}(X, Y) < 0$, we say that X and Y are *negatively correlated;* and when cov $(X, Y) = 0$, we say that they are *uncorrelated.* The covariance and its sign are a way of determining whether a type of relationship exists between the random variables, and in what direction (positive or negative) this relationship goes.

We have now considered another way (besides independence) of investigating relationships between random variables. How do the concepts of independence and the covariance compare as different ways of investigating whether and to what extent such relationships exist? One difference is that the investigation of whether random variables are independent or not tells us only whether or not there is a relationship. The covariance, on the other hand, tells us the direction of that relationship. The covariance and independence, however, seek to uncover somewhat different relationships between random variables; let us next move to a discussion of this issue.

PROBLEMS

8.11 For the joint distribution $p(X, Y)$ of problem 8.4, calculate cov (X, Y).

8.12 For the joint distribution $p(X, Y)$ of problem 8.5, calculate cov (X, Y).

8.13 For the joint distribution $p(X, Y)$ of problem 8.9, calculate cov (X, Y).

8.14 For the joint distribution $p(X, Y)$ of problem 8.10, calculate cov (X, Y).

8.15 Consider the joint distribution

X \ Y	1	2
1	.2	.3
2	.3	.2

and suppose that $W = [X - E(X)][Y - E(Y)]$.

(a) Find $p(W)$.
(b) Determine $E(W)$ from $p(W)$.
(c) Find cov (X, Y) from $p(X, Y)$.
(d) Compare your answers to parts (b) and (c) and explain.

8.16 For the joint distribution of problem 8.15, suppose that $T = X^2 + Y^2$.

(a) Find $\Sigma_X \Sigma_Y (X^2 + Y^2) p(X, Y)$.
(b) Find $p(T)$.

(c) Find $E(T)$ from $p(T)$.

(d) Compare your answers to parts (a) and (c) and explain.

*8.17 In the text we suggested that

$$\text{cov}\,(X,\,Y) = E\{[X - E(X)][Y - E(Y)]\}$$

may be rewritten as

$$\text{cov}\,(X,\,Y) = E(XY) - E(X)\,E(Y)$$

Prove this result.

RELATIONSHIP BETWEEN INDEPENDENCE AND UNCORRELATEDNESS

Consider the following distribution:

X \ Y	-1	0	1
-1	0	1/4	0
0	1/4	0	1/4
1	0	1/4	0

X and Y are clearly not independent random variables; intuitively, we can see this by noting that if $Y = -1$ we *know* that $X = 0$, but if $Y = 0$ we *know* that $X \neq 0$. (Can you formally show that these random variables are not independent?) If you were to calculate the covariance, however, you would find it to be zero (verify this result). Therefore, although X and Y are not independent, they are uncorrelated!

This example is sufficient to illustrate that the concept of independence is in some sense different from that of uncorrelatedness, since variables can be uncorrelated and not be independent. More formally, the fact that random variables are uncorrelated does not imply that they are independent.

To explore this more fully let us think about, at an intuitive level, what the concepts of independence and uncorrelatedness are based on. With independence, we are saying that knowing the value of Y does not *in any way* influence our evaluation of the probabilities regarding X; we do not ask as to *how* the probabilities regarding X are affected, just whether or not they are. With the covariance, we can best see what we are seeking to uncover by referring back to Figure 8.1. The key to variables that are positively correlated is that when one is above (below) average it is then relatively more likely that the other variable will be above (below) average.[6] Thus a *particular type of relationship* is involved; there are other types of relationships, however, as we shall suggest below.

Consider the above joint probability distribution. X and Y both have means of 0. Regardless of the value of Y, the conditional distribution of X (given that value of Y) has a mean of 0. Similarly, regardless of the value of X, the conditional distribution of Y (given that value of X) has a mean of 0. This finding suggests that no matter whether Y is -1, 0, or 1 (that is, below, at, or above average) we expect X to be 0 (its average). Similarly, no matter whether X is -1, 0, or 1, we expect Y to be 0 (its average). *There is no tendency for one variable's being above average to be associated with the other's being above (or below) average.* We should then expect the covariance to be zero, as it is.

When we say that random variables are independent, we are saying that there is no relationship of any form between the two variables. On the other hand, when we say that random variables are uncorrelated we are saying that there is no relationship of a particular form—for example, one being above average suggests that the other is more likely than otherwise to be above (or below) average. This statement does not rule out other types of relationships that are not expressible, at least partially, in this form. It is then suggested, correctly, that if random variables are independent, then they must also be uncorrelated. (A proof of this proposition, for the discrete case, is contained in Appendix 1 to this chapter.) Uncorrelatedness, however, does not imply independence.

We can formalize this discussion a bit by saying that cov $(X, Y) = 0$ implies a lack of a relationship between X and Y expressible in the context of a linear form, but it says nothing about relationships not expressible in such a context. Clearly, if $Y = a + bX$ (a linear function) and (say) $b > 0$, then when X is above average Y is above average, suggesting a nonzero covariance. And, if $b = 0$ (that is, the relationship is not expressible as a linear function), then X being above average is not linked with Y being above or below average.

---

[6] A comparable statement applies for the case of negatively correlated random variables.

Consider now the joint distribution above; here

$$X^2 + Y^2 = 1$$

for each of the possible pairs of values of X and Y that occur with nonzero probability. That is, there is a perfect *nonlinear* relationship between X and Y! Certainly, then, they are not independent.

The covariance then "buys" something (showing whether the relationship is positive or negative) at the cost of limiting the type of relationship we are looking for. This is generally a satisfactory trade-off in empirical work because, after all, the kind of relationship we are most interested in is one in which it is true that when one variable is above average ("big"), there is a tendency for the other to be above average ("big") or below average ("small").

The covariance, however, cannot tell us the strength of the relationship—that is, whether the variables are strongly or weakly correlated. Why not? We shall next consider this problem and suggest a solution.

PROBLEMS

8.18 In the text we argued that independence implies uncorrelatedness, but that uncorrelatedness does not imply independence. Suppose that we are considering two random variables X and Y that have a covariance of 1. Can you determine whether or not X and Y are independent? Rigorously justify your answer, and generalize this result.

8.19 Consider the $p(X, Y)$

X \ Y	0	1	2
0	.1	0	.1
1	0	.3	0
2	0	.3	0
3	.1	0	.1

(a) Find cov (X, Y).

(b) Are X and Y independent? Fully justify your answer.

(c) How does $E(X|Y)$ change as Y changes? How does $E(Y|X)$ change as X changes? Use your responses to these questions as part of an explanation for your answers to parts (a) and (b).

CORRELATION COEFFICIENT

Suppose that X represents educational attainment (in years) and Y represents monthly income, for a group of individuals. The covariance between X and Y can be written as

$$\text{cov}(X, Y) = E(XY) - E(X)E(Y)$$

If the covariance is positive, we know that above-average values of X have at least some tendency to be associated with above-average values of Y. What can we say about the strength of that tendency? That is, what can we say about how closely X and Y are related?

It is tempting to suspect that the (absolute) magnitude of the covariance is an indicator of the strength of a relationship; we might suspect that the greater the covariance (in absolute value) the closer the relationship between the two random variables. It turns out, however, that this suspicion, though on the right track, fails because of a hidden problem. To investigate this, let us suppose that in the above problem we found $\text{cov}(X, Y)$ to be Φ. This finding was based upon X measured as years of education and Y measured as monthly income. Suppose that we changed the units of measurement so that education (call it X^*) was measured in months of educational attainment and income (call it Y^*) was measured as annual income.

Substantively, there should be no difference; the relationship between X and Y is the same as the relationship between X^* and Y^*. But it turns out that $\text{cov}(X, Y)$ is different from $\text{cov}(X^*, Y^*)$! To see this, note that $X^* = 12X$ and $Y^* = 12Y$. Therefore,

$$\begin{aligned}
\text{cov}(X^*, Y^*) &= E(X^*Y^*) - E(X^*)E(Y^*) \\
&= E[(12X)(12Y)] - E(12X)E(12Y) \\
&= 144\,E(XY) - 144\,E(X)E(Y) \\
&= 144[E(XY) - E(X)E(Y)] \\
&= 144\,\text{cov}(X, Y) \\
&= 144\Phi
\end{aligned}$$

That is, though $\text{cov}(X, Y)$ and $\text{cov}(X^*, Y^*)$ have the same sign, the latter is larger by a factor of 144—all because we have changed the units of measurement! Since the magnitude of the covariance depends on the units of measurement, we cannot depend on the *size* of the covariance to indicate the strength of the relationship (although we can depend on its *sign* to indicate the direction of the relationship).

But how can we then measure the strength of a relationship? The key is to develop a measure of correlation that is unit-free, for then changes in the units of measurement will not change the magnitude of the statistic. In the problem at hand the units of X are years of education and the units of Y are dollars per month. The covariance between X and Y is then measured in units of (years of education) $\times$ (dollars per month). The covariance between X^* and Y^*, however, is measured in units of (months of education) $\times$ (dollars per year).

The standard way to make our measure of correlation unit-free is to compute a statistic known as the correlation coefficient ρ, which is defined (for both the continuous and discrete cases) as

$$\rho = \frac{\text{cov}(X, Y)}{\sigma_X \sigma_Y}$$

where σ_X and σ_Y are the standard deviations of the two random variables. Since σ_X is measured in years of education, and σ_Y is measured in dollars per month, the units of ρ (for the X and Y variables from above) are

$$\frac{(\text{years of education}) \times (\text{dollars per month})}{(\text{years of education}) \times (\text{dollars per month})}$$

That is, ρ is unit-free. If we were to change the units of measurement of either X or Y or both, the value of ρ would not change.

It turns out that it will always be the case that $-1 \le \rho \le 1$. If $\rho = 1$ we have a perfect positive linear relationship, and if $\rho = -1$ we have a perfect negative linear relationship. We know that the type of relationship that the covariance seeks to uncover is a linear relationship; the same must be true for the correlation coefficient since they are directly related. To prove the above assertion all we need to do is show that if there is a perfect linear relationship ($Y_i = a + bX_i$ for each i), then $|\rho| = 1$; after all, a perfect linear relationship is the closest possible type of linear relationship. Appendix 2 to this chapter proves this proposition.

Therefore, we can say that the closer ρ is to 1 in absolute value, the closer is the relationship which we have uncovered. When ρ is .9 we have the suggestion of a relatively strong positive relationship, whereas when $\rho = -.1$ we have the suggestion of a relatively weak negative relationship.

The correlation coefficient then gives us two types of information regarding the type of relationship that exists between two random variables; it tells us not only the direction but also the strength of that relationship. But what does it say about causality? That is, what does it say about the underlying *processes* at work linking the two random variables? This leads us to a discussion of the relationship between *correlation* on the one hand and *causality* on the other.

PROBLEMS

8.20 For the joint distribution $p(X, Y)$ in problem 8.4, find the correlation coefficient. (See also problem 8.11.)

8.21 For the joint distribution $p(X, Y)$ in problem 8.5, find the correlation coefficient. (See also problem 8.12.)

8.22 For the $p(X, Y)$

X \ Y	1	2	3
2	.2	0	0
4	.1	.1	.2
6	0	.1	.3

find the correlation coefficient.

8.23 For the joint distribution

X \ Y	−1	0	1
−1	0	0	.3
0	0	.4	0
1	.3	0	0

find the correlation coefficient. Can you provide an intuitive explanation for this result?

8.24 Suppose that X is a random variable with the distribution

X	p(X)
1	.3
2	.3
3	.4

and suppose that $Y = X + 1$.

(a) Find $p(X, Y)$.
(b) Find the correlation coefficient between X and Y.
(c) Provide an intuitive explanation for your answer to part (b).

CORRELATION VERSUS CAUSALITY

We begin this section with a statement, and we shall seek to illustrate its correctness and importance in the discussion that follows. The statement is as follows: Determining that X and Y are correlated does not imply that changes in X will *cause* changes in Y, or that changes in Y will *cause* changes in X. Typically, this statement is expressed as "correlation does not imply causality" (although of course the two *may* be associated).

Correlation exists when changes in X are associated in a probabilistic sense with changes in Y. Causality exists when the changes in X are the motivating force behind the changes in Y. Let us pursue this matter further.

Suppose that X is some measure of a monetary aggregate (e.g., the rate of growth of the money supply) and Y is some measure of aggregate economic performance (e.g., the rate of growth of GNP). Suppose (as the historical record suggests for the United States) that there is a fairly strong correlation between X and Y. What can we say about the *causal* nature of the relationship between X and Y?

One possibility is that changes in X induce changes in Y (a type of monetarist argument) and the strong correlation that we have observed between these random variables is simply a manifestation of that underlying relationship.

Although some economists are persuaded that this explanation is in fact a correct one, it is important to note that the observed correlation between X and Y does *not* imply this conclusion. It is possible, for example, that the causality in fact goes from Y to X and not the other way around! One can imagine a scenario in which increases in Y lead the Federal Reserve Board to pursue a policy of monetary expansion so as to provide the additional liquidity necessitated by economic growth.[7] In this case we would still see a

---

[7] This additional liquidity ameliorates the upward pressure on interest rates that would otherwise occur.

strong correlation between the two variables, even though there is no presumed causality from X to Y.

The important question of "which way is the causality" must be answered in addressing the interpretation of an observed correlation, but it is only the simplest manifestation of the problem of inferring causality. To illustrate the potential for more complicated processes, let us suppose that X (some measure of an individual's educational attainment) and Y (some measure of this individual's income) are positively correlated. Certainly one possibility is that there is causality from X to $Y;$ more schooling leads to a higher skill level, greater on-the-job productivity, and therefore a higher wage.

For a second possibility, let us consider the possible impact of a third variable measuring the income of the parents of the individual in question (let us call this P). Individuals whose parents have relatively high incomes tend to earn high incomes themselves because of the "inheritability" of economic class (i.e., there is causality from P to Y). On the other hand, individuals whose parents have relatively high incomes tend to acquire relatively large amounts of education (i.e., there is causality from P to X). So for an individual whose parents have a high income (P is above average) there is a tendency for both X and Y to be above average. Thus, X and Y would be positively correlated even if there was no causality between X and Y (in either direction)!

Any judgment regarding causality must ultimately come from sources other than the observed correlation between the variables. This statement suggests that correlations can be properly interpreted only in the context of some appeal to a theoretical or other type of belief system that suggests the appropriate causal mechanisms relevant to the problem at hand. In making these cautionary remarks we do not mean to downplay the importance of the statistical techniques we have presented. Rather, we wish to emphasize that their interpretation is something to be done with care, and that we are never completely removed from the theoretical perspective in which we are viewing the situation.

WHAT IF THE POPULATIONS ARE NOT COMPLETELY OBSERVED?

The mean and variance of a random variable are *population* characteristics; they may be calculated from an observed population, or from a probability distribution (which, given the link between relative frequency and probability, is essentially a way of characterizing a population). Similarly, the covariance is a population characteristic. But what do we do in the typical case of an observational study where we observe only a random sample of pairs of values of X and Y? We have earlier addressed this issue in the case of sample means and sample variances; we now turn to a discussion of the *sample covariance*.

Suppose that we observe the entire population of (pairs of values of) X and Y. (We shall here think of the problem as based upon observation as opposed to as characterized by a joint distribution.) As discussed above, we could calculate the covariance as

$$\text{cov}(X, Y) = \frac{1}{m} \sum_{i=1}^{m} [X_i - E(X)][Y_i - E(Y)] \qquad (8.6)$$

where m is the size of the population. This expression is the same as Eq. (8.1); we have simply pulled the $(1/m)$ factor outside the summation sign. The sum is over the individual pairs of values of X and Y that comprise the population.

If the entire population is not observed, we cannot of course perform the calculation embodied in Eq. (8.6). Our job, then, is to come up with a sample statistic as an "estimate" of the true population covariance. We take a random sample of pairs of values of X and Y and record the observations (X_i, Y_i). We can substitute the sample means for the population means, and we can just add up values over the sample (of size n) that we have observed. This procedure suggests a possible estimate of the covariance

$$\frac{1}{n} \sum_{i=1}^{n} (X_i - \bar{X})(Y_i - \bar{Y})$$

where now, of course, the sum is just over the random sample we have observed. This technique would be one good way of estimating the covariance for either discrete or continuous random variables; there is, however, a slightly better way. Remember that when we calculated the sample variance of an individual random variable we said there was a technical reason by which it was "better" to divide by $(n - 1)$ instead of n? For the same technical reason (which will be discussed in Chapter 10) it is also better in this case to divide by $(n - 1)$. Our usual estimate of the covariance is then

$$\widehat{\text{cov}(X, Y)} = \frac{1}{n - 1} \sum_{i=1}^{n} (X_i - \bar{X})(Y_i - \bar{Y}) \qquad (8.7)$$

where the "hat" over cov (X, Y) denotes an estimate of this parameter. Equation (8.7) is applicable to either discrete or continuous random variables.

For calculation purposes, it is often easier to rewrite Eq. (8.7) as

$$\overline{\text{cov} (X, Y)} = \frac{1}{n-1}\left[\left(\sum_{i=1}^{n} X_iY_i\right) - n(\bar{X})(\bar{Y})\right] \quad (8.8)$$

You are asked to prove this equivalence in a problem following this section.

The sample correlation coefficient is then

$$\hat{\rho} = \frac{\overline{\text{cov} (X, Y)}}{s_X s_Y}$$

where s_X and s_Y are, respectively, the sample standard deviations of X and Y. It will still be the case that this estimated correlation coefficient will be between -1 and 1, with values closer to one in absolute value suggesting stronger relationships.

PROBLEMS

√ **8.25** Consider the following six observations on X and Y:

X_i	Y_i
2	3
3	3
5	4
4	3
4	4
6	7

Calculate $\hat{\rho}$, the estimated correlation coefficient.

8.26 Consider the following five observations on X and Y:

X_i	Y_i
2	6
3	5
4	5
5	2
6	2

Calculate $\hat{\rho}$, the estimated correlation coefficient.

*8.27 Prove that

$$\overline{\text{cov}(X, Y)} = \frac{1}{n-1} \sum_{i=1}^{n} (X_i - \bar{X})(Y_i - \bar{Y})$$

may be rewritten as

$$\overline{\text{cov}(X, Y)} = \frac{1}{n-1} \left[\left(\sum_{i=1}^{n} X_i Y_i \right) - n(\bar{X})(\bar{Y}) \right]$$

Illustrate this equivalence using the data from problem 8.25.

FUNCTIONS OF TWO (OR MORE) RANDOM VARIABLES

To close this chapter we address some issues of a rather more technical nature that will be quite useful to us in the future.

Suppose that X and Y are random variables, and that we consider a linear function of them. That is, suppose that

$$W_i = aX_i + bY_i + c$$

for each observation i, where a, b, and c are fixed constants. (Any function of two random variables that can be written in this form is a linear function.)

Suppose that X_i and Y_i are two characteristics of the observations in a population of size m, and that we observe this entire population. We then observe the entire population of W_i's. Now,

$$E(W) = \frac{1}{m} \sum_{i=1}^{m} W_i$$

$$= \frac{1}{m} \sum_{i=1}^{m} (aX_i + bY_i + c)$$

where the sum is over the individual observations of the random variables that comprise the population. If we push the summation sign through, we get

$$E(W) = \frac{1}{m} \left(a \sum_{i=1}^{m} X_i + b \sum_{i=1}^{m} Y_i + mc \right)$$

and thus

$$E(W) = a \left(\frac{1}{m} \sum_{i=1}^{m} X_i \right) + b \left(\frac{1}{m} \sum_{i=1}^{m} Y_i \right) + c$$

$$= a E(X) + b E(Y) + c$$

This result is very important because it suggests that the linearity property of E can be extended to the case of two (and, in fact, more than two, can you see why?) random variables. Since we have achieved this result in the context of an observational approach (rather than through the joint distribution) the result is equally valid for discrete and continuous random variables.

Suppose now that we consider the determination of σ_W^2. Our discussion from Chapter 7 suggests that

$$\sigma_W^2 = E\{[W - E(W)]^2\}$$

But, since W can be rewritten in terms of X and Y, we can say that

$$\begin{aligned}
\sigma_W^2 &= E\{[aX + bY + c - [aE(X) + bE(Y) + c]]^2\} \\
&= E\{(a[X - E(X)] + b[Y - E(Y)])^2\} \\
&= E\{a^2[X - E(X)]^2 + b^2[Y - E(Y)]^2 \\
&\quad + 2ab[X - E(X)][Y - E(Y)]\}
\end{aligned}$$

The linearity property of E allows us to then "push" it through until we hit the random variables:

$$\sigma_W^2 = a^2 E\{[X - E(X)]^2\} + b^2 E\{[Y - E(Y)]^2\} \\
+ 2ab E\{[X - E(X)][Y - E(Y)]\}$$

In essence, we have treated $[X - E(X)]^2$, $[Y - E(Y)]^2$, and $[X - E(X)][Y - E(Y)]$ as new random variables, and then exploited the linearity property of E with regard to these newly defined random variables. Notice that

$$E\{[X - E(X)]^2\} = \sigma_X^2$$
$$E\{[Y - E(Y)]^2\} = \sigma_Y^2$$
$$E\{[X - E(X)][Y - E(Y)]\} = \text{cov}\,(X, Y)$$

Thus

$$\sigma_W^2 = a^2\sigma_X^2 + b^2\sigma_Y^2 + 2ab\,\text{cov}\,(X, Y)$$

EXAMPLE ■

Suppose that X has a mean of 3 and a variance of 4, Y has a mean of 5 and a variance of 6, and the covariance between X and Y is 7. If $W = 2X + 3Y + 4$, what are the mean and variance of W?

$$E(W) = 2E(X) + 3E(Y) + 4 = 2(3) + 3(5) + 4 = 25$$

$$\sigma_W^2 = (2)^2\sigma_X^2 + (3)^2\sigma_Y^2 + 2(2)(3)\,\text{cov}\,(X,\,Y)$$

$$= 4(4) + 9(6) + 12(7) = 154$$

Note that if the X and Y variables are independent, then

$$\sigma_W^2 = a^2\sigma_X^2 + b^2\sigma_Y^2$$

since cov $(X,\,Y) = 0$ if X and Y are independent. (Notice the similarity of this result to that regarding the variance of a function of one random variable.) ■

These results can be generalized to linear functions of many random variables. If $X_1, \ldots, X_n$ are random variables and if

$$Y = a_0 + a_1 X_1 + \cdots + a_n X_n$$

(where the a_i's are constants), then

$$E(Y) = a_0 + a_1\,E(X_1) + \cdots + a_n\,E(X_n)$$

In addition, if cov $(X_i, X_j) = 0$ for all $i \neq j$ (that is, if the X's are mutually uncorrelated), then

$$\sigma_Y^2 = a_1^2\sigma_1^2 + a_2^2\sigma_2^2 + \cdots + a_n^2\sigma_n^2$$

where σ_i^2 is the variance of X_i and σ_Y^2 is the variance of Y. Notice the similarity between these results and those achieved earlier regarding functions of one random variable (Chapter 3); in particular, notice that the intercept a_0 does not affect the variance (after all, it is the same for each observation) and that in the variance formula the coefficients a_i enter as squares (as in the case of a function of a single random variable). (If the X_i's were not mutually uncorrelated, then the variance of Y would depend upon the covariances between the various possible pairs of X's.) These results are of particular importance in the work we begin in Chapter 9 because the sample statistics on which we base inferences can often be viewed as linear functions of many random variables. Note that if the X_i's are mutually independent, then they will also be mutually uncorrelated, since independence implies uncorrelatedness.[8]

---

[8] See the earlier discussion in Chapter 5 in which the concept of mutually independent events was presented in more detail.

PROBLEMS

8.28 Suppose that X is a random variable with a mean of 10 and a variance of 9, Y is a random variable with a mean of 5 and a variance of 25, and cov $(X, Y) = 2$. If $W = X + 2Y + 3$, find the mean and variance of W.

8.29 X is a random variable with a mean of 5 and a variance of 4, Y is a random variable with a mean of 6 and a variance of 9, and the covariance between X and Y is -1. If $W = 2X - 3Y + 1$, find the mean and variance of W.

8.30 X is a random variable with a mean of 1 and a standard deviation of 2, and Y is a random variable with a mean of 2 and a standard deviation of 3. The correlation coefficient between X and Y is .3. If $W = 2X + 2Y + 7$, find the mean and standard deviation of W.

8.31 In a large population of individuals, wage income has a mean of $20,000 per year and a standard deviation of $5000 per year. Asset income has a mean of $3000 per year and a standard deviation of $2500 per year. The correlation coefficient between wage income and asset income is .4. Total income is the sum of wage and asset income. Find the mean and standard deviation of total income for this population.

8.32 Suppose that X and Y are independent random variables, each having a mean of 2 and a variance of 9. If $S = X + Y$, find the mean and variance of S. How are the mean and variance of S related to the means and variances of X and Y? How are the standard deviations related?

8.33 The random variables X_1, X_2, and X_3 are mutually independent; each has a mean of 10 and a variance of 5. If we construct the weighted average

$$A = \frac{1}{6} X_1 + \frac{2}{6} X_2 + \frac{3}{6} X_3$$

find the mean and variance of A.

*8.34 In a large lecture course there are three exams. On the first exam (E_1) there was a mean of 80 and a standard deviation of 10. On the second exam (E_2) there was a mean of 75 and a standard deviation of 15. On the third exam (E_3) there was a mean of 82 and a standard deviation of 8.

E_1 counts 20%, E_2 counts 30%, and E_3 counts 50% of a student's course grade. That is, a student's grade for the course is

$$\text{grade} = .2E_1 + .3E_2 + .5E_3$$

(a) Find $E(\text{grade})$ and σ_{grade}.

(b) On what assumption is your calculation of σ_{grade} based?

(c) If $Y = a_0 + a_1X_1 + a_2X_2 + a_3X_3$, where the X_i's are mutually correlated, derive a general formula for calculating the variance of Y.

(d) Apply your result from part (c) to the determination of the standard deviation of the random variable *grade* considered above if the correlation coefficients are as follows: .6 between E_1 and E_2, .6 between E_2 and E_3, and .4 between E_1 and E_3.

SUMMARY

In this chapter we have discussed what probability theory suggests about the relationships between random variables and how such relationships might be quantified and interpreted. Although we have generally limited ourselves to the context of two random variables, many of the concepts can be extended to more than two random variables (for example, the results just considered regarding linear functions).

Sample statistics (such as the sample mean) are functions of many random variables. The behavior of such sample statistics is then closely linked to issues regarding joint distributions. We now turn to a discussion of such statistics.

REVIEW PROBLEMS FOR CHAPTER 8

8.35 Suppose that $E(X) = 2$, $E(Y) = 3$, and cov $(X, Y) = 1$. Find $E(XY)$.

8.36 For X and Y from problem 8.35, if $W = 2XY + 5$, find $E(W)$.

8.37 X has a normal distribution with mean 5 and variance 25. Y has a normal distribution with mean 3 and variance 16. In addition, X and Y are independent. If $W = X + Y$, find

$\Pr(W > 9)$. (*Hint:* Linear functions of independent normals are normal.)

8.38 Consider the $p(X, Y)$

X \ Y	1	2	3
1	.3	.1	0
2	.4	.1	.1

Find $\Pr(X + Y \geq 4)$.

*8.39 Suppose that we have two hats, A and B. A contains five red and five blue marbles. B contains seven red and three blue marbles. I pick three marbles with replacement from each hat. What is the probability that among the six marbles chosen I will have exactly three reds? At least three reds?

*8.40 Suppose that there are two stages in preparing a paper: outlining and writing. Among a group of students, the time spent outlining has a mean of 2 hours and a standard deviation of 1 hour. The time spent writing has a mean of 5 hours and a standard deviation of 3 hours. The correlation coefficient between outlining time and writing time is $-.5$ (because a better outline reduces writing time). Find the mean and standard deviation of total time preparing this paper.

APPENDIX 1 TO CHAPTER 8

In this appendix we offer a proof (for the case of discrete random variables) that independence implies uncorrelatedness, that is, that if X and Y are independent random variables, then their covariance is 0.[9]

From Eq. (8.5) the covariance between X and Y can be written as

$$\text{cov}(X, Y) = \left[\sum_X \sum_Y XY\, p(X, Y) \right] - E(X)\, E(Y)$$

[9] This proof is a modified version of that offered in Thomas H. Wonnacott and Ronald J. Wonnacott, *Introductory Statistics for Business and Economics,* 4th ed. (New York: John Wiley & Sons, 1990), p. 742.

Since X and Y are independent, we know that $p(X, Y) = p(X) \times p(Y)$ for all (X, Y). The covariance formula, in this case, becomes

$$\mathrm{cov}\,(X, Y) = \left[\sum_X \sum_Y XY\, p(X)\, p(Y) \right] - E(X)\, E(Y)$$

Consider the double summation. As far as the sum over Y is concerned, X and $p(X)$ are fixed numbers. This suggests that X and $p(X)$ may be moved outside the Σ_Y sign; that is,

$$\mathrm{cov}\,(X, Y) = \left[\sum_X X\, p(X) \sum_Y Y\, p(Y) \right] - E(X)\, E(Y)$$

Since

$$\sum_X X\, p(X) = E(X) \quad \text{and} \quad \sum_Y Y\, p(Y) = E(Y)$$

we can then say that

$$\mathrm{cov}\,(X, Y) = E(X)\, E(Y) - E(X)\, E(Y) = 0$$

This expression proves that, for the case of discrete random variables, independence implies uncorrelatedness. This result also applies to continuous random variables, but the proof in that case requires the use of calculus.

APPENDIX 2 TO CHAPTER 8

In this appendix we offer a proof to show that if there is a perfect linear relationship between X and Y then the correlation coefficient ρ will equal one in absolute value ($|\rho| = 1$).

If there is a perfect linear relationship between X and Y then

$$Y_i = a + bX_i$$

for each i—that is, for each observation within the population. (Or we can simply write $Y = a + bX$ as a general way of characterizing X and Y as random variables.)

Let us first determine the covariance. We know that

$$E(Y) = a + b\, E(X)$$

and therefore

$$Y - E(Y) = a + bX - [a + b\, E(X)] = b[X - E(X)]$$

Thus

$$\begin{aligned}
\text{cov}\,(X,\,Y) &= E\{[X - E(X)][Y - E(Y)]\} \\
&= E\{[X - E(X)]b[X - E(X)]\} \\
&= bE\{[X - E(X)]^2\} \\
&= b\sigma_X^2
\end{aligned}$$

(2A8.1)

We also know that

$$\sigma_Y^2 = b^2\sigma_X^2$$

and therefore

$$\sigma_Y = |b|\sigma_X$$

(2A8.2)

(Remember, we say $|b|$, since a square root is always positive, but b may be negative.) Since

$$\rho = \frac{\text{cov}\,(X,\,Y)}{\sigma_X\sigma_Y}$$

we know that

$$\rho = \frac{b\sigma_X^2}{\sigma_X|b|\sigma_X} = \frac{b}{|b|}$$

by Eqs. (2A8.1) and (2A8.2).

Clearly,

$$|\rho| = \frac{|b|}{|b|} = 1$$

To elaborate, if $b > 0$, we know that $\rho = 1$, which is the case of a perfect positive linear relationship. If $b < 0$, then $|b| = -b$ and hence $\rho = -1$, which is the case of a perfect negative linear relationship.

DISCUSSION QUESTIONS FOR CHAPTER 8

1. Construct a hypothetical process to explain how, although the amount of time students study and the grades they get are positively correlated, additional study time does not *cause* a student to get higher grades.

2. Suppose that we determine that the sample correlation coefficient between X and Y is positive. Explain how this might occur even if X and Y (as populations) are uncorrelated.

3. If $W = aX + bY + c$, then $\sigma_W^2 = a^2\sigma_X^2 + b^2\sigma_Y^2 + 2ab \operatorname{cov}(X, Y)$. Provide intuition as to why the variance of W is increased (decreased) because of a positive (negative) correlation between X and Y, in the case where a and b are both positive numbers.

Chapter *9*

Sampling Theory

The fundamental statistical problem faced in applied research is that of using the information in a sample to infer the characteristics of a larger population. Because the elements of the population that appear in our sample are the result of a random process (picking the sample), sample statistics (such as $\bar{X}$) are random variables. We then face the problem of exploiting concepts from probability theory (the study of randomness) to enhance our ability to make inferences.

In this chapter we begin to use probability theory to explore sample statistics as random variables. This technique will serve as a foundation for the third main section of this book, in which we attempt to better understand the process of inference and increase the sophistication of the inferences we can make.

THE NATURE OF SAMPLING

In Chapter 4 we defined the concept of a random sample. Whereas a sample is any subset of the population, a random sample is a sample chosen in a manner such that each element of the population has the same chance of being selected. With a random sample, then, our observations are not "biased" toward one (or more) type(s) of observations. We can conceptually view the process of choosing a random sample as one in which we put each observation on a piece of paper (where all these pieces of paper are identical), put these pieces of paper into a large drum, thoroughly mix them, and then blindly choose from the drum.

When discussing the random characteristics of samples, it is easiest to make one further assumption concerning the nature of the sample. The process of picking a sample from a population is essentially one of sampling without replacement; after all, we generally do not return earlier observations to the population for future draws. In the case of sampling without replacement, successive draws are approximately independent if we are sampling a relatively small number from a relatively large population; we shall generally use this assumption because (1) it is often accurate (in economics we are often sampling from a large population) and (2) it simplifies the analysis that follows.

Note that if we have a random sample wherein the individual observations are independent random variables, then these independent random variables will all have the same probability distribution. This conclusion is demonstrated by considering an example.

Suppose that the population X from which we are sampling (hereinafter referred to as the *parent population*) is a very large population that is normally distributed (by assumption) with an unknown mean μ and variance σ^2. Since the population is very large, the individual observations (chosen without replacement) will be independent. If X_1 is the result of our first draw from the population (i.e., the first element in our sample) then what is the distribution of X_1? Clearly, it must be normal with mean μ and variance σ^2. The same can be said for $X_2, \ldots, X_n$. Thus, in this scenario the successive observations in our sample $(X_1, \ldots, X_n)$ are identically distributed with a distribution equal to that of the parent population X.

This result suggests a definition of a special type of random sample, which will be the basis for the analysis that follows.[1]

Definition of a "Simple Random Sample"   A simple random sample from a population is a random sample in which the indi-

---

[1] We shall work only in the context of a simple random sample, as defined below. There are, however, other types of samples that are at times encountered (e.g., samples in which the individual observations are not independent because of small population sampling).

vidual observations are mutually independent and identically distributed random variables.

Given this background, we are now ready to discuss the characteristics of the sample mean $\bar{X}$ as a random variable.

DISTRIBUTION OF THE SAMPLE MEAN

If we are interested in making inferences regarding the mean of a population μ, our general method is to begin with the sample mean $\bar{X}$. In this section we shall focus on determining the characteristics of $\bar{X}$ as a random variable so as to lay a foundation from which we can eventually develop more sophisticated inferences (that is, inferences above and beyond a mere "guess").

Suppose that X is a characteristic of a population and that we observe a simple random sample of size n ($X_1, \ldots, X_n$). The sample mean $\bar{X}$ is calculated as

$$\bar{X} = \frac{1}{n} \sum_{i=1}^{n} X_i$$

It is helpful to rewrite this equation as

$$\bar{X} = \frac{1}{n} X_1 + \frac{1}{n} X_2 + \cdots + \frac{1}{n} X_n \tag{9.1}$$

Equation (9.1) makes it clear that $\bar{X}$ is a linear function of n random variables and therefore is itself a random variable. What is more, the individual X_i's are independent (thus the importance of a simple random sample); therefore we can use our results from Chapter 8 regarding linear functions of independent random variables to determine the properties of $\bar{X}$.

Let us suppose that the parent population X has mean μ and variance σ^2. First we determine the mean and variance of $\bar{X}$; to do this we simply determine the mean and variance of a linear function of n independent random variables that are identically distributed, in accordance with Eq. (9.1):

$$E(\bar{X}) = \frac{1}{n} E(X_1) + \frac{1}{n} E(X_2) + \cdots + \frac{1}{n} E(X_n)$$

$$= \frac{1}{n}\mu + \frac{1}{n}\mu + \cdots + \frac{1}{n}\mu \quad (n \text{ times})$$

$$= \frac{1}{n}\mu(n)$$

$$= \mu$$

It is important to interpret this result carefully. We know that $\bar{X}$ is a random variable because it is a function of the individual X_i's (which are random variables). The derivation above tells us that the mean of the probability distribution of the random variable $\bar{X}$ is μ (the mean of the parent population). That is, on average $\bar{X}$ equals μ.

Do not be confused by thinking of a "mean of a sample mean," although this value is what we have just determined. There is no need for confusion if you keep it clear that the sample mean is a random variable; it therefore has a probability distribution characterized by a mean. And we should be comforted by the fact that on average $\bar{X}$ equals μ for, after all, $\bar{X}$ is our "guess" of μ!

What is the intuition behind this result? Since we base our inference on a simple random sample, we suspect that on average this sample is representative of the population as a whole (of course, it could by chance not be—we could pick five 7-footers when we sample heights), and therefore that the sample mean on average takes on a value equal to the population mean. That is, if we were to take thousands and thousands of simple random samples from this population and calculate $\bar{X}$ for each one of them, the average of these $\bar{X}$'s should be just about μ. Of course, in practice we only have one observation on $\bar{X}$, and we therefore must consider the variability of this random variable.

What then can we determine regarding the variance of $\bar{X}$? Since $\bar{X}$ is a linear function of independent random variables, and since each X_i has a variance of σ^2, we know from Eq. (9.1) that the variance of $\bar{X}$ is

$$\left(\frac{1}{n}\right)^2\sigma^2 + \left(\frac{1}{n}\right)^2\sigma^2 + \cdots + \left(\frac{1}{n}\right)^2\sigma^2 = \left(\frac{1}{n}\right)^2\sigma^2(n)$$

$$= \frac{\sigma^2}{n}$$

It is important here to distinguish between the variance of $\bar{X}$ and the variance of X; the former is the variance of the sample mean, whereas the latter is the variance of the parent population. Clearly, the variance of $\bar{X}$ is only a fraction of the variance of any individual observation.

When we calculate a sample mean, there is some tendency for high values to be offset by low values, and consequently there is less variability in the sample mean than in any individual observation. In order to get an extreme value of $\bar{X}$, we must get several extreme X's as opposed to just one. Certainly, it is far more likely to pick an individual whose height exceeds 7 feet than it is to pick a sample of individuals in which the average height exceeds 7 feet. This result suggests that the variance of $\bar{X}$ is less than that of X and, in fact, that this variance decreases as the sample size increases. This conclusion is confirmed by the result above.

When we say that the variance of $\bar{X}$ is σ^2/n, it is important to realize that σ^2 refers not to the variance of the random variable in question ($\bar{X}$) but rather to the variance of some other random variable (X). Because we often work with different random variables with different variances, we often use the notation var (W) to denote the variance of the random variable W. In the problem at hand, var $(X) = \sigma^2$ and var $(\bar{X}) = \sigma^2/n$, where σ^2 is some number.

We now know the mean and variance of $\bar{X}$. What about the shape of its distribution?

THE PROBABILITY DISTRIBUTION OF THE SAMPLE MEAN

In Chapter 6 we argued that a linear function of a normally distributed random variable is itself normal; that is, if $Y = a + bX$ and X is normal, then Y is also normal. This result can be extended to a linear function of n mutually independent random variables; that is, if $X_1, \ldots, X_n$ are mutually independent and normally distributed random variables, and if

$$Y = a_0 + a_1 X_1 + \cdots + a_n X_n$$

then Y will have a normal distribution (for which the mean and variance may be determined using our results on linear functions).[2]

---

[2] The best intuition for this result comes from viewing it as an extension of the earlier result for a function of one normal random variable.

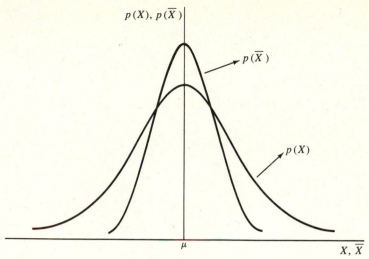

Figure 9.1 The sampling distribution of $\bar{X}$.

This result suggests that if the parent population is normally distributed, that is, if $X \sim N(\mu, \sigma^2)$, then $\bar{X}$ (from a simple random sample) will also have a normal distribution [since $\bar{X}$ is a linear function of the normal X_i's; see Eq. (9.1)]. Given $E(\bar{X})$ and var $(\bar{X})$ as derived above, then $\bar{X} \sim N(\mu, \sigma^2/n)$. In Figure 9.1 we illustrate this situation.

But what about the case where the parent population X has a distribution that is not normal? It turns out that (except in some odd cases that are generally not relevant for applied purposes) as long as the sample size is reasonably large, the distribution of $\bar{X}$ may be reasonably well approximated as normal. This result is the basis for the central role played by the normal distribution in applied statistical analysis, and it has to do with a formal result known as the *central limit theorem*. We do not state the central limit theorem here because there are some sophisticated theoretical issues involved in presenting it properly; we instead state what, for our purposes, is its main implication.

Implication of the Central Limit Theorem   If $X_1, \ldots, X_n$ is a simple random sample from a population with mean μ and variance σ^2, then as n gets large it will generally be the case that $p(\bar{X})$ can be reasonably approximated as a normal distribution with mean μ

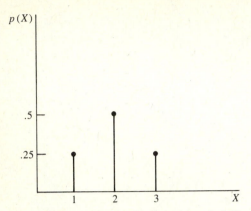

Figure 9.2   A symmetric discrete probability distribution.

and variance σ^2/n; that is, $\bar{X}$ is approximately distributed as $N(\mu, \sigma^2/n)$ when n is large.[3]

In general, as n "gets large," the distribution of $\bar{X}$ becomes "close" to a normal and may be reasonably approximated as such. How large the sample must be for the distribution of $\bar{X}$ to become "close" to a normal varies depending on the parent population, but it is often the case, even with a moderately sized n of 30 or so, that the distribution of $\bar{X}$ is fairly close to that of a normal distribution.

How can we motivate this result? Let us consider some examples. In Figure 9.2 we present a discrete probability distribution that has a symmetric shape; the mean of this distribution is 2. It should not be surprising that if we were to take increasingly larger samples from this population, the distribution of $\bar{X}$ would quickly look much like a normal distribution; $\bar{X}$ becomes more nearly continuous, and from the beginning there is a general symmetric shape. But what about a situation such as that given in Figure 9.3, wherein $p(1) = .4$, $p(2) = .3$, $p(3) = .2$, and $p(4) = .1$?

In Figure 9.3 we show a probability distribution very different from that of a normal distribution, primarily because of its lack of

---

[3] Technically, we should say that the central limit theorem suggests that $(\bar{X} - \mu)/(\sigma/\sqrt{n})$ has a distribution that approaches the standard normal as n approaches infinity. Since the variance of $\bar{X}$ is σ^2/n, we know that the variance of $\bar{X}$ approaches 0 as n approaches infinity, making it problematic to refer to the limiting distribution of $\bar{X}$ as normal; see, for example, John E. Freund and Ronald E. Walpole, *Mathematical Statistics,* 4th ed. (Englewood Cliffs, NJ: Prentice-Hall, 1987), pp. 274–277. In practice, however, the two views are indistinguishable since the sample size is always finite.

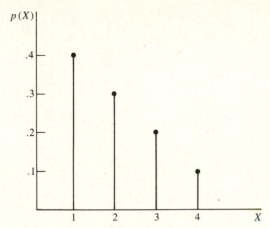

Figure 9.3 An asymmetric discrete probability distribution.

symmetry. The central limit theorem tells us that as the sample size n increases, the probability distribution of $\bar{X}$ approaches a normal distribution. To begin to illustrate this finding consider the distribution of $\bar{X}$ when we pick a simple random sample of size 2 from this population.

There will be 16 (that is, 4×4) possible ordered pairs (X_1, X_2), where X_1 represents the result of the first draw and X_2 the result of the second. We below list these 16 possibilities and the probabilities with which they occur [due to independence, $p(X_1, X_2) = p(X_1)p(X_2)$]. We also list the value of $\bar{X}$ associated with each outcome.

(X_1, X_2)	$p(X_1, X_2)$	$\bar{X}$
(1, 1)	.4(.4) = .16	1
(1, 2)	.4(.3) = .12	1.5
(1, 3)	.4(.2) = .08	2
(1, 4)	.4(.1) = .04	2.5
(2, 1)	.3(.4) = .12	1.5
(2, 2)	.3(.3) = .09	2
(2, 3)	.3(.2) = .06	2.5
(2, 4)	.3(.1) = .03	3
(3, 1)	.2(.4) = .08	2
(3, 2)	.2(.3) = .06	2.5
(3, 3)	.2(.2) = .04	3
(3, 4)	.2(.1) = .02	3.5
(4, 1)	.1(.4) = .04	2.5
(4, 2)	.1(.3) = .03	3
(4, 3)	.1(.2) = .02	3.5
(4, 4)	.1(.1) = .01	4

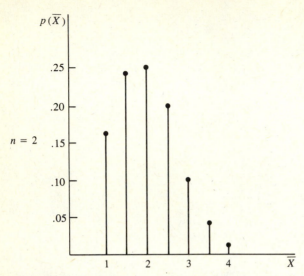

Figure 9.4   Illustrating the central limit theorem for $n = 2$.

Note that the 16 possible outcomes are mutually exclusive. Remembering that the probability of the union of mutually exclusive events is the sum of their individual probabilities, we can then determine from the above chart the possible values of the sample mean $\bar{X}$ and the probabilities $p(\bar{X})$ with which they occur:

$\bar{X}$	$p(\bar{X})$ (for $n = 2$)
1	.16
1.5	.24
2	.25
2.5	.20
3	.10
3.5	.04
4	.01

Notice that the distribution when $n = 2$ (plotted in Figure 9.4) reaches a peak at $\bar{X} = 2$ with some tapering off in each direction, although it is certainly not yet a symmetric distribution.

EXERCISE ■
Determine the mean and variance of $p(\bar{X})$ as given above, and confirm that this result is consistent with the results presented earlier concerning the characteristics of $\bar{X}$ as they relate to those of the parent population.   ■

Now, let us consider what happens when the sample is of size 3. In this case there are 64 possible outcomes ($4 \times 4 \times 4$), and if you were to derive the distribution of $\bar{X}$, you would find it to be as follows:

$\bar{X}$	$p(\bar{X})$ (for n = 3)
1	.064
4/3	.144
5/3	.204
2	.219
7/3	.174
8/3	.111
3	.056
10/3	.021
11/3	.006
4	.001

(As an exercise, derive these probabilities.) Now consider Figure 9.5; even with a sample of only size 3 we begin to see the distribution moving toward the general shape of the normal distribution! We also include here a graph of $p(\bar{X})$ when the sample is of size 4.[4] With each additional observation, it becomes more nearly continuous and more nearly symmetric.

Variations on the central limit theorem play a key role in much of the statistical analysis used in economics, and this theorem is the fundamental reason for the attention paid to the normal distribution. This theorem suggests that the work we did in Chapter 6 on the normal distribution should be relevant (at least as an approximation) for studying the behavior of $\bar{X}$ as a random variable.

PROBLEMS

9.1 We have a large bowl containing many chips. One-third of these chips are marked with a 1, one-third are marked with a 2, and one-third are marked with a 3.

---

[4] The probability distribution $p(\bar{X})$ when $n = 4$ is: $p(1) = .0256$, $p(1.25) = .0768$, $p(1.5) = .1376$, $p(1.75) = .1840$, $p(2) = .1905$, $p(2.25) = .1608$, $p(2.5) = .1124$, $p(2.75) = .0648$, $p(3) = .0310$, $p(3.25) = .0120$, $p(3.5) = .0036$, $p(3.75) = .0008$, $p(4) = .0001$.

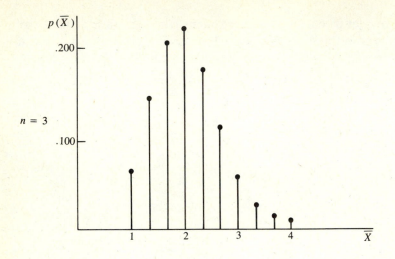

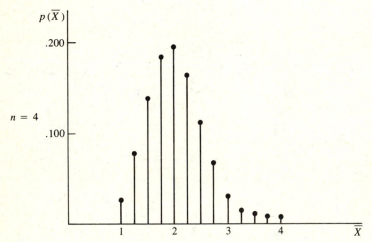

Figure 9.5 Illustrating the central limit theorem for $n = 3$ and $n = 4$.

(a) You randomly choose one chip. Find $p(X)$, $E(X)$, and var (X).

(b) You randomly choose two chips. Find $p(\bar{X})$, and then find $E(\bar{X})$ and var $(\bar{X})$ from $p(\bar{X})$. Verify that $E(\bar{X}) = E(X)$ and var $(\bar{X}) = $ var $(X)/n$ (i.e., $E(\bar{X}) = \mu$ and var $(\bar{X}) = \sigma^2/n$.)

(c) Repeat part (b) for a sample of three chips.

9.2 Redo problem 9.1, except now with a bowl in which 50% of the chips are marked 1, 30% are marked 2, and 20% are marked 3.

9.3 Suppose that X has a normal distribution with mean 5 and variance 16, and that you take a simple random sample of size 9 from this population. Find $E(\bar{X})$ and var $(\bar{X})$:

(**a**) Treating $\bar{X}$ as a linear function of the X's and using the known properties of the X population.

(**b**) Using the relationship between $p(X)$ and $p(\bar{X})$ discussed in the text.

9.4 Suppose that $E(X) = \mu$ and var $(X) = \sigma^2$, but μ and σ^2 are unknown. Describe how our knowledge concerning $E(\bar{X})$ and var $(\bar{X})$ is affected.

APPLICATIONS OF THE CENTRAL LIMIT THEOREM

From our discussions above, we know that the probability distribution of $\bar{X}$ can be at least approximated as $N(\mu, \sigma^2/n)$, where μ and σ^2 are, respectively, the mean and variance of the parent population. This statement suggests that probabilities describing the behavior of $\bar{X}$ can be determined presuming we have certain pieces of information regarding the parent population. Let us consider an example.

A store owner knows from past experience that daily receipts have a mean of $200 and a standard deviation of $125. During a given month (30 business days), what is the probability that average daily receipts are at least $250?

In this problem we can view the parent population as characterized by a mean $\mu = \$200$ with standard deviation $\sigma = \$125$. (That is, we are viewing the population parameters as known.) We have a sample of size $n = 30$. The question is: What is $\Pr(\bar{X} \geq 250)$? If we assume that the sample is a simple random sample,[5] then we know that $\bar{X}$ is at least approximately normal with mean μ (=200) and variance σ^2/n (=(125)^2/30).

Since

$$\bar{X} \sim N\left(\mu, \frac{\sigma^2}{n}\right)$$

---

[5] The simple random sample assumption may be problematic in this case (see discussion question 3 at the end of this chapter), but we use it for illustrative purposes.

we know that

$$\frac{\bar{X} - \mu}{\sigma/\sqrt{n}} \sim N(0, 1)$$

Why? The quantity $(\bar{X} - \mu)/(\sigma/\sqrt{n})$ is a normal random variable $(\bar{X})$ minus its mean (μ), divided by its standard deviation $(\sigma/\sqrt{n})$. That is,

$$\frac{\bar{X} - \mu}{\sigma/\sqrt{n}} = \frac{\bar{X} - E(\bar{X})}{\sqrt{\text{var}(\bar{X})}}$$

Probabilities regarding $\bar{X}$ may then be determined by an application of the standardization transformation first considered in Chapter 6. For the problem at hand,

$$\Pr(\bar{X} \geq 250) = \Pr\left(\frac{\bar{X} - \mu}{\sigma/\sqrt{n}} \geq \frac{250 - 200}{125/\sqrt{30}}\right)$$

$$= \Pr(Z \geq 2.19) = .0143$$

Note that, if the parent population were normal, the probability that receipts were at least $250 *on some particular day* would be

$$\Pr(X \geq 250) = \Pr\left(\frac{X - \mu}{\sigma} \geq \frac{250 - 200}{125}\right) = \Pr(Z \geq .4) = .3446$$

That is, the probability that a particular observation is extreme by a certain amount (or more) is larger than the same probability for an average of observations in a sample. Why? The key is that var $(\bar{X}) <$ var (X), and thus extreme values are less likely with $\bar{X}$.

To investigate a related question, suppose that the shopkeeper needed $5400 in revenues during a month (30 days) to meet some payments. What is the probability of doing this?

The question involves the probability that

$$\sum_{i=1}^{30} X_i \geq 5400$$

But this situation is true if and only if

$$\frac{1}{30} \sum_{i=1}^{30} X_i \geq \frac{5400}{30}$$

that is, if

$$\bar{X} \geq 180$$

We know that

$$\Pr(\bar{X} \geq 180) = \Pr\left(\frac{\bar{X} - \mu}{\sigma/\sqrt{n}} \geq \frac{180 - 200}{125/\sqrt{30}}\right)$$

$$= \Pr(Z \geq -.88) = 1 - \Pr(Z < -.88) = .8106$$

Since a value regarding a sum can be reexpressed as a corresponding value regarding a sample mean, the central limit theorem is more broadly applicable than it might appear to be at first glance.

PROBLEMS

9.5 For the probability distribution given in Figure 9.3, we derived the distribution $p(\bar{X})$ for the $n = 3$ case in the text. If $n = 3$, find $\Pr(\bar{X} \geq 3)$

(a) using $p(\bar{X})$ as derived in the text;
(b) using the central limit theorem.
Why are these two answers different?

9.6 What would the relationship be between the two answers suggested by the different approaches in problem 9.5 if the parent population were normally distributed?

9.7 Suppose that jelly beans have a mean weight of 5 grams with a standard deviation of 1 gram. A box contains 25 jelly beans. Find $\Pr(\bar{X} \geq 4.7)$.

9.8 In problem 9.7, to what value must we reduce the standard deviation of jelly bean weight so as to make the probability of interest equal to .95?

9.9 Suppose that over a long historical period the mean inflation rate has been 4% with standard deviation 2%. In a particular decade, what is the probability that the average inflation rate is at least 3%? Comment on the role played in your answer by the assumption of a simple random sample, and on the appropriateness of this assumption.

9.10 Using the data from problem 9.9, how many years of observations would be necessary so as to make the probability of interest equal to .99?

9.11 For a given year, the rate of return on all stocks was normally distributed, with a mean of 10% and a standard deviation of 8%.

(a) An investor randomly chose one stock at the beginning of the year to hold for the course of the year. What is the probability that the rate of return on this stock was at least 8%?

(b) Another investor randomly chose 20 stocks at the beginning of the year to hold for the course of the year. What is the probability that the average rate of return on these stocks was at least 8%?

(c) What do your answers to parts (a) and (b) suggest about the benefits of portfolio diversification?

9.12 Asset income across individuals has a mean (per year) of $500 with standard deviation $400. In a simple random sample of 30 individuals, what is the probability that the total amount of asset income will exceed $18,000?

9.13 Mary is a construction worker who is irregularly employed. Her wages per month have a mean of $2000 but a standard deviation of $1000. Find the probability that her income in a given year will exceed $20,000, assuming that her amounts of income across successive months are independent.

9.14 In problem 9.13, what would Mary's mean monthly income need to be so as to make the probability of interest equal to 90%?

9.15 A wall is going to be built by stacking nine bricks. Bricks have a height that is random, with mean 3 inches and standard deviation one-half inch. Find the probability that the height of the wall exceeds 29 inches. (Assume that there is no mortar between bricks.)

SUMMARY

The sample mean $\bar{X}$ is our basis for inference regarding the population mean μ. But remember that μ is a constant, whereas $\bar{X}$ is a random variable. Since $\bar{X}$ is a random variable, μ is but one of its possible values, and there is then the possibility (if not the likelihood) that $\bar{X}$ will differ from μ. That is, we must concern ourselves with the possibility that, given the random nature of $\bar{X}$, the estimate of μ we make based upon $\bar{X}$ will be incorrect (i.e., $\bar{X} \neq \mu$). But what are the possibilities and likelihoods in this respect? That is, how likely are we to be wrong? By how much are we likely to be wrong?

The key to these questions is the probability distribution of $\bar{X}$. But that is what we have just derived! In other words, our work concerning $p(\bar{X})$ will have direct implications for issues regarding the quality of the estimate, likely margins of error, and sophisticated inferences in general.

A similar analysis could be undertaken for the sample variance as compared with the population variance (or, for that matter, any sample statistic as compared with the respective population parameter); although the particulars are different in the case of the sample variance (in fact, the normal distribution is less important), the method of analysis is similar. The distribution of the sample variance is discussed in the appendix to this chapter.

In Section III we shall use the characteristics of sample statistics as random variables to make more specific, and to expand upon, the process of inference. To this problem we now turn.

REVIEW PROBLEMS FOR CHAPTER 9

9.16 An investor has three independent sources of income: wages, rents, and interest. Wages are normally distributed, with mean \$15,000 and standard deviation \$2000. Rents are normally distributed, with mean \$2000 and standard deviation \$500. Interest is normally distributed, with mean \$500 and standard deviation \$200. Find the probability that the investor's total income exceeds \$16,000.

9.17 Suppose that we take a simple random sample of 15 individuals with wages, rents, and interest as described in problem 9.16. Find the probability that the total income of these 15 individuals exceeds \$275,000.

9.18 Suppose that X is a normally distributed random variable, with mean 10 and variance 25. We take a simple random sample of size 4—that is, $X_1, \ldots, X_4$. If

$$W = .4X_1 + .3X_2 + .2X_3 + .1X_4$$

find $\Pr(W > 12)$.

*9.19 Suppose that there are 13 people (nine men and four women) standing on a platform, which has a weight capacity of 2100 pounds. Weights of men are normally distributed, with mean 160 pounds and standard deviation 20 pounds. Weights of women are normally distributed, with mean 125 pounds and standard deviation 15 pounds. What is the probability that the platform will collapse?

*9.20 Suppose that X has a normal distribution, with mean 6 and variance 16. We take a simple random sample of size 25 from this population. If

$$\Pr[(6 - \Psi) < \bar{X} < (6 + \Psi)] = .95$$

then what is Ψ? What does this result tell you regarding a likely margin of error for $\bar{X}$ as compared to the true population mean of 6?

*9.21 Suppose that X has a normal distribution with unknown mean μ and unknown variance σ^2. We take a simple random sample of size n from this population. If

$$\Pr[(\mu - \Psi) < \bar{X} < (\mu + \Psi)] = .95$$

then what is Ψ? What does this result tell you regarding a likely margin of error for $\bar{X}$ as a source of inference regarding the true but unknown population mean μ?

APPENDIX TO CHAPTER 9

In this appendix we shall discuss the distribution of the sample variance s^2, where

$$s^2 = \frac{1}{n - 1} \sum_{i=1}^{n} (X_i - \bar{X})^2$$

Let us first discuss this presuming that our sample is a simple random sample from a normally distributed parent population with mean μ and variance σ^2. [This discussion is based in part on Freund and Walpole (see note 3), pp. 285–287.] In this case, we know that $(X_i - \mu)/\sigma$ is a random variable that is standard normal, and the $(X_i - \mu)/\sigma$'s (across i) are independent. Thus,

$$\sum_{i=1}^{n} \left(\frac{X_i - \mu}{\sigma} \right)^2 \sim \chi^2_{(n)}$$

that is, this expression has a chi-square distribution with n degrees of freedom (see the appendix to Chapter 6).

It turns out that

$$\sum_{i=1}^{n} \left(\frac{X_i - \mu}{\sigma} \right)^2 = \left(\frac{\bar{X} - \mu}{\sigma/\sqrt{n}} \right)^2 + \frac{\sum_{i=1}^{n} (X_i - \bar{X})^2}{\sigma^2} \qquad \text{(A9.1)}$$

This equation is the key to determining the distribution of the sample variance, as we shall now see.

Note that $[(\bar{X} - \mu)/(\sigma/\sqrt{n})]^2$ is the square of a standard normal random variable, and thus it has a chi-square distribution with one degree of freedom. Equation (A9.1) is then

$$\chi^2_{(n)} = \chi^2_{(1)} + \frac{\sum_{i=1}^{n} (X_i - \bar{X})^2}{\sigma^2}$$

A random variable that has a chi-square distribution with n degrees of freedom can be written as the sum of squares of n independent standard normal random variables. Since a random variable that has a chi-square distribution with 1 degree of freedom can be written as the square of a single standard normal, we have the suggestion that

$$\frac{\sum_{i=1}^{n} (X_i - \bar{X})^2}{\sigma^2}$$

can be thought of as the sum of squares of $(n - 1)$ independent standard normals if these right-hand-side terms (including the $\chi^2_{(1)}$ variable) are independent. (It turns out that they are, but we shall just assert this.) This result suggests that

$$\frac{\sum_{i=1}^{n} (X_i - \bar{X})^2}{\sigma^2} \sim \chi^2_{(n-1)}$$

and since

$$\sum_{i=1}^{n} (X_i - \bar{X})^2 = (n - 1)s^2$$

that

$$\frac{(n - 1)s^2}{\sigma^2} \sim \chi^2_{(n-1)}$$

(Remember that the parent population is presumed to be normal.) This conclusion is our primary result concerning the distribution of s^2. For a given n, $p(s^2)$ then has the general chi-square shape, as seen in Figure A9.1 (although s^2 itself is *not* a chi-square). This

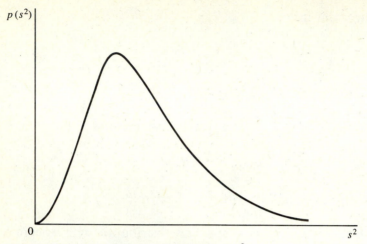

Figure A9.1 The sampling distribution of s^2.

result should be intuitively pleasing, since we know that s^2 cannot be negative (why?) but can be large.

Since

$$\chi^2_{(a)} = Z^2_1 + Z^2_2 + \cdots + Z^2_a$$

(where the Z_i's are a set of independent standard normals) we know that

$$E(\chi^2_{(a)}) = E(Z^2_1) + E(Z^2_2) + \cdots + E(Z^2_a)$$

$$= a$$

[since $E(Z^2_i) = 1$; can you show this?]. That is, the expected value of a chi-square is its degrees of freedom. It also turns out that the variance of a chi-square is twice its degrees of freedom—that is, var $[\chi^2_{(a)}] = 2a$. (The proof requires techniques beyond this course.) We can use these facts to determine the mean and variance of the sample variance s^2 when the parent population is normal.

We know that if

$$W = \frac{(n-1)s^2}{\sigma^2}$$

then $W \sim \chi^2_{(n-1)}$ and thus $E(W) = (n-1)$ and var $(W) = 2(n-1)$. But

$$s^2 = \left(\frac{\sigma^2}{n-1}\right)W$$

that is, s^2 is a linear (and in fact a proportional) function of W. Thus

$$E(s^2) = \left(\frac{\sigma^2}{n-1}\right)E(W) = \left(\frac{\sigma^2}{n-1}\right)(n-1) = \sigma^2$$

$$\text{var}(s^2) = \left(\frac{\sigma^2}{n-1}\right)^2 \text{var}(W) = \frac{\sigma^4}{(n-1)^2}2(n-1) = \frac{2\sigma^4}{n-1}$$

are properties of s^2 (when the parent population is normal).

  If the parent population is not normal, it is still the case that $E(s^2) = \sigma^2$. The variance of s^2, however, depends on the distribution of the parent population. We shall omit a discussion of these more complicated issues.

DISCUSSION QUESTIONS FOR CHAPTER 9

1. Given what you now know about the distribution of $\bar{X}$, what would you say happens to the probability that $\bar{X}$ lies within some fixed range of μ as the sample size n increases? Justify your answer both mathematically and intuitively.

2. Suppose you thought that $\mu = \mu_0$ (where μ_0 is some specific number you have specified) but you were not sure. In a simple random sample you find $\bar{X}$ to be quite a bit above μ_0. How might you use this information to evaluate the validity of your original position ($\mu = \mu_0$)?

3. Consider the text example of the shopkeeper who is interested in whether average revenues for the week are at least $250 (this appears in the section on applications of the central limit theorem). In this problem we presumed that the 30 days of the business month were a simple random sample. Was this assumption reasonable? Why might it not have been a reasonable assumption?

SOPHISTICATED STATISTICAL INFERENCE

Chapter *10*

Estimation

The simplest type of inference is a guess—an estimate; for example, we estimate the population mean μ with the sample mean $\bar{X}$. Although this approach might on the surface appear quite simple, lurking underneath is an as-yet-unanswered question: Why do we estimate this parameter this way? In fact, there are in general many ways of estimating parameters, and for us to make a judgment as to how to proceed we must make it clear what we are looking for in an estimator. That is, what makes a good estimator and what makes a not-so-good estimator?

The process of inference, as far as we have taken it, is essentially that of estimation. Our first step in developing more sophisticated inferences is to formalize the process of estimation, inasmuch as estimates are starting points for more sophisticated inferences. In this chapter, therefore, we develop the criteria by which we choose a process of estimation, and we apply them in a number of relatively familiar situations.

THE CRITERION OF UNBIASEDNESS

Recall from Chapter 9 that a sample statistic (such as $\bar{X}$) is a random variable; the implication is that there are multiple possible values

187

that the statistic can take on. Given that the population parameter of interest is a fixed number, the estimate we make using the sample statistic may be right or wrong. And given that sample statistics typically have probability distributions that are continuous (e.g., $\bar{X}$ is at least approximately normal) the probability that the statistic is equal to the parameter of interest is typically 0 [e.g., $\Pr(\bar{X} = \mu) = 0$ if $\bar{X}$ is normal]. This finding suggests, clearly, that we need to be somewhat humble regarding the reliability of simple estimates based on sample statistics.[1] It also suggests that asking the question "How likely is it that the estimator gives us the right answer?" is not going to be a particularly fruitful way of proceeding in our effort to develop criteria for evaluating estimators.

We must then turn to characteristics of the probability distribution of the estimator for determining its desirability or undesirability. The expected value of an estimator can be determined from knowledge concerning its probability distribution. In Chapter 9, for example, we determined that $E(\bar{X}) = \mu$. If you think about it, it should be apparent that this is an example of a highly desirable characteristic for an estimator; we know that $\bar{X}$ may differ from μ, but we also know that on average it does not. Certainly, it seems reasonable to avoid estimators that on average give you the wrong answer (i.e., the expected value is different from the parameter of interest).[2] Our first criterion for estimators is then that on average they should equal the parameter which we are trying to estimate. Estimators meeting this criterion are said to be *unbiased*.

Definition of Unbiasedness   Suppose that θ is a parameter characterizing a population, and that $\hat{\theta}$ is a sample statistic used to estimate θ. Then $\hat{\theta}$ is said to be an unbiased estimator of θ if $E(\hat{\theta}) = \theta$.

As an example, suppose that we have a simple random sample from a population with unknown mean μ and that we use the sample mean $\bar{X}$ to estimate μ. We know from our work in Chapter 9 that $E(\bar{X}) = \mu$, and we can therefore say that $\bar{X}$ is an unbiased estimator

---

[1] Remember that a zero probability does not imply that the event is impossible; rather, it is a statement of the long-run relative frequency with which we would expect the event to occur. See the discussion of probability density functions in Chapter 6.

[2] This conclusion is actually not as clear as we have just suggested. Later in this chapter we shall discuss its limitations.

of μ. That is, there is no tendency toward over- or underestimating μ when we use $\bar{X}$; given that $\bar{X}$ has a symmetric probability density function (the normal distribution) we know that $\bar{X}$ is just as likely to be too big as it is to be too small. (Note that here μ was an example of a θ and $\bar{X}$ was an example of a $\hat{\theta}$.)

For another example, let us suppose that we are interested in estimating the mean height of U.S. males. For our sample, we go to a gym and randomly select 10 male basketball players and then calculate the sample mean of their heights. We use this sample mean to estimate the population mean. In this case we have an example of a biased estimator, for it is more likely than not that the sample mean will exceed the population mean; that is, $E(\hat{\theta}) > \theta$. (After all, the mean of interest is for *all* U.S. males, and it is likely that basketball players are on average taller than nonplayers.) In this example the bias was really the result of our method of sampling; our sample was not a random sample because taller individuals had a greater chance of being selected than did shorter individuals. Nevertheless, the end result was still a biased estimate.

Finally, consider the sample variance. Suppose there is a population with variance σ^2, and we estimate this parameter by the sample variance

$$s^2 = \frac{1}{n-1} \sum_{i=1}^{n} (X_i - \bar{X})^2$$

In the appendix to Chapter 9 we discussed the fact that $E(s^2) = \sigma^2$, which implies that s^2 is an unbiased estimator of σ^2. This conclusion is, by the way, the justification for dividing by $(n-1)$ instead of n; when we divide by $(n-1)$, we have an unbiased estimator.

We can attempt to provide some intuition for this result by noting that it is only to the extent that n exceeds one that we even have the ability to measure dispersion; one observation is "used" in forming $\bar{X}$. There are then only $(n-1)$ observations contributing to dispersion about the sample mean, and therefore we divide by $(n-1)$.

Note that if we were to estimate σ^2 with division by n, that is, by forming

$$\hat{\theta} = \frac{1}{n} \sum_{i=1}^{n} (X_i - \bar{X})^2$$

it would be the case that $E(\hat{\theta}) \neq \sigma^2$ and that $\hat{\theta}$ is an example of a biased estimator. In particular, notice that

$$\hat{\theta} = \left(\frac{n-1}{n}\right)\left(\frac{1}{n-1}\right) \sum_{i=1}^{n} (X_i - \bar{X})^2$$

$$= \left(\frac{n-1}{n}\right)s^2$$

which suggests that

$$E(\hat{\theta}) = \left(\frac{n-1}{n}\right)E(s^2) = \left(\frac{n-1}{n}\right)\sigma^2 < \sigma^2$$

(Notice that $\hat{\theta}$ is a linear function of s^2, and therefore the expectation operator can be applied as above.) Hence if we were to use $\hat{\theta}$, there would be a tendency for us to underestimate the variance.

We define the *bias* of an estimator as the difference between its expected value and the parameter we are trying to estimate. That is,

$$\text{bias } (\hat{\theta}) = E(\hat{\theta}) - \theta$$

In the example above, with $\hat{\theta} = \dfrac{1}{n} \sum_{i=1}^{n} (X_i - \bar{X})^2$ as an estimator of σ^2,

$$\text{bias } (\hat{\theta}) = \left(\frac{n-1}{n}\right)\sigma^2 - \sigma^2 = \frac{-\sigma^2}{n}$$

The concept of unbiasedness is fairly straightforward; nevertheless, there are some possible limitations to this criterion for evaluating estimators that need attention. In particular,

1. There are often many unbiased estimators (so how do we choose among them?).
2. There may be situations in which a biased estimator is preferred.
3. There may be situations in which no unbiased estimator is available.

These issues will be discussed below, but they do not change the fact that unbiasedness is a fundamentally important property that we generally desire.

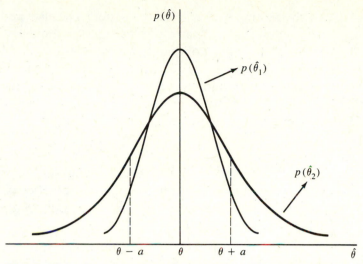

Figure 10.1   Comparing the efficiencies of unbiased estimators.

THE CRITERION OF EFFICIENCY

Estimators are random variables, and they therefore have probability distributions. The criterion of unbiasedness suggests that we should use an estimator that on average gives us the right answer even though for a particular sample it might be in error. But how do we choose between unbiased estimators, since often more than one will be available?

Consider Figure 10.1, within which the probability distributions of two different estimators ($\hat{\theta}_1$ and $\hat{\theta}_2$) of the parameter θ are drawn. Both estimators are unbiased estimators. Which should we use?

Unbiasedness tells us that on average we get the right answer, even though an individual estimate may be wrong. This is true for both $\hat{\theta}_1$ and $\hat{\theta}_2$. The two estimators differ, however, regarding their typical errors. Suppose that we formed some interval around the true parameter θ, from $\theta - a$ to $\theta + a$, where a has been chosen such that if we are outside this range we have made a "big" mistake. Clearly, the probability of a "big" mistake is smaller with $\hat{\theta}_1$ than with $\hat{\theta}_2$; formally,

$$\Pr[(\theta - a) < \hat{\theta}_1 < (\theta + a)] > \Pr[(\theta - a) < \hat{\theta}_2 < (\theta + a)]$$

In this case, $\hat{\theta}_1$ appears to be a preferable estimator.

What drives this result, of course, is the fact that $\hat{\theta}_1$ has a smaller variance than $\hat{\theta}_2$ does. In general, when comparing unbiased estimators, the one with the smaller variance is "better" because we are more likely to be within a given "narrow" range of the parameter of interest.[3] We then say that when comparing unbiased estimators, the one with the smaller variance is *more efficient* (in that it somehow uses the given information in a manner that creates a higher-quality estimate; that is, it uses this information more efficiently). Let us pause here for an example.

EXAMPLE ∎

Consider a population with unknown mean μ and variance σ^2. We observe a simple random sample of four observations, X_1, X_2, X_3, X_4 and consider the following two estimators of μ:

$$\hat{\mu}_1 = \frac{1}{4} X_1 + \frac{1}{4} X_2 + \frac{1}{4} X_3 + \frac{1}{4} X_4$$

$$\hat{\mu}_2 = \frac{1}{3} X_1 + \frac{1}{3} X_2 + \frac{1}{6} X_3 + \frac{1}{6} X_4$$

Since $\hat{\mu}_1 = \bar{X}$, and since $E(\bar{X}) = \mu$, we know that $\hat{\mu}_1$ is an unbiased estimator of μ. To determine that $\hat{\mu}_2$ is unbiased, we take its expected value (the key here is finding the expected value of a linear function):

$$E(\hat{\mu}_2) = \frac{1}{3} E(X_1) + \frac{1}{3} E(X_2) + \frac{1}{6} E(X_3) + \frac{1}{6} E(X_4)$$

$$= \frac{1}{3} \mu + \frac{1}{3} \mu + \frac{1}{6} \mu + \frac{1}{6} \mu = \mu$$

We now know that we are comparing two unbiased estimators. To determine which is better, we determine their variances. This procedure involves determining the variance of a linear function of four independent random variables. Since $\hat{\mu}_1 = \bar{X}$, we know that

$$\text{var}\,(\hat{\mu}_1) = \frac{\sigma^2}{4} = .25\sigma^2$$

---

[3] Strictly speaking, this statement pertains only to estimators with well-behaved and relatively symmetric distributions. That is, because of asymmetries the variance might not fully characterize the dispersion of interest in the estimator under consideration. But for the situations most commonly considered in economics it is a good working rule.

[Remember that var $(\bar{X}) = \sigma^2/n$, where σ^2 is the variance of the parent population.] For $\hat{\mu}_2$,

$$\text{var}\,(\hat{\mu}_2) = \left(\frac{1}{3}\right)^2 \text{var}\,(X_1) + \left(\frac{1}{3}\right)^2 \text{var}\,(X_2)$$

$$+ \left(\frac{1}{6}\right)^2 \text{var}\,(X_3) + \left(\frac{1}{6}\right)^2 \text{var}\,(X_4)$$

$$= \frac{1}{9}\,\sigma^2 + \frac{1}{9}\,\sigma^2 + \frac{1}{36}\,\sigma^2 + \frac{1}{36}\,\sigma^2$$

$$= \frac{10}{36}\,\sigma^2 = .278\sigma^2$$

Since

$$.25\sigma^2 < .278\sigma^2$$

regardless of the value of σ^2, we know that

$$\text{var}\,(\hat{\mu}_1) < \text{var}\,(\hat{\mu}_2)$$

Therefore $\hat{\mu}_1$ is a more efficient estimator. ∎

In many cases we can determine the unbiased estimator with the smallest possible variance (of any unbiased estimator); such an estimator would be called the *minimum-variance unbiased estimator* and is more efficient than any other unbiased estimator. For example, in the case of sampling from a normal population, it can be shown that the sample mean $\bar{X}$ is a minimum-variance unbiased estimator; this fact, in conjunction with the simplicity of calculating $\bar{X}$, is the typical justification for its use.[4]

PROBLEMS

10.1 Suppose that we have a simple random sample of size 3 from a population with mean μ. Our estimator of μ is $\hat{\mu}$, defined as follows:

---

[4] When sampling from a nonnormal population, there may be unbiased estimators of μ that are more efficient than $\bar{X}$. These estimators, however, would be nonlinear functions of the X_i's; that is, $\bar{X}$ is a minimum-variance *linear* unbiased estimator regardless of the form of the parent population.

$$\hat{\mu} = \frac{1}{6} X_1 + \frac{1}{6} X_2 + \frac{2}{3} X_3$$

Is $\hat{\mu}$ an unbiased estimator of μ? If not, what is its bias? Fully justify your answer.

10.2 For the sample in problem 10.1, consider another estimator of μ (call it $\hat{\theta}$) that is defined as follows:

$$\hat{\theta} = \frac{1}{2} X_1 + \frac{1}{2} X_2 + \frac{1}{2} X_3$$

Is $\hat{\theta}$ an unbiased estimator of μ? If not, what is its bias? Fully justify your answer.

10.3 We have a simple random sample of size 4 from a population with mean μ. Consider the following two estimators of μ:

$$\hat{\mu}_1 = \frac{1}{10} X_1 + \frac{1}{10} X_2 + \frac{2}{5} X_3 + \frac{2}{5} X_4$$

$$\hat{\mu}_2 = \frac{1}{9} X_1 + \frac{1}{9} X_2 + \frac{1}{9} X_3 + \frac{2}{3} X_4$$

(a) Prove that both of these estimators are unbiased estimators of μ.

(b) Which is better? Fully justify your answer.

10.4 Suppose that we have independent samples of size 10 and 15, respectively, from a population with mean μ. We calculate the sample mean for each sample and refer to them as $\bar{X}_{10}$ and $\bar{X}_{15}$.

(a) Is $\bar{X}_{10}$ or $\bar{X}_{15}$ a better estimator of μ? Fully justify your answer.

(b) Consider the following ways of combining $\bar{X}_{10}$ and $\bar{X}_{15}$ to form a single estimate of μ:

$$\hat{\theta}_1 = \frac{1}{2} \bar{X}_{10} + \frac{1}{2} \bar{X}_{15}$$

$$\hat{\theta}_2 = \frac{10}{25} \bar{X}_{10} + \frac{15}{25} \bar{X}_{15}$$

The difference has to do with whether the sample means are equally weighted or given differential weights based upon the different sample sizes. Which estimator is better? Fully justify your answer.

10.5 Suppose that we have a simple random sample of size n from a population with mean μ. Consider the following estimator of μ:

$$\hat{\mu} = \frac{1}{n-1} \sum_{i=1}^{n} X_i$$

Is $\hat{\mu}$ an unbiased estimator of μ? If not, what is its bias?

10.6 Suppose that we have a simple random sample of size n from a population with variance σ^2. We estimate this variance as follows:

$$\hat{\sigma}^2 = \frac{1}{n+1} \sum_{i=1}^{n} (X_i - \bar{X})^2$$

What is the bias of $\hat{\sigma}^2$ as an estimator of σ^2?

10.7 Suppose that $E(X) = E(Y) = \mu$ but that var $(Y) = .5$ var (X). We have independent simple random samples of size 2 each from these populations. Consider the following two estimators of μ:

$$\hat{\mu}_1 = \frac{1}{4} X_1 + \frac{1}{4} X_2 + \frac{1}{4} Y_1 + \frac{1}{4} Y_2$$

$$\hat{\mu}_2 = \frac{1}{6} X_1 + \frac{1}{6} X_2 + \frac{1}{3} Y_1 + \frac{1}{3} Y_2$$

Which is better? Fully and rigorously justify your answer.

10.8 Suppose that $E(X) = \mu$ and $E(Y) = 2\mu$. We have independent simple random samples of size n each from these two populations. Suppose that var $(X) =$ var (Y). Consider the following estimators of μ:

$$\hat{\mu}_1 = \bar{X}$$

$$\hat{\mu}_2 = \frac{1}{2} \bar{Y}$$

$$\hat{\mu}_3 = \frac{1}{4} \bar{X} + \frac{3}{8} \bar{Y}$$

$$\hat{\mu}_4 = \frac{1}{5} \bar{X} + \frac{2}{5} \bar{Y}$$

Which estimator should we use? Fully and rigorously justify your answer.

*10.9 Redo problem 10.8, except now assume that var (Y) = 4 var (X).

*10.10 We know that $E(s^2) = \sigma^2$—that is, that the sample variance is an unbiased estimator of the population variance. We also typically estimate the population standard deviation σ with the sample standard deviation s. Is s an unbiased estimator of σ? If not, is its bias positive or negative? (*Hint:* For any random variable X, var $(X) = E(X^2) - [E(X)]^2$.)

COMPARING UNBIASED AND BIASED (OR TWO BIASED) ESTIMATORS

The discussion above suggests that unbiased estimators should be chosen above biased estimators. Although this framework is the one that is usually applied, there are, at least in principle, situations in which a biased estimator *might* be viewed as preferable to an unbiased estimator. Figure 10.2 depicts such a situation.

The parameter of interest is θ, and we plot the distributions of two estimators $\hat{\theta}_1$ and $\hat{\theta}_2$, where the former is an unbiased estimator but the latter is not. Clearly, the advantage to $\hat{\theta}_1$ is that on average we get the right answer, whereas with $\hat{\theta}_2$ on average we underestimate θ. On the other hand, if we were to construct a range of values from $\theta - a$ to $\theta + a$ (i.e., within a "narrow" range around θ), it would be the case that

$$\Pr[(\theta - a) < \hat{\theta}_2 < (\theta + a)] > [\Pr(\theta - a) < \hat{\theta}_1 < (\theta + a)]$$

Although $\hat{\theta}_2$ is biased, it has a larger probability of taking on a value within some "narrow" range of θ; if a mistake of a or less is considered "acceptable" and of more than a "unacceptable," it might be sensible to rely on $\hat{\theta}_2$ instead of $\hat{\theta}_1$!

Suppose that θ represents the actual rate of return on a stock over a given period, and that $\hat{\theta}_1$ and $\hat{\theta}_2$ are different ways of estimating (forecasting) this rate. It is likely that for some individuals $\hat{\theta}_2$ would be the preferable way of forming the forecast; the chance of a big mistake is small, and if anything our bias is toward underestimating performance (i.e., we are "too conservative"). Of course there are other situations in which one might imagine that $\hat{\theta}_1$ is the preferable

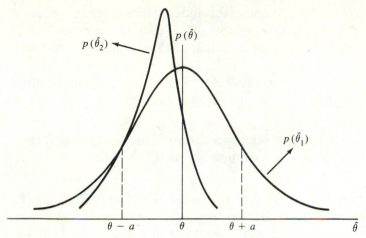

Figure 10.2 Comparing a biased estimator ($\hat{\theta}_2$) and an unbiased estimator ($\hat{\theta}_1$).

method of estimation (if this were one of many stocks in a portfolio, for example, so that risk was lessened through diversification).

How might we quantitatively compare biased versus unbiased (or two biased) estimators? Our discussion suggests that there are two negative characteristics of estimators that we should like to minimize: (1) amount of bias and (2) variance. When comparing unbiased estimators, it is clear that the comparison is then among the variances (as above). But what about estimators with different biases?

We might form an index that combines the two undesirable characteristics of estimators (variance and bias) into one number. One way would be simply to add the variance to the bias, although there would be a problem since the bias might be negative and therefore offset (at least in part) the variance. To correct this situation we might form

$$\text{var}\,(\hat{\theta}) + [\text{bias}\,(\hat{\theta})]^2$$

where we square the bias so as to ensure that the second term is nonnegative.[5] We might then adopt the rule of choosing the esti-

---

[5] Note that the two terms on the right-hand side will be measured in the same units. These units are the square of the units in which the observations are measured; if, for example, our observations on X are in inches, then the variance is in square inches and the square of the bias is in square inches as well (since the bias would be in inches).

mator with the smaller value of this index. (In the case of unbiased estimators the bias is zero, and this rule is then the same as the one suggested earlier in this chapter; that is, choose the estimator with the smaller variance.)

The index referred to above is generally referred to as the *mean squared error*. Formally, we define it as follows

Definition of the Mean Squared Error   The mean squared error of an estimator $\hat{\theta}$ of θ, written mse $(\hat{\theta})$, is $E[(\hat{\theta} - \theta)^2]$.

[Note that this value would be the variance of $\hat{\theta}$ only if $E(\hat{\theta}) = \theta$.] This definition appears different from the index suggested above, but it turns out that the expressions are identical. That is, it will always be the case that[6]

$$\text{mse } (\hat{\theta}) = E[(\hat{\theta} - \theta)^2] = \text{var } (\hat{\theta}) + [\text{bias } (\hat{\theta})]^2$$

Although comparing the mean squared errors is a possible decision-making rule when deciding between biased and unbiased (or two biased) estimators, there are other possible decision-making rules. We suggest the mean squared error rule as one possibility.[7]

---

[6] To prove the result that follows, note that the mean squared error may be rewritten as

$$\text{mse } (\hat{\theta}) = E\{[\hat{\theta} - E(\hat{\theta})] + [E(\hat{\theta}) - \theta]\}^2$$

where we have subtracted and added $E(\hat{\theta})$. Squaring the term gives us

$$\text{mse } (\hat{\theta}) = E\{[\hat{\theta} - E(\hat{\theta})]^2 + [E(\hat{\theta}) - \theta]^2 + 2[E(\hat{\theta}) - \theta][\hat{\theta} - E(\hat{\theta})]\}$$

If we push the E through, remembering that $E(\hat{\theta})$ and θ are constants and that the expected value of a constant is itself, we get

$$\text{mse } (\hat{\theta}) = E\{[\hat{\theta} - E(\hat{\theta})]^2\} + [E(\hat{\theta}) - \theta]^2 + 2[E(\hat{\theta}) - \theta][E(\hat{\theta}) - E(\hat{\theta})]$$

The final term on the right side is zero since $E(\hat{\theta}) - E(\hat{\theta}) = 0$. Then, since

$$E\{[\hat{\theta} - E(\hat{\theta})]^2\} = \text{var } (\hat{\theta}) \quad \text{and} \quad E(\hat{\theta}) - \theta = \text{bias } (\hat{\theta})$$

we get

$$\text{mse } (\hat{\theta}) = \text{var } (\hat{\theta}) + [\text{bias } (\hat{\theta})]^2$$

which proves the result.

[7] Another possibility is to use the so-called "maximum likelihood" approach; this approach is discussed in advanced treatments of statistics and we shall not pursue it further here.

EXAMPLE ■

In this example we show how the mean squared errors of two biased estimators can be compared. Suppose that we have a simple random sample of size 3 from a population with mean μ and variance σ^2. Consider the following two estimators of μ:

$$\hat{\theta}_1 = \frac{1}{4} X_1 + \frac{1}{4} X_2 + \frac{1}{4} X_3$$

$$\hat{\theta}_2 = \frac{1}{7} X_1 + \frac{1}{7} X_2 + \frac{3}{7} X_3$$

It can be determined that

$$E(\hat{\theta}_1) = \frac{3}{4} \mu \quad \text{and} \quad \text{bias} (\hat{\theta}_1) = -.25\mu$$

$$E(\hat{\theta}_2) = \frac{5}{7} \mu \quad \text{and} \quad \text{bias} (\hat{\theta}_2) = -.29\mu$$

$$\text{var} (\hat{\theta}_1) = \frac{3}{16} \sigma^2 = .19\sigma^2$$

$$\text{var} (\hat{\theta}_2) = \frac{11}{49} \sigma^2 = .22\sigma^2$$

(Verify these computations.) Thus

$$\text{mse} (\hat{\theta}_1) = .19\sigma^2 + (-.25\mu)^2 = .19\sigma^2 + .06\mu^2$$

$$\text{mse} (\hat{\theta}_2) = .22\sigma^2 + (-.29\mu)^2 = .22\sigma^2 + .08\mu^2$$

Since $.22\sigma^2 > .19\sigma^2$ and $.08\mu^2 > .06\mu^2$ (regardless of the values of σ^2 and μ) we know that mse $(\hat{\theta}_1) <$ mse $(\hat{\theta}_2)$.   ■

In economics we generally use unbiased estimators when they are available, and in fact the search for unbiased estimators is an important process. There are, however, some situations in which no unbiased estimators are available; these situations appear frequently in regression analysis beyond the simplest cases. In the appendix to this chapter we shall describe a somewhat more general concept than unbiasedness, which is often used in such cases.

PROBLEMS

10.11 Consider the sample mean $\bar{X}$ as an estimator of the population mean μ. Find the mean squared error of $\bar{X}$.

10.12 A simple random sample of size 4 has been taken from a population with unknown mean μ. Consider the following estimator of μ:

$$\hat{\mu} = \frac{1}{3} X_1 + \frac{1}{3} X_2 + \frac{1}{3} X_3 + \frac{1}{3} X_4$$

Find the mean squared error of $\hat{\mu}$.

10.13 Consider the following two estimators of the population mean μ based upon the sample described in problem 10.12:

$$\hat{\mu}_1 = \frac{1}{8} X_1 + \frac{1}{8} X_2 + \frac{3}{8} X_3 + \frac{3}{8} X_4$$

$$\hat{\mu}_2 = \frac{1}{5} X_1 + \frac{1}{5} X_2 + \frac{1}{5} X_3 + \frac{1}{5} X_4$$

Calculate the mean squared error of each estimator, and then discuss how these mean squared errors might be related to each other in terms of their relative sizes.

10.14 Continuing problem 10.13, suppose that

$$\hat{\mu}_3 = \frac{1}{7} X_1 + \frac{1}{7} X_2 + \frac{3}{8} X_3 + \frac{5}{8} X_4$$

What is the mean squared error of $\hat{\mu}_3$? What can you say about how this value compares to the mean squared errors of $\hat{\mu}_1$ and $\hat{\mu}_2$?

SUMMARY

In this chapter we have discussed more specifically than before the process of inference called estimation and have presented some criteria for use in determining the appropriate method of estimation. Regardless of the estimator we choose, however, we are left with the powerful fact that we are likely to wind up with an incorrect estimate (since estimators are random variables). In the next chapter we specifically acknowledge this fact and explore an approach to inference that is based on it.

REVIEW PROBLEMS FOR CHAPTER 10

10.15 Suppose that we take a simple random sample of size 3 from a population with mean μ. Consider the following estimators of μ:

$$\hat{\mu}_1 = \frac{1}{3} X_1 + \frac{1}{3} X_2 + \frac{1}{3} X_3$$

$$\hat{\mu}_2 = \frac{1}{6} X_1 + \frac{1}{6} X_2 + \frac{4}{6} X_3$$

$$\hat{\mu}_3 = \frac{1}{7} X_1 + \frac{2}{7} X_2 + \frac{3}{7} X_3$$

$$\hat{\mu}_4 = \frac{1}{5} X_1 + \frac{2}{5} X_2 + \frac{3}{5} X_3$$

Which estimators are unbiased? Fully justify your answer. Do you see the general rule that is suggested?

10.16 Suppose that X and Y represent two populations with the following characteristics:

$$E(X) = E(Y) = \mu \qquad \text{var}\,(X) = 5 \qquad \text{var}\,(Y) = 10$$

You have independent simple random samples of size 2 from each population. Consider the following two estimators of μ:

$$\hat{\theta}_1 = \frac{1}{4} X_1 + \frac{1}{4} X_2 + \frac{1}{4} Y_1 + \frac{1}{4} Y_2$$

$$\hat{\theta}_2 = \frac{2}{5} X_1 + \frac{2}{5} X_2 + \frac{1}{10} Y_1 + \frac{1}{10} Y_2$$

(a) Prove that these are both unbiased estimators of μ.
(b) Which is better? Fully justify your answer.

10.17 Suppose that we have two unbiased and independent estimates $\hat{\theta}_1$ and $\hat{\theta}_2$ of a parameter θ. However, $\hat{\theta}_2$ is less reliable than $\hat{\theta}_1$ is; the standard deviation of $\hat{\theta}_2$ is twice the standard deviation of $\hat{\theta}_1$. Consider the following methods of combining $\hat{\theta}_1$ and $\hat{\theta}_2$ so as to form a single estimate of θ:

$$\hat{W}_1 = \frac{1}{2} \hat{\theta}_1 + \frac{1}{2} \hat{\theta}_2$$

$$\hat{W}_2 = \frac{4}{5}\hat{\theta}_1 + \frac{1}{5}\hat{\theta}_2$$

$$\hat{W}_3 = \frac{5}{6}\hat{\theta}_1 + \frac{1}{6}\hat{\theta}_2$$

$$\hat{W}_4 = \hat{\theta}_1$$

(a) Which are unbiased? Fully justify your answer.

(b) Rank the combined estimators in terms of efficiency.

10.18 Suppose that we have two types of observations from a parent population X having mean μ and variance σ^2. X_1 and X_2 are from a simple random sample of size 2, and are accurately measured. X_3 and X_4 are from an independent simple random sample in which the observations are subject to measurement error. That is,

$$\text{var}(X_1) = \text{var}(X_2) = \sigma^2$$

$$\text{var}(X_3) = \text{var}(X_4) = \sigma^2 + k$$

where $k > 0$ is the measurement error factor. Consider the following two estimators of μ:

$$\hat{\theta}_1 = \frac{1}{4}X_1 + \frac{1}{4}X_2 + \frac{1}{4}X_3 + \frac{1}{4}X_4$$

$$\hat{\theta}_2 = \frac{1}{3}X_1 + \frac{1}{3}X_2 + \frac{1}{6}X_3 + \frac{1}{6}X_4$$

Suppose that $k = .5\sigma^2$. Which estimator is better? Fully justify your answer.

*10.19 Ms. Jones is a developer who owns a square plot of land. To estimate its area, she measures its length. In doing this she makes a random error; her measure X of the length is a random variable with mean μ (the true length) and variance σ^2. Aware of the possible error, she takes a second independent measurement Y, where Y is a random variable with the same probability distribution as X. Consider the following two estimators of the area of the plot of land (μ^2):

$$\hat{A}_1 = \left(\frac{X + Y}{2}\right)^2$$

$$\hat{A}_2 = \frac{X^2 + Y^2}{2}$$

Which estimator has a smaller bias?

*10.20 Ms. Jones owns another plot of land that is rectangular. Its true width is W and its true length is L. To estimate its area she measures its width and length, but in doing so she makes random errors. Her measure of the width is X_1, which is a random variable with mean W and variance σ^2. Her measure of the length is X_2, which is a random variable with mean L and variance σ^2. The correlation coefficient between X_1 and X_2 is .5. Her estimate of the area is $\hat{A} = X_1 X_2$. What is the bias of $\hat{A}$ as an estimate of the true area?

APPENDIX TO CHAPTER 10

In this appendix we discuss the concept of *consistency* as a property an estimator may have. This property is a key element in evaluating the performance of estimators in relatively sophisticated econometric situations. In Figure A10.1 we have drawn various probability distributions (associated with increasing sample sizes, where $n_1 < n_2 < n_3 < n_4$) for some estimator $\hat{\theta}$ of θ.

The distribution of $\hat{\theta}$ when $n = n_1$ suggests a biased estimator; however, as the sample size increases, we find that (1) the difference between $E(\hat{\theta})$ and θ falls (the bias decreases) and (2) the variance decreases. There is a class of estimators for which this general pattern emerges; that is, both the bias and variance approach zero as the sample size increases.

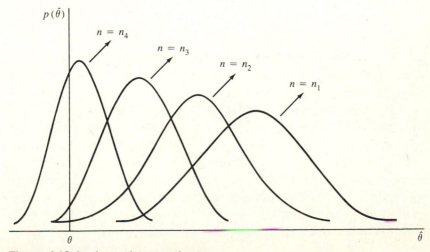

Figure A10.1 A consistent estimator.

This pattern of behavior is very desirable for it suggests that the probability that $\hat{\theta}$ falls within some (narrow) range of θ (from $\theta - k$ to $\theta + k$, where k is some number) is increasing as n rises, and it becomes close to one once the sample is of sufficient size. More formally,

$$\Pr(|\hat{\theta} - \theta| < k)$$

rises as n increases; and as n gets very large, this probability gets close to one.

This explanation of the property of consistency is an intuitive one. Both unbiased and biased estimators can have this property, but it is also possible for both unbiased and biased estimators not to have it.

Formally, an estimator $\hat{\theta}$ of θ is consistent if and only if

$$\lim_{n \to \infty} \Pr(|\hat{\theta} - \theta| < k) = 1$$

for any $k > 0$. [The parameter k is viewed as some arbitrarily small positive number; see, for example, Jan Kmenta, *Elements of Econometrics,* 2d ed. (New York: Macmillan, 1986), p. 166.] Although this definition is difficult to use in practice, it turns out that there is a condition for consistency that is often easy to use and is in fact a common way of determining the presence of this property. If both the variance and the bias of an estimator approach zero as n approaches infinity, then the distribution of $\hat{\theta}$ "collapses" over θ, which is exactly what happens in Figure A10.1 where a consistent estimator is depicted. Since the mean squared error is related to the variance and the bias by

$$\text{mse}\,(\hat{\theta}) = \text{var}\,(\hat{\theta}) + [\text{bias}\,(\hat{\theta})]^2$$

the suggestion is that if the mean squared error of $\hat{\theta}$ approaches zero as n approaches infinity, then $\hat{\theta}$ is a consistent estimator of θ. This result is commonly referred to as the mean squared error condition.

It is important to note that, although the mean squared error condition being satisfied implies that $\hat{\theta}$ is consistent, a $\hat{\theta}$ that does not satisfy this condition may still be consistent. That is, the mean squared error condition is a sufficient but not a necessary condition for consistency. This statement suggests that showing that the mean squared error does not approach zero does not then prove that the estimator is inconsistent. In general, we would show that an estimator is inconsistent through the use of the concept of a *probability limit,* which is discussed in courses in econometrics; this approach would often involve showing that the distribution of the estimator collapses over a number other than θ.

As an example, consider $\bar{X}$ as an estimator of μ. We know that

$$\text{mse }(\bar{X}) = \frac{\text{var }(X)}{n}$$

[since bias $(\bar{X}) = 0$] and therefore that mse $(\bar{X}) \to 0$ as $n \to \infty$. The sample mean $\bar{X}$ is then an example of a consistent estimator, which suggests that if we have a large enough sample its value is very likely to be very close to μ.

Consistency is an example of what we call a *large-sample* property; that is, it says nothing about the behavior of the estimator in a small sample but rather refers to how the pattern of behavior of the estimator changes as the sample size becomes large. Consistency is then most appropriate when we have a sample that is of at least a reasonably large size, but how many observations are necessary depends on the problem at hand.

The concept of consistency is important in econometrics because there are many situations where, although there are no unbiased estimators available, there are consistent estimators available. The search for acceptable estimators then becomes a search for consistent ones. For a further discussion of the property of consistency, see an econometrics text.

DISCUSSION QUESTIONS FOR CHAPTER 10

1. What information would be helpful in determining the likely magnitude of a mistake one makes when using $\bar{X}$ to estimate μ?

2. How might the information involved in determining the relative efficiency of unbiased estimators be used to make a more sophisticated inference regarding the parameter of interest?

3. Discuss the sense in which the mean squared error decision rule (i.e., pick the estimator with the smallest mean squared error) is a generalization of our discussion regarding the relative efficiency of unbiased estimators.

4. The mean squared error might be described as "the expected square of our mistake." Explain.

Chapter *11*

Confidence Intervals

Chapter 9 was concerned with the process of sampling and its general implications for sample statistics. Chapter 10 was concerned with using sample statistics as estimators of population parameters. In this chapter we acknowledge the limitations of the estimation processes described in Chapter 10 and consider a more general framework for making inferences concerning population parameters. We shall now begin to develop and use more sophisticated inferences to learn more about the characteristics of populations that are not completely observed.

MOTIVATION

We begin with an observation concerning the process of estimation: Even if we are using an estimator that is unbiased and efficient, for all intents and purposes the answer we get is bound to be wrong. Why?

Consider the simple case in which we estimate an unknown population mean μ of a population that is known to be normally distributed by taking a simple random sample and calculating $\bar{X}$. We know that the distribution of $\bar{X}$ is normal; since the normal distribution is a continuous probability distribution, we know that the probability that $\bar{X}$ takes on any specific value is zero. That suggests that $\Pr(\bar{X} = \mu) = 0$; hence, the probability that we get the right answer is zero! Although this statement does not imply that it is impossible for us to get the right answer, it suggests that we should generally assume that $\bar{X} \neq \mu$.

One way to interpret this result is to say that since we are likely to be wrong, why do anything at all? The problem with this response is that it ignores the fact that there are different ways of being wrong, and that we in fact can say something about the likelihoods associated with various ways of being wrong! To be more precise, this conclusion takes into account the fact that sample statistics are random variables but ignores the fact that they often have probability distributions that we can determine and exploit in an effort to broaden the context within which we make inferences. Let us now pursue this observation.

MARGINS OF ERROR

Let us exploit what we know about probability distributions of sample statistics to try to develop methods leading to inferences that have a good chance of being correct. Consider the example of the sample mean $\bar{X}$ calculated from a simple random sample from a normally distributed population with mean μ; its probability distribution is shown in Figure 11.1. The parameter k is the number such that

$$\Pr[(\mu - k) < \bar{X} < (\mu + k)] = .95 \qquad (11.1)$$

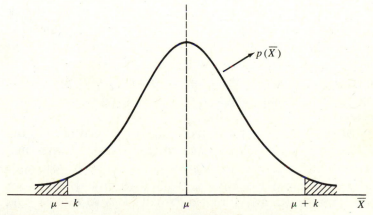

Figure 11.1 The sampling distribution of $\bar{X}$.

The value of k can be determined through our knowledge regarding the probability distribution of $\bar{X}$; given the symmetric nature of this distribution, the probability in each of the two tails of $\bar{X}$'s distribution (i.e., below $\mu - k$ and above $\mu + k$) is .025.

Earlier in this chapter we noted that $\Pr(\bar{X} = \mu) = 0$ due to the fact that $\bar{X}$ is a continuous random variable. Equation (11.1), on the other hand, tells us that $\bar{X}$ is "off" by k or less 95% of the time; that is, there is a 95% probability that $\bar{X}$ lies within k units of μ.

We can exploit this fact to develop a more sophisticated inference regarding μ. Consider the interval

$$(\bar{X} - k) \text{ to } (\bar{X} + k) \tag{11.2}$$

If $\bar{X}$ is in the interval $(\mu - k, \mu + k)$ then the interval in Eq. (11.2) will contain the value of μ. (If $\bar{X}$ is between μ and $\mu + k$, then the lower bound $\bar{X} - k$ will be less than μ, and if $\bar{X}$ is between μ and $\mu - k$, then the upper bound $\bar{X} + k$ will be more than μ; in both cases μ is within the interval in Eq. (11.2).) If, however, $\bar{X}$ is not in the interval $(\mu - k, \mu + k)$, then the Eq. (11.2) interval will not contain μ.

Since Eq. (11.2) is an interval that contains μ whenever $\bar{X}$ is in the interval $(\mu - k, \mu + k)$, and since $\Pr[(\mu - k) < \bar{X} < (\mu + k)] = .95$, we then conclude that the interval in Eq. (11.2) is one that 95% of the time contains μ.

This result is very important, and we should at this point make sure we see its significance. We know that $\Pr(\bar{X} = \mu) = 0$. On the other hand, if we say that μ is somewhere within the interval given by Eq. (11.2) without saying exactly where within that interval it is, then we are making a statement regarding μ that is true 95% of the time! In some sense we have "traded" the precision of a simple estimate (which is true 0% of the time) for the high likelihood of saying something that is correct. But perhaps the more constructive way to view this situation is to say that the precision of a simple estimate is really an illusion; the simple estimate $\bar{X}$ of μ is bound to be wrong because of sampling variability, so what is the point of the precision?

The value k is essentially a "likely margin of error," and our approach here has then led to a "sophisticated" type of inference. Sometimes k will be rather small, in which case we have a good degree of precision in our inference, but this will not always be the case. Often, the key to arriving at a "small" k is to have a large sample, as we shall see below.

As an example, the value of k we are discussing is often seen in public opinion polls; if a poll suggests that 60% of the population thinks the president is doing a good job and the poll has a sampling error of plus or minus three percentage points, then the three percentage points is the value of k.

The interval in Eq. (11.2) is generally referred to as a *95% confidence interval.* But where does the value of k come from? Refer back to Figure 11.1; the value of k is chosen such that $\Pr[(\mu - k) < \bar{X} < (\mu + k)] = .95$, which implies that k's value is determined from the characteristics of the distribution of $\bar{X}$. Since we know a great many things about the distribution of $\bar{X}$, we should be able to determine the value of k. First, however, let us consider some interpretive issues.

What is the sense in which Eq. (11.2) is a 95% confidence interval? There are some subtle issues here that deserve exploration. First, notice that once the Eq. (11.2) interval has been calculated, μ is either in the interval or not, and there is no element of randomness in this determination. Once the interval has been calculated as (a, b) where a and b are numbers, it is fixed (i.e., not random), as is μ. Thus μ is either in the interval or not. We do not say that the *probability* that μ is in the fixed interval is 95%, since neither μ nor the fixed interval is random.[1]

So where does the 95% come in? Our knowledge of the distribution of $\bar{X}$ tells us that $\bar{X}$ will be between $\mu - k$ and $\mu + k$ with probability .95; given that probability is a long-run relative frequency, this observation suggests that if we were to take thousands and thousands of simple random samples (of a given size) and for each one of them calculate $\bar{X}$, just about 95% of these $\bar{X}$'s would be between $\mu - k$ and $\mu + k$. We do not know whether the one $\bar{X}$ we actually have is in that range or not, but since 95% of all $\bar{X}$'s are, we are 95% *confident* that ours is. Since $(\mu - k) < \bar{X} < (\mu + k)$ implies $(\bar{X} - k) < \mu < (\bar{X} + k)$, we are 95% confident that the interval $(\bar{X} - k, \bar{X} + k)$ contains μ. We therefore refer to Eq. (11.2) as a *95% confidence interval.* Although we do not know whether our particular confidence interval contains μ or not, we are 95% confident that it does since 95% of all such confidence intervals contain μ.

Why do we choose 95% as the level of confidence? This choice is admittedly arbitrary, but we can motivate it a bit by referring

---

[1] In actuality, the probability that μ is in the fixed interval is either 0 or 1 but we do not know which.

back to Figure 11.1. The value of k was chosen so that $\Pr[(\mu - k)$ $< \bar{X} < (\mu + k)] = .95$. This choice suggests that in order to increase the level of confidence we would have to increase the size of k; our confidence interval would then be wider (making our inference less precise). On the other hand, decreasing the value of k (making our inference more precise) decreases the level of confidence. So we have a trade-off between precision and confidence, and 95% is often seen as an appropriate choice to make regarding this trade-off. But this choice is in some sense arbitrary, and in some situations alternative levels of confidence are seen as more appropriate (90% and 99% are common alternatives).

Let us now begin to address the most obvious unsettled issue, which is the determination of the value of k. We stick for the moment to the case of $\bar{X}$ as a sample statistic for inferring the value of μ, although the concept of a confidence interval is clearly applicable to any parameter.

DETERMINING THE VALUE OF k—A FIRST STEP

The determination of the value of k is based upon what we know concerning the distribution of $\bar{X}$. Let us continue with the above assumption that the parent population X is normal. This implies

$$\bar{X} \sim N\left(\mu, \frac{\sigma^2}{n}\right)$$

where μ and σ^2 are, respectively, the mean and variance of the parent population. If we take a normally distributed random variable and subtract its mean and divide by its standard deviation the result will be a random variable that has a standard normal distribution. That is,

$$\frac{\bar{X} - \mu}{\sigma/\sqrt{n}} \sim N(0, 1)$$

We can use this fact to determine the value of k. In Figure 11.2 we draw a standard normal distribution. The $Z_{.025}$ value is defined such that $\Pr(Z > Z_{.025}) = .025$; that is, the area under the curve to the right of this point is .025. In view of the symmetry of the standard

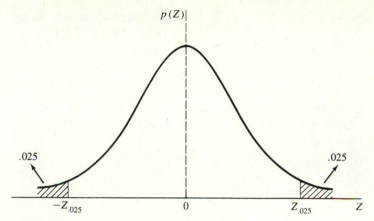

$p(Z)|$

.025

.025

$-Z_{.025}$ 0 $Z_{.025}$ Z

Figure 11.2 Interpreting the $Z_{.025}$ value.

normal distribution about zero we know that $\Pr(Z < -Z_{.025}) = .025$ (as shown in the figure) and therefore that

$$\Pr(-Z_{.025} < Z < Z_{.025}) = .95$$

This result suggests that

$$\Pr\left[-Z_{.025} < \left(\frac{\bar{X} - \mu}{\sigma/\sqrt{n}}\right) < Z_{.025}\right] = .95$$

The above event describes a situation in which two things are simultaneously true. If we perform the same operation on each part of the inequality, we do not change the nature of the event it describes and therefore do not change the probability. What we want to do is to isolate $\bar{X}$ in the middle by itself and hope that the expressions on the two ends will be of the form $\mu - k$ and $\mu + k$; if we can, we will have an expression like Eq. (11.1) in which the value of k is specified, and we can then determine the 95% confidence interval following the logic described above.

To accomplish this goal we first multiply through by $\sigma/\sqrt{n}$:

$$\Pr\left[-Z_{.025} \frac{\sigma}{\sqrt{n}} < (\bar{X} - \mu) < Z_{.025} \frac{\sigma}{\sqrt{n}}\right] = .95$$

Finally, we add μ to each part of the inequality:

$$\Pr\left[\left(\mu - Z_{.025} \frac{\sigma}{\sqrt{n}}\right) < \bar{X} < \left(\mu + Z_{.025} \frac{\sigma}{\sqrt{n}}\right)\right] = .95 \quad (11.3)$$

If we compare Eqs. (11.3) and (11.1), we see that they are the same given that

$$k = Z_{.025} \frac{\sigma}{\sqrt{n}}$$

We have therefore determined the value of k. How did we do it? By exploiting what we had learned in Chapter 9 concerning the distribution of $\bar{X}$ as a random variable.

Recall that the discussion above suggested that a 95% confidence interval for μ is

$$(\bar{X} - k \, , \, \bar{X} + k)$$

We conclude then by saying that a 95% confidence interval for μ is

$$\bar{X} - Z_{.025} \frac{\sigma}{\sqrt{n}} \text{ (lower bound)} \quad \text{and} \quad \bar{X} + Z_{.025} \frac{\sigma}{\sqrt{n}} \text{ (upper bound)}$$

$$(11.4)$$

that is,

$$\bar{X} \pm Z_{.025} \frac{\sigma}{\sqrt{n}}$$

Table II of the standard normal distribution can be used to determine that $Z_{.025} = 1.96$, as $\Pr(Z > 1.96) = .025$.

Strictly speaking, Eq. (11.4) describes a 95% confidence interval for μ only in the case where the parent population X is normal. But it would be a good approximation of the 95% confidence interval if this were not the case (as long as n is at least moderately large) because of the central limit theorem and the approximate normality of $\bar{X}$.

A confidence interval at a different level of confidence can be analogously constructed; all that would change is the Z value. A 99% confidence interval, for example, would be

$$\bar{X} \pm Z_{.005} \frac{\sigma}{\sqrt{n}}$$

where $\Pr(Z > Z_{.005}) = .005$. The level of the Z statistic is then half the difference between 1 and the level of confidence (expressed as a decimal).

There is a problem with the interval given by Eq. (11.4), however, because in order to calculate it we must know the population variance σ^2. But how can we? Presumably, we are inferring the characteristics of μ because we are not observing the entire population of X's. But if we do not know the population of X's, how can we know σ^2? We now address this problem through an approach that builds upon the conceptual framework just developed but substitutes an estimate of σ^2 for its true (and unknown) value.

PROBLEMS

11.1 Consider a population with unknown mean μ but a known variance $\sigma^2 = 25$. In a simple random sample of size 36 we find $\bar{X} = 8$.

(a) Build a 95% confidence interval for μ.
(b) Build a 90% confidence interval for μ.
(c) Build a 99% confidence interval for μ.

11.2 Consider a population with unknown mean μ but a known standard deviation $\sigma = 4$. In a simple random sample of size 15 we find $\bar{X} = 3$.

(a) Build a 95% confidence interval for μ.
(b) Build a 90% confidence interval for μ.
(c) Build a 99% confidence interval for μ.

11.3 Using the data of problem 11.2, to what value must n be increased (presuming that $\bar{X}$ stayed the same) so as to make the 95% confidence interval extend from 2 to 4?

11.4 Using the data of problem 11.2, but keeping $n = 15$, determine the level of confidence that is associated with a confidence interval extending from 2 to 4.

CONFIDENCE INTERVAL FOR μ
WHEN THE VARIANCE IS UNKNOWN

In general, if we are trying to infer the characteristics of μ (in view of the fact that we do not observe the entire population), then it will be the case that σ^2 is unknown. This situation suggests that, although Eq. (11.4) is still the correct formula for a 95% confidence interval, it is not empirically useful since σ^2 is unknown. What do we do in such a situation?

The confidence interval (11.4) is based upon the distribution theory

$$\frac{\bar{X} - \mu}{\sigma/\sqrt{n}} \sim N(0, 1)$$

When σ^2 is unknown we must estimate it; our usual approach is to construct the sample variance

$$s^2 = \frac{1}{n-1} \sum_{i=1}^{n} (X_i - \bar{X})^2$$

We might then think in terms of basing our confidence interval on the expression

$$\frac{\bar{X} - \mu}{s/\sqrt{n}} \qquad (11.5)$$

and its distribution. Notice that this expression does not have a standard normal distribution; by substituting s^2 for σ^2 we have introduced a new random variable into the expression and have therefore changed its probability distribution. If we can determine the distribution of this expression, however, we will then be able to derive a confidence interval using a strategy similar to that exploited above (this interval will just be based upon a different probability distribution). But what is the distribution of Eq. (11.5)?

Since $E(\bar{X}) = \mu$, the numerator in Eq. (11.5) is on average zero; this result suggests (but does not prove)[2] the possibility that the distribution of Eq. (11.5) is centered about zero. Given that the distribution of $\bar{X}$ is symmetric about μ, there is also a suggestion that Eq. (11.5) might have a distribution that is symmetric about zero.

How does the distribution of Eq. (11.5) then in fact compare with that of the standard normal? Since we have s^2 instead of σ^2 in the denominator, and since the former is a random variable while the latter is a constant, we should expect that the distribution of Eq. (11.5) is more disperse than the standard normal distribution.

---

[2] Since $(\bar{X} - \mu)/(s/\sqrt{n})$ is a nonlinear function of the two random variables $\bar{X}$ and s, we *cannot* simply say that

$$E\left(\frac{\bar{X} - \mu}{s/\sqrt{n}}\right) = \frac{E(\bar{X} - \mu)}{E(s)/\sqrt{n}}$$

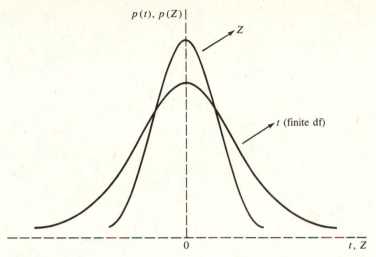

Figure 11.3 The *t* distribution compared with the standard normal distribution.

The expression given in Eq. (11.5) is generally modeled as having a probability distribution known as the *t* distribution with $n - 1$ *degrees of freedom*. The degrees of freedom is a parameter characterizing the *t* distribution, and it can range over the positive integers from 1 to ∞. We lose a degree of freedom (degree of variation) from the sample size n because it is only to the extent that n exceeds one that we can even form s—and therefore the entire expression given in Eq. (11.5).[3]

In Figure 11.3 we draw the probability density function for a random variable having a *t* distribution, and also that of the standard normal as a reference. Notice the greater degree of dispersion in the *t*. Notice also that the *t* distribution is in fact centered and symmetric about zero.

What role does the degrees of freedom play in the determination of the probability distribution of a *t*-distributed random variable? The *t* distribution is more disperse than the standard normal because it has an additional source of random variability (s^2). Since s^2 is a sample statistic used to estimate σ^2, we would generally expect that its variance decreases as the sample size increases, in much the

[3] Strictly speaking, Eq. (11.5) has exactly a *t* distribution only if the parent population is normal; however, the *t* distribution is often used as an approximation even when such is not the case. The *t* distribution's origins are discussed in more detail in Appendix 1 to this chapter.

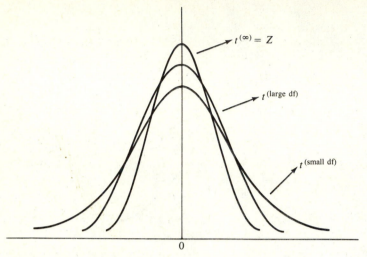

Figure 11.4 Degrees of freedom and the *t* distribution.

same way that the variance of $\bar{X}$ decreases as the sample size increases.[4] This result, in conjunction with the fact that s^2 is an unbiased estimator of σ^2, suggests that the difference between the *t*
distribution and the standard normal should get smaller as the sample size increases. In fact, as the degrees of freedom $(n-1)$ approaches infinity, the *t* distribution approaches the standard normal;
Figure 11.4 illustrates the evolution of the *t* distribution as the degrees
of freedom $(n-1)$ increases.

Formally, we say

$$\frac{\bar{X}-\mu}{s/\sqrt{n}} \sim t^{(n-1)}$$

where the superscript on the *t* gives the degrees of freedom. We have
not given the explicit formula for the probability density function
of the *t* distribution, but we know that by determining the appropriate areas under the curve we can determine any probability regarding a *t* distribution in which we might be interested. The calculations have already been done for us for a wide range of problems

---

[4] To be more precise, if the parent population is normal, the variance of s^2
is $2\sigma^4/(n-1)$, as discussed in the appendix to Chapter 9. (Discussion of the
variance of s^2 becomes more complicated, however, once we consider nonnormal
parent populations.)

that we typically encounter; these results are contained within Table IV (page 390) in this text.

This table contains one row for each value of the degrees of freedom (up to 30). The number (call it θ) in the t_α column in the β degrees of freedom row represents that value of that t distribution such that $\Pr(t^{(\beta)} > \theta) = \alpha$. For example, $\Pr(t^{(10)} > 2.23) = .025$. Once the degrees of freedom exceeds 30, the t distribution is very close to the standard normal, and for our purposes we may approximate such a t distribution with the standard normal distribution.[5] Formally, the t distribution with infinite degrees of freedom is the standard normal distribution. (See the bottom row of Table IV.)

Let us now consider how we can use this background on the t distribution to determine a formula for a 95% confidence interval for μ when the population variance is unknown. Here, we follow the same general strategy as in the case where the population variance is known. We know that

$$\frac{\bar{X} - \mu}{s/\sqrt{n}} \sim t^{(n-1)}$$

which suggests that

$$\Pr\left(-t_{.025}^{(n-1)} < \frac{\bar{X} - \mu}{s/\sqrt{n}} < t_{.025}^{(n-1)}\right) = .95$$

By multiplying through by $s/\sqrt{n}$ and then adding μ, we arrive at

$$\Pr\left[\left(\mu - t_{.025}^{(n-1)}\frac{s}{\sqrt{n}}\right) < \bar{X} < \left(\mu + t_{.025}^{(n-1)}\frac{s}{\sqrt{n}}\right)\right] = .95$$

This expression is of the form $\Pr[(\mu - k) < \bar{X} < (\mu + k)] = .95$; we have determined the value of $k!$ Thus, a 95% confidence interval for μ is

$$\bar{X} - t_{.025}^{(n-1)}\frac{s}{\sqrt{n}} \text{ (lower bound)} \quad \text{and} \quad \bar{X} + t_{.025}^{(n-1)}\frac{s}{\sqrt{n}} \text{ (upper bound)}$$

that is,

$$\bar{X} \pm t_{.025}^{(n-1)}\frac{s}{\sqrt{n}}$$

[5] For some research problems this approximation is unsatisfactory, in which case a more complete t table should be consulted. But for our purposes this approximation is adequate.

This interval should be interpreted in the same way as discussed above; since 95% of all such intervals would contain μ, we are 95% confident that ours does.[6]

As was earlier the case, for a different level of confidence we simply change the level of the t statistic. For example, a 99% confidence interval for μ would be

$$\bar{X} \pm t_{.005}^{(n-1)} \frac{s}{\sqrt{n}}$$

EXAMPLE ■

Suppose that we are interested in the mean annual income of the families residing in a particular city. In a simple random sample of 25 such families we find $\bar{X} = \$25,000$ and $s = \$8000$. A 95% confidence interval for μ would then be

$$\bar{X} \pm t_{.025}^{(24)} \frac{s}{\sqrt{n}} = 25000 \pm (2.06) \frac{8000}{\sqrt{25}} = 25000 \pm 3296$$

$$= (21704, 28296)$$

That is, we are 95% confident that the true mean family income for this city is between \$21,704 and \$28,296 per year. A 99% confidence interval would be

$$\bar{X} \pm t_{.005}^{(24)} \frac{s}{\sqrt{n}} = 25000 \pm (2.80) \frac{8000}{\sqrt{25}} = 25000 \pm 4480$$

That is, we are 99% confident that the true mean income is between \$20,520 and \$29,480 per year. (Notice that the 99% confidence interval is wider than the 95% confidence interval is. Why?)   ■

EXAMPLE ■

Appendix 2 to this chapter contains data on selected variables for a sample of 50 metropolitan areas in the United States. We might consider using these data to build confidence intervals for the value of the variable of interest in the mean metropolitan

---

[6] Strictly speaking, this expression is a 95% confidence interval for μ only if the parent population is normal; however, it is often used as an approximation when such is not the case.

area (within the population of all metropolitan areas in the United States).

This work is greatly facilitated by having access to a statistical program on a computer (either a mainframe or a personal computer), inasmuch as the calculation of the sample mean and sample variance here requires a large number of repetitive calculations. There are many such programs available, and we assume that the reader has access to one.

A 95% confidence interval for the poverty rate in the mean metropolitan area can be calculated using this sample of 50 observations. The sample statistics are

$$\bar{X} = 11.41 \qquad s = 2.99 \qquad n = 50$$

and the 95% confidence interval is then

$$\bar{X} \pm t_{.025}^{(49)} \frac{s}{\sqrt{n}} = 11.41 \pm 1.96 \frac{2.99}{\sqrt{50}}$$

$$= 11.41 \pm .83 = (10.58, 12.24)$$

where we have approximated the t with the standard normal ($Z_{.025} = 1.96$) since the degrees of freedom exceeds 30. [Many statistical programs that calculate confidence intervals do not use this approximation but use the precise value of the t distribution. In this case such a confidence interval would be (10.564, 12.264), which is very close to the result calculated above.] ∎

The data in Appendix 2 will be the source for some examples and problems that follow (and also some examples and problems in Chapter 12), and in general provide a data set that can be used at an instructor's or student's discretion to make up further problems.

Here we have been concerned with the notion of a confidence interval as it applies to a particular individual parameter (a population mean). Clearly, this conceptual framework can be applied to any population parameter; the specifics (i.e., estimator, distribution theory) would change, but the general strategy would remain the same. We shall now turn our discussion to a particularly interesting example of another "parameter" for which a confidence interval is often interesting and informative.

PROBLEMS

11.5 We take a simple random sample of size 25 from a population with unknown mean and variance. We find $\bar{X} = 10$ and $s^2 = 20$.

(a) Build a 95% confidence interval for the population mean μ.

(b) Build a 90% confidence interval for μ.

(c) Build a 99% confidence interval for μ.

11.6 We take a simple random sample of size 50 from a population with unknown mean and variance. We find $\bar{X} = 23$ and $s = 15$.

(a) Build a 95% confidence interval for the population mean μ.

(b) Build a 90% confidence interval for μ.

(c) Build a 99% confidence interval for μ.

11.7 As a financial aid officer at a large university, you are interested in the mean summer income of students. To investigate this you take a simple random sample of 100 students and find $\bar{X} = \$2000$ with $s = \$500$. Build a 95% confidence interval for the mean summer income of all students within this population.

11.8 A simple random sample of 25 recent graduates from a particular college gives a sample mean starting salary of $24,000 with sample standard deviation $3000. Build a 95% confidence interval for the mean starting salary of all graduates from this college.

11.9 For the following sample of states, we record the number of persons (in thousands) who were union members in 1978:

Connecticut	296	Indiana	643
Massachusetts	611	Nevada	80
Ohio	1294	Tennessee	303
Wisconsin	522		

Build a 95% confidence interval for total union membership *in the nation as a whole* (i.e., all 50 states). [*Source of data:* U.S. Bureau of Labor Statistics. *Directory of National Unions and Employee Associations,* 1979. Bulletin 2079 (Washington, DC: U.S. Government Printing Office, 1980), Appendix J.]

11.10 Continuing problem 11.9, the actual total union membership (in thousands) for the nation as a whole in 1978 was
20,459. Comment on your answer to problem 11.9, given
this new information.

11.11 For the following sample of ten states, we observe the percent
of bachelor's degrees in 1981–82 conferred upon women:

Rhode Island 51.1%	Ohio 49.7%
Minnesota 51.8%	Nebraska 50.8%
Virginia 54.3%	Florida 46.3%
Arkansas 51.4%	Idaho 44.7%
Utah 41.4%	Alaska 58.1%

Build a 95% confidence interval for the national mean of
percent of bachelor's degrees conferred upon women in
1981–82. [*Source of data:* U.S. Bureau of the Census, *State
and Metropolitan Area Data Book,* 1986 (Washington, DC:
U.S. Government Printing Office, 1986), p. 539.]

11.12 Consider the following data on the U.S. economy:

Year	Nominal interest rate, 3-month Treasury bills	Inflation rate from previous year
1980	11.5%	12.5%
1981	14.0	8.9
1982	10.7	3.8
1983	8.6	3.8
1984	9.6	3.9
1985	7.5	3.8
1986	6.0	1.1
1987	5.8	4.4
1988	6.7	4.4
1989	8.1	4.6

[*Source: Economic Report of the President: 1990* (Washington, DC: U.S. Government Printing Office, 1990), pp.
376, 363.] Defining the real interest rate as the difference
between the nominal interest rate and the inflation rate,
build a 95% confidence interval for the mean U.S. real interest rate. Comment on the appropriateness of the assumption that the data are a simple random sample in this
problem.

11.13 (Requires computer.) Using the data in Appendix 2 to this
chapter, build a 95% confidence interval for the percent of

the population that was of Spanish origin in the mean U.S. metropolitan area in 1980.

11.14 (Requires computer.) Using the data in Appendix 2 to this chapter, build a 95% confidence interval for income per capita in the mean U.S. metropolitan area in 1983.

CONFIDENCE INTERVALS FOR DIFFERENCES IN MEANS

We are often interested in the relationship between two populations, exemplified in questions such as the following:

Do individuals with a disease who take a particular drug live on average longer than those with the disease who do not take the drug?

How does the typical male's wage compare to that of the typical female?

Are individuals with high blood pressure more likely (than those without high blood pressure) to have heart attacks?

Do students who graduate from college *summa cum laude*[7] earn on average higher salaries (than those who don't)?

In fact, many of the interesting and important questions that we confront in applied research involve comparing the characteristics of two (or sometimes more than two) populations and then trying to explain whatever differences we observe. Let us analyze and illustrate some of the basic issues involved here in the context of the last example suggested above, that of comparing the salaries of college graduates who received the *summa cum laude* distinction and the salaries of college graduates who did not receive this distinction.

We visualize this problem as one in which there are two populations: those who were *summa cum laude* college graduates and those who were college graduates without this distinction. We are interested in comparing the characteristics of these two populations.

---

[7] *Summa cum laude* is a distinction earned by some college graduates for outstanding performance. Although the determination of this distinction varies across institutions, it typically involves having a GPA above a particular number (often on the order of 3.8 or 3.9 out of a possible 4.0), with perhaps certain other requirements.

Let X_d (d for distinction) represent the salaries of those who were *summa cum laude;* $X_{d,i}$ then represents the salary of the ith person who was *summa cum laude.* Similarly, $X_{n,i}$ (n for no distinction) represents the salary of the ith person who was not *summa cum laude.* These populations have means μ_d and μ_n, respectively. The question of interest is then as to the relationship between these means; in particular, does μ_d exceed μ_n, and, if so, by how much?

We are in this case trying to investigate the role played by the *summa cum laude* distinction from an empirical perspective; presumably, this distinction is evidence of superior performance, which we might then expect to be associated with higher future wages. Note that the relationship between μ_d and μ_n is only a relationship between means; it is quite likely that even if $\mu_d > \mu_n$ there would still be some individuals who were *summa cum laude* who earn less than some who did not attain this distinction.

Since we do not observe the entire populations, we must use samples to infer the characteristics of μ_d and μ_n. We shall below discuss in more detail issues regarding the sampling technique and the general framework within which such questions might be studied, but for now we presume that we have simple random samples of size n_d and n_n respectively from the two populations. These samples yield sample means $\bar{X}_d$ and $\bar{X}_n$. We also assume for the moment that the parent populations (X_d and X_n) both have normal distributions.

Given that the question is whether μ_d exceeds μ_n, we often "redefine" the problem in terms of a "new" parameter $\mu_d - \mu_n$, which is just the difference in means. Of course, the main question of interest is then: Is $\mu_d - \mu_n > 0$ (i.e., is $\mu_d > \mu_n$)? The advantage of this step is that we now have the question defined in terms of a single parameter that is a function of the two original means.

What have we done in the past when trying to infer the characteristics of an unknown parameter? Think about the case of a single mean. First, we estimate it. Then, we exploit what we can determine regarding the distribution of the estimator to build a confidence interval. Let us follow this strategy in the problem at hand.

To estimate $\mu_d - \mu_n$ we calculate the sample means $\bar{X}_d$ and $\bar{X}_n$ and form $\bar{X}_d - \bar{X}_n$. What are the characteristics of $\bar{X}_d - \bar{X}_n$ as an estimator? First, we determine its expected value:

$$E(\bar{X}_d - \bar{X}_n) = E(\bar{X}_d) - E(\bar{X}_n)$$

$$= \mu_d - \mu_n$$

This expression tells us that $\bar{X}_d - \bar{X}_n$ is an unbiased estimator of $\mu_d - \mu_n$.

Since the parent populations are normal, we know that $\bar{X}_d$ and $\bar{X}_n$ both have normal distributions. The fact that there are different individuals in the two samples suggests that these samples are independent, and therefore that $\bar{X}_d$ and $\bar{X}_n$ are independent random variables.[8] As we have seen before, a linear function of independent normally distributed random variables will itself be normal, suggesting that our estimator $\bar{X}_d - \bar{X}_n$ is normally distributed. We then need only to determine its variance in order to have a complete characterization of its probability distribution.

Since $\bar{X}_d - \bar{X}_n$ is a linear function of independent random variables, we can easily determine its variance using results from Chapters 8 and 9:

$$\text{var}\,(\bar{X}_d - \bar{X}_n) = (1)^2\,\text{var}\,(\bar{X}_d) + (-1)^2\,\text{var}\,(\bar{X}_n)$$

$$= \frac{\sigma_d^2}{n_d} + \frac{\sigma_n^2}{n_n}$$

where σ_d^2 and σ_n^2 are the variances of the respective parent populations, and n_d and n_n are the respective sample sizes. We can then summarize our findings by saying that

$$\bar{X}_d - \bar{X}_n \sim N\left(\mu_d - \mu_n, \quad \frac{\sigma_d^2}{n_d} + \frac{\sigma_n^2}{n_n}\right) \qquad (11.6)$$

which is the key piece of distribution theory from which a formula for a 95% confidence interval can be derived. From Eq. (11.6) we could use the general strategy outlined earlier in this chapter to derive a formula for a 95% confidence interval for $\mu_d - \mu_n$. (Can you do this?)

---

[8] It will not always be the case that the samples are independent. For example, if we were investigating the effectiveness of a weight reduction program we might observe the *same* group of individuals before and after the program and compare the mean weights at these two times. Here we clearly do not have independent samples because we have the same individuals observed at two different points in time; if we have heavier-than-average individuals (as compared to all those who undertake the program) at the start we are likely to have heavier-than-average individuals (as compared with all those who finish the program) at the end. There are alternative techniques for dealing with these types of situations that exploit the fact that we can uniquely match individuals at two different points in time. But the situations most commonly confronted in economics are in the framework of independent samples.

At this point, a word regarding the relationship between the variances of the two populations is in order. The population of *summa cum laude* individuals (X_d) has variance σ_d^2, and the population of individuals without this distinction (X_n) has variance σ_n^2. It will generally be the case that both of these variances are unknown, since we do not observe the entire populations. The implication is that there are then two variances that we must estimate. Although there is no major problem with this, often an assumption is made in order to simplify this estimation; this assumption is that the population variances, though unknown, are equal. We shall make this assumption in the analysis that follows and shall base our confidence interval formula upon it.[9] (Note that our assumption is that the population, not sample, variances are equal.)

We then have one parameter σ^2 ($= \sigma_d^2 = \sigma_n^2$) to estimate, and we have two separate samples (from two populations having that variance) available for that estimation. But we only want one estimate of σ^2! That is, although we can calculate the respective sample variances s_d^2 and s_n^2, since generally they will be different, which one do we use? Our solution to this problem is to combine these variance estimates into a single estimate.

The standard way of doing this is to construct an estimator that "pools" together the information from the two samples. This estimator, s_p^2 (where the p is for "pooled"), is a weighted average of the two individual sample variances, where the weights are based on the respective sample sizes.

Definition of the Pooled Variance Estimate   Suppose that X_1 and X_2 represent two populations with equal but unknown variances. We have independent simple random samples from these two populations. We estimate the common variance σ^2 as

$$s_p^2 = \left(\frac{n_1 - 1}{n_1 + n_2 - 2} \right) s_1^2 + \left(\frac{n_2 - 1}{n_1 + n_2 - 2} \right) s_2^2$$

[9] Obviously, there will be some situations in which this assumption is simply not appropriate. There are techniques available for cases in which we do not assume that the population variances are equal, but they are a bit messier than the one we shall develop, and often it makes surprisingly little difference whether we assume the population variances are the same or not. For a further discussion see, for example, Gerald Keller et al., *Statistics for Management and Economics: A Systematic Approach* (Belmont, CA: Wadsworth Publishing Company, 1988), pp. 304–306.

where n_1 and n_2 are the respective sample sizes, and s_1^2 and s_2^2 are the respective sample variances. s_p^2 is the so-called *pooled variance estimate.*

Notice that $(n_1 - 1)$ is the degrees of freedom in the sample from the X_1 population, $(n_2 - 1)$ is the degrees of freedom in the sample from the X_2 population, and $(n_1 + n_2 - 2)$ is the sum of these two values. The weights in the s_p^2 formula are then based upon the relative sizes of the degrees of freedom. In the problem at hand:

$$s_p^2 = \left(\frac{n_d - 1}{n_d + n_n - 2}\right)s_d^2 + \left(\frac{n_n - 1}{n_d + n_n - 2}\right)s_n^2$$

We shall now investigate how we can integrate this material to develop an operational formula for a confidence interval.

Equation (11.6) suggests that

$$\frac{\bar{X}_d - \bar{X}_n - (\mu_d - \mu_n)}{\sqrt{\dfrac{\sigma^2}{n_d} + \dfrac{\sigma^2}{n_n}}} = \frac{\bar{X}_d - \bar{X}_n - (\mu_d - \mu_n)}{\sigma\sqrt{\dfrac{1}{n_d} + \dfrac{1}{n_n}}} \sim N(0, 1)$$

(given that $\sigma_d^2 = \sigma_n^2 = \sigma^2$), inasmuch as this expression involves just a normal minus its mean divided by its standard deviation. This result is not yet operational because we do not know σ^2. But we can estimate it with s_p^2. If we substitute s_p for σ in this expression we change the relevant distribution theory; it should not be surprising that the expression that results turns out to have a t distribution:

$$\frac{\bar{X}_d - \bar{X}_n - (\mu_d - \mu_n)}{s_p\sqrt{\dfrac{1}{n_d} + \dfrac{1}{n_n}}} \sim t^{(n_d + n_n - 2)}$$

(Essentially, we have substituted an estimate of the variance for its true value, and then used the t distribution to account for the added variability associated with the additional random variable.[10])

---

[10] Formally, this expression is exactly a t distribution only if the underlying parent populations are both normal; in reality, however, it is often used as an approximation even when such is not the case. Notice that the degrees of freedom of the t is the combined degrees of freedom of the two samples.

Therefore

$$\Pr\left\{-t_{.025}^{(n_d+n_n-2)} < \left[\frac{\bar{X}_d - \bar{X}_n - (\mu_d - \mu_n)}{S_p\sqrt{\dfrac{1}{n_d} + \dfrac{1}{n_n}}}\right] < t_{.025}^{(n_d+n_n-2)}\right\} = .95$$

and

$$\Pr\left\{\left[(\mu_d - \mu_n) - t_{.025}^{(n_d+n_n-2)} S_p\sqrt{\frac{1}{n_d} + \frac{1}{n_n}}\right] < [\bar{X}_d - \bar{X}_n]\right.$$

$$\left. < \left[(\mu_d - \mu_n) + t_{.025}^{(n_d+n_n-2)} S_p\sqrt{\frac{1}{n_d} + \frac{1}{n_n}}\right]\right\} = .95$$

This expression is of the form seen in Eq. (11.1), with the parameter μ replaced by $(\mu_d - \mu_n)$ and the estimator $\bar{X}$ replaced by $(\bar{X}_d - \bar{X}_n)$. The term

$$t_{.025}^{(n_d+n_n-2)} S_p\sqrt{\frac{1}{n_d} + \frac{1}{n_n}}$$

is then analogous to the k in Eq. (11.1), implying that a 95% confidence interval for $(\mu_d - \mu_n)$ is

$$\bar{X}_d - \bar{X}_n \pm t_{.025}^{(n_d+n_n-2)} S_p\sqrt{\frac{1}{n_d} + \frac{1}{n_n}} \qquad (11.7)$$

(As before, if the degrees of freedom exceeds 30 we approximate the t with the standard normal.) Let us use and interpret this result in a numerical example.[11]

EXAMPLE ■

Suppose that we observe a simple random sample of 13 individuals who are *summa cum laude* college graduates and an independent simple random sample of 16 individuals who are college graduates but did not receive the *summa cum laude* distinction. We find that the sample mean salary of *summa*

---

[11] Strictly speaking, this expression is a 95% confidence interval for $\mu_d - \mu_n$ only if the parent populations are normal; often, however, it is used as an approximation when such is not the case.

cum laude graduates is $30,000 per year and that the sample mean salary of graduates who did not receive the *summa cum laude* distinction is $28,000 per year. The sample standard deviations are $s_d = \$8000$ and $s_n = \$7000$. What can we infer about the relationship between μ_d and μ_n?

We have some evidence to suggest that those who are *summa cum laude* earn more, on average, than those who are not; in our samples, the mean salary among those college graduates who received the *summa cum laude* distinction exceeds the mean salary of other college graduates. But is this persuasive evidence? One possibility is that the difference in sample means is indicative of a difference in the underlying population means. But another possibility is that the underlying population means are equal and the difference in sample means is due simply to random variability.

To investigate this, we pursue the calculation of a likely margin of error, so as to determine the maximum plausible variation in the difference in sample means. That is, we build a 95% confidence interval for the difference in population means:

$$\bar{X}_d - \bar{X}_n \pm t_{.025}^{(n_d + n_n - 2)} s_p \sqrt{\frac{1}{n_d} + \frac{1}{n_n}}$$

First, we must calculate s_p^2 (and therefore implicitly s_p):

$$s_p^2 = \left(\frac{n_d - 1}{n_d + n_n - 2}\right) s_d^2 + \left(\frac{n_n - 1}{n_d + n_n - 2}\right) s_n^2$$

$$= \left(\frac{13 - 1}{13 + 16 - 2}\right)(8000)^2 + \left(\frac{16 - 1}{13 + 16 - 2}\right)(7000)^2$$

$$= 55,666,667$$

implying that $s_p = \$7461$. The 95% confidence interval is then

$$\mu_d - \mu_n = \bar{X}_d - \bar{X}_n \pm t_{.025}^{(n_d + n_n - 2)} s_p \sqrt{\frac{1}{n_d} + \frac{1}{n_n}}$$

$$= 30000 - 28000 \pm t_{.025}^{(27)}(7461) \sqrt{\frac{1}{13} + \frac{1}{16}}$$

$$= 30000 - 28000 \pm (2.05)(7461) \sqrt{\frac{1}{13} + \frac{1}{16}}$$

$$= 2000 \pm 5711$$

$$= (-3711, 7711)$$

That is, a 95% confidence interval for the difference in means is ($-$3711, \$7711). The fact that zero is within this interval suggests that it is plausible that such a difference in sample means could have occurred even if the underlying population means were equal. That is, zero is a plausible value of the difference in mean salaries between college graduates who received the *summa cum laude* distinction and college graduates who did not receive this distinction. This realization is important because it suggests that we should not be overconfident regarding the reliability of the difference in sample means observed within our samples.[12] ∎

EXAMPLE ∎

This example relies on the data contained in Appendix 2 to this chapter. Suppose that we are interested in the relationship between the number of social security recipients (per 1000 population) in southern as compared to northeastern metropolitan areas. In this sample we observe 20 southern and 10 northeastern metropolitan areas. If we let μ_s be the mean value of social security recipients (per 1000 population) in southern metropolitan areas, and μ_{ne} be the analogous parameter for northeastern metropolitan areas, we might form a 95% confidence interval for $\mu_s - \mu_{ne}$. Our sample statistics are

$$\bar{X}_s = 178.5 \qquad s_s = 68.0 \qquad n_s = 20$$

$$\bar{X}_{ne} = 156.0 \qquad s_{ne} = 19.5 \qquad n_{ne} = 10$$

Assuming that the population variances are equal, $s_p = 57.1$ and a 95% confidence interval for $\mu_s - \mu_{ne}$ is

---

[12] This conclusion is the first hint of the procedure known as hypothesis testing, which will be discussed from a different and more formal perspective in Chapter 12.

$$\bar{X}_s - \bar{X}_{ne} \pm t_{.025}^{(28)} s_p \sqrt{\frac{1}{n_s} + \frac{1}{n_{ne}}}$$

$$= 178.5 - 156.0 \pm (2.05)(57.1)\sqrt{\frac{1}{20} + \frac{1}{10}}$$

$$= 22.5 \pm 45.3$$

$$= (-22.8, 67.8)$$

The sample means suggest that southern metropolitan areas have more social security recipients (per 1000 population) than northeastern metropolitan areas do, but the confidence interval suggests that we cannot rule out the possibility that this difference is due simply to random variability. ∎

We now turn to a further discussion of interpretive issues involving confidence intervals for differences in means.

PROBLEMS

11.15 In independent samples of 25 male and 18 female workers, we find weekly wages as follows:

$$\bar{X}_m = \$400 \qquad \bar{X}_f = \$370 \qquad n_m = 25$$
$$s_m = \$150 \qquad s_f = \$120 \qquad n_f = 18$$

where the m subscript represents the males and the f subscript represents the females. Assuming that the population variances are equal, build a 95% confidence interval for $\mu_m - \mu_f$ and carefully interpret your result.

11.16 Suppose that high school students taking the SAT have the option of taking a particular preparatory course. Let us take independent simple random samples of 20 students each who have taken and not taken the preparatory course. We find

$$\bar{X}_t = 1060 \qquad s_t = 120 \qquad n_t = 20$$
$$\bar{X}_{nt} = 990 \qquad s_{nt} = 70 \qquad n_{nt} = 20$$

where $\bar{X}_t$ is the sample mean combined score (verbal plus math) for those who have taken the preparatory course, $\bar{X}_{nt}$

is the sample mean combined score for those who have not taken the preparatory course, etc. Assuming equal population variances, build a 95% confidence interval for $\mu_t - \mu_{nt}$ and carefully interpret your results.

11.17 In independent simple random samples of male athletes and male nonathletes at a large university, we find the following sample characteristics for their GPAs:

Athletes	Nonathletes
$\bar{X} = 2.8$	$\bar{X} = 2.9$
$s = .4$	$s = .5$
$n = 15$	$n = 25$

Build a 95% confidence interval for the difference in population means, assuming that the population variances are equal.

11.18 At a large university, you take independent simple random samples of in-state students and out-of-state students, and find the following sample characteristics of their GPAs:

In-state	Out-of-state
$\bar{X} = 2.6$	$\bar{X} = 2.9$
$s = .5$	$s = .3$
$n = 8$	$n = 11$

Build a 95% confidence interval for the difference in population means, assuming that the population variances are equal.

11.19 Suppose that you are interested in the relationship between mean annual incomes of workers in two different countries. You have independent simple random samples of workers and their annual incomes from the two countries, as follows:

Country *A*	Country *B*
$40,000	$37,283
27,328	41,164
16,414	7,315
15,307	21,465
61,245	29,307
9,059	33,049
	13,085
	22,123

Build a 95% confidence interval for $\mu_A - \mu_B$, assuming equal population variances.

11.20 Investigating the difference between per capita incomes in western states and southern states, you take a sample of four western states and six southern states and note their per capita incomes in 1983:

West	South
Montana $7971	Delaware $9963
New Mexico $7877	Virginia $10,136
Washington $9915	South Carolina $7830
Hawaii $9776	Kentucky $7724
	Mississippi $6801
	Oklahoma $9092

This listing is based upon the U.S. Census definitions of southern and western states. [*Source:* U.S. Bureau of the Census, *State and Metropolitan Area Data Book, 1986* (Washington, DC: U.S. Government Printing Office, 1986), p. 560.] Build a 95% confidence interval for $\mu_W - \mu_S$, assuming that the population variances are equal.

11.21 Using the same states as in problem 11.20 we note the poverty rates for all persons in 1979:

West	South
Montana 12.3%	Delaware 11.9%
New Mexico 17.6%	Virginia 11.8%
Washington 9.8%	South Carolina 16.6%
Hawaii 9.9%	Kentucky 17.6%
	Mississippi 23.9%
	Oklahoma 13.4%

Build a 95% confidence interval for the difference in mean poverty rates between southern and western states. [*Source:* U.S. Bureau of the Census, *State and Metropolitan Area Data Book, 1986* (Washington, DC: U.S. Government Printing Office, 1986), p. 561.]

11.22 (Requires computer.) Using the data in Appendix 2 to this chapter, build a 95% confidence interval for $\mu_L - \mu_S$, where μ_L is the mean percent of persons age 25 plus with 16 or more years of school in 1980 in metropolitan areas with populations of at least 500,000 persons, and μ_S is the mean percent of persons age 25 plus with 16 or more years of school in 1980 in metropolitan areas with populations under 500,000. (Assume equal population variances.)

11.23 (Requires computer.) Using the data in Appendix 2 to this chapter, build a 95% confidence interval for $\mu_L - \mu_S$, where μ_L is the mean poverty rate in 1979 for metropolitan areas with populations of at least 500,000 persons, and μ_S is the mean poverty rate in 1979 for metropolitan areas with populations under 500,000. (Assume equal population variances.)

INTERPRETING CONFIDENCE INTERVALS FOR DIFFERENCES IN MEANS

Consider a situation in which we are interested in whether or not (and the extent to which) there is a difference in the mean earnings of male versus female workers. If we let μ_m equal the male mean and μ_f equal the female mean, the parameter of interest is then $\mu_m - \mu_f$. To investigate this issue we might build a 95% confidence interval for $\mu_m - \mu_f$ using the method described above. Suppose that we find that the 95% confidence interval for this parameter ranges only over the positive numbers, suggesting that we are at least 95% confident that μ_m exceeds μ_f (i.e., the male mean exceeds the female mean).

It is very important to notice that the finding above tells us only that the male mean exceeds the female mean; *it does not tell us why this is the case.* There are, in fact, several possible causes for this difference that are consistent with the above conclusion regarding the relationship between these means:

1. There might be some attributes of individuals that affect wages (educational attainment, amount of job experience, etc.) and are systematically different between male and female workers, and the difference in earnings is due to the differential attainment of these attributes.
2. There is some process of discrimination by which employers systematically pay female workers less than their male counterparts.

3. For some reason, female workers are in occupations that, due to market forces of supply and demand, have lower wages than the occupations frequented by their male counterparts.

Unfortunately, the above-referenced confidence interval gives us no way of knowing which of the three (or which combination of these three) explanations accounts for the observed difference.

It is very important to keep this point in mind, and the problem is due largely to the nature of the data that are the basis of our analysis. We have here imagined that we have an *observational* study, in which we go out and observe a group of persons and things and one or more of their characteristics. In such a study we might determine that there is some relationship between two variables (e.g., gender and earnings) but we do not know from the data the cause of this relationship. It is possible, in fact, that the relationship is really due to some third variable driving the process that we have not incorporated into our analysis. (See, for example, the education and experience variables suggested in explanation 1 above.)

How do we then go about the process of trying to determine causes? Ultimately, the answer must come from some appeal to a theoretical perspective as to the underlying process. But we might be able to make further progress empirically. One thing we might be able to do is to attempt to rule out certain possible explanations. For example, in the above problem we might consider male and female workers with identical educational characteristics; any difference in their earnings is then presumably not due to differences in education. By this technique we attempt to *control* for one or more factors. The process might involve selective sampling (e.g., female versus male college graduates) or, more generally, some multivariate statistical procedure that can "separate out" the influences of one or more other variables. Multiple regression, in fact, is one way of attempting such a procedure; it will be the topic of Chapter 14.

An observational study can be contrasted with what is commonly called an *experimental* study, in which the values of the variable under investigation (in the above example this variable is gender) are determined by the researcher. In other words, the values of this variable are controlled within the context of the experiment, as opposed to simply observed. An example of an experimental study would be one in which we take a group of individuals having a terminal disease and randomly give some of them a drug and others a placebo and observe whether there are any differences on average in the survival times between members of the two groups. If there are, it is likely that the drug is the reason, inasmuch as the random distribution of the drug (say by a coin toss) typically rules out the possibility that it will be correlated with some other causal factor not under consideration.[13]

In economics we typically operate in the context of observational studies because we find it difficult to create controlled experimental situations in which to investigate economic questions. (Obviously, it would be difficult to create an experimental study to investigate gender!) Although there are cases of experimental studies in economics, we most commonly are in an observational context and thus have a heightened ambiguity concerning causal mechanisms.

At this point it is important to recognize, then, that interpretive difficulties are often present in observational studies. Statistical procedures tell us *what* the relationship is between two or more variables, but they do not tell us *why*. The "why" part must in general ultimately come from some theoretical understanding of the underlying processes driving the relationships, and there is then potentially room for conflict as a result of disagreements about these underlying processes. This situation does not imply, however, that there is no room for substantial progress from an empirical perspective, especially in terms of ruling out certain possibilities.

SUMMARY

In this chapter we have developed the concept of a confidence interval as a first way of making relatively sophisticated inferences about parameters of populations. In the next chapter we shall take many of the ideas here presented and discuss them in a slightly different context, in which we attempt to "test" the validity of various hypotheses.

[13] On the other hand, if we simply observe a group of patients, some of whom have taken the drug and others of whom have not, where this determination is not due to a purely random process, seeing a difference in survival times does not prove that the drug is the reason. Perhaps the drug is expensive so that only the rich can afford it, but the rich might live longer with the disease due to their superior access to health care. It is then possible that the latter fact is the determinant of the observed difference in survival times. In this observational study, then, reaching a conclusion as to causality is much more difficult.

REVIEW PROBLEMS FOR CHAPTER 11

11.24 Consider the effects of a pre–freshman year program—that is, a tutorial program some students take before they start their freshman year in college. Discuss the issues you would need to address in investigating the effectiveness of this program and interpreting your findings.

11.25 A story noted that there were more cases of smallpox reported in the third world in areas that had hospitals than in areas that did not. The conclusion was then drawn that the hospitals and the accompanying medical personnel must have been the source of the infection. Comment on this conclusion.

11.26 Suppose that we divide time periods into two types: periods of rapid monetary growth (say, the money supply grows at 5% or more per year) and periods of low monetary growth (the money supply grows at less than 5% per year). We are interested in how inflation rates compare between these two types of time periods, so we build a 95% confidence interval for $\mu_R - \mu_L$ (where μ_R is the mean inflation rate during periods of rapid monetary growth and μ_L is the mean inflation rate during periods of low monetary growth) and find that its total range is over positive numbers. Evaluate the claim that this result proves that monetary expansion causes inflation.

11.27 Mary has a job in city 1 that pays $2000 per month. She is considering moving to city 2, but does not know what her earnings will be. She does know, however, that her living expenses will be $300 per month higher in city 2. Mary knows that there are many jobs open in her field in city 2 but does not know, of course, which one she will be able to get. In a sample of ten of these jobs, she finds the average monthly wage to be $3000, with a sample standard deviation of $500. What would you advise her about the likely change in her financial situation if she were to move to city 2?

11.28 Continuing problem 11.27, Mary is also considering moving to city 3. She knows that her living expenses in city 3 will be $500 per month higher than in city 1 ($200 per month higher than in city 2). Mary knows that there are many jobs open in her field in city 3 but again does not know which one she will be able to get. In a sample of 20 of these jobs, she finds the average monthly wage to be $3600, with a sample standard deviation of $400. What would you advise her about the likely change in her financial situation if she were to move to city 3 as opposed to city 2?

11.29 The pooled variance estimate s_p^2 is defined as

$$s_p^2 = \left(\frac{n_1 - 1}{n_1 + n_2 - 2}\right)s_1^2 + \left(\frac{n_2 - 1}{n_1 + n_2 - 2}\right)s_2^2$$

Show that an equivalent formula for s_p^2 is

$$s_p^2 = \frac{\sum_{i=1}^{n_1} (X_{1i} - \bar{X}_1)^2 + \sum_{i=1}^{n_2} (X_{2i} - \bar{X}_2)^2}{n_1 + n_2 - 2}$$

where X_{1i}, $i = 1, \ldots, n_1$, is a simple random sample from the X_1 population that has a sample mean $\bar{X}_1$, and X_{2i}, $i = 1, \ldots, n_2$, is a simple random sample from the X_2 population that has sample mean $\bar{X}_2$.

*11.30 Assuming that $\sigma_1^2 = \sigma_2^2 = \sigma^2$, we estimate the common variance σ^2 as

$$s_p^2 = \left(\frac{n_1 - 1}{n_1 + n_2 - 2}\right)s_1^2 + \left(\frac{n_2 - 1}{n_1 + n_2 - 2}\right)s_2^2$$

where n_1 and n_2 are the respective sample sizes, and s_1^2 and s_2^2 are the respective sample variances.

(a) Prove that s_p^2 is an unbiased estimator of σ^2.
(b) If the parent populations are both normal, what is the variance of s_p^2?

*11.31 In a sample of ten states, we find the following percentages of persons 25 years of age or older who are high school graduates in 1970 and 1980:

State	1980	1970	State	1980	1970
Maine	68.7	54.7	West Virginia	56.0	41.6
Connecticut	70.3	56.0	Kentucky	53.1	38.5
Indiana	66.4	52.9	Louisiana	57.7	42.2
Iowa	71.5	59.0	Wyoming	77.9	62.8
Kansas	73.3	59.9	Nevada	75.5	65.2

Notice that these samples are not independent because we observe the same state at two different time periods (e.g., the 1980 figure for Maine is related to the 1970 figure for Maine); our normal procedure for a confidence interval for a difference in means presumes independent samples. Build a 95% confidence interval for $\mu_{1980}-\mu_{1970}$. (*Hint:* Construct a new random variable equal to the change from 1970 to 1980, and then build a 95% confidence interval for the mean of this *single* random variable.) [*Source for data:* U.S. Bureau of the Census. *State and Metropolitan Area Data Book, 1986* (Washington, DC: U.S. Government Printing Office, 1986), p. 539.]

*11.32 Suppose that we have a simple random sample of size 5 from a normal population with unknown mean μ but known variance $\sigma^2 = 9$. We estimate μ with

$$\hat{\mu} = \frac{1}{10} X_1 + \frac{1}{10} X_2 + \frac{2}{10} X_3 + \frac{3}{10} X_4 + \frac{3}{10} X_5$$

and find $\hat{\mu} = 5$. Use this information to build a 95% confidence interval for μ. (*Hint:* What is the probability distribution of $\hat{\mu}$?)

APPENDIX 1 TO CHAPTER 11

In this appendix we discuss how the t distribution has its foundations in the normal and chi-square distributions. [For a more detailed discussion, see, for example, John E. Freund and Ronald E. Walpole, *Mathematical Statistics,* 4th ed. (Englewood Cliffs, NJ: Prentice-Hall, 1987), pp. 289–292.]

Suppose that Z is a random variable having a standard normal distribution. Suppose that Y is a random variable that is independent of Z and has a chi-square distribution with k degrees of freedom. Then, if

$$T = \frac{Z}{\sqrt{Y/k}}$$

then $T \sim t^{(k)}$.

In the case of a single mean, if the parent population is normal, we know that

$$\frac{\bar{X} - \mu}{\sigma/\sqrt{n}} \sim N(0, 1)$$

(see Chapter 9) and

$$\frac{(n-1)s^2}{\sigma^2} \sim \chi^2_{(n-1)}$$

(see the appendix to Chapter 9). It is then suggested that

$$\frac{\dfrac{\bar{X} - \mu}{\sigma/\sqrt{n}}}{\sqrt{\dfrac{(n-1)s^2}{\sigma^2} \Big/ (n-1)}} \sim t^{(n-1)}$$

if $\bar{X}$ and s^2 are independent, which in this case they are. However,

$$\frac{\dfrac{\bar{X} - \mu}{\sigma/\sqrt{n}}}{\sqrt{\dfrac{(n-1)s^2}{\sigma^2} \Big/ (n-1)}} = \left(\frac{\bar{X} - \mu}{\sigma/\sqrt{n}}\right)\left(\frac{\sigma}{s}\right) = \frac{\bar{X} - \mu}{s/\sqrt{n}}$$

It is thus implied that, in the case of a simple random sample from a normal parent population,

$$\frac{\bar{X} - \mu}{s/\sqrt{n}} \sim t^{(n-1)}$$

This result is, of course, the one presented in the text. Note how it gives us $(n-1)$ as the degrees of freedom of the t.

It turns out that the distribution of $(\bar{X} - \mu)/(s/\sqrt{n})$ is often approximated by $t^{(n-1)}$ even if the parent population is nonnormal, and we generally adopt this procedure.

APPENDIX 2 TO CHAPTER 11

The following data are from U.S. Bureau of the Census, *State and Metropolitan Area Data Book, 1986* (Washington, DC: U.S. Government Printing Office, 1986), pp. XXXIV–XLIX.

(1) Metr. area	(2) % black (1980)	(3) % Spanish (1980)	(4) AFDC recip. per 1000 (1984)	(5) Soc. sec. recip. per 1000 (1982)	(6) % with 16+ yrs school (1980)	(7) Income per cap. (1983)	(8) Poverty rate (1979)	(9) Unemp. rate (1984)	(10) Region	(11) Size
1	6.05	11.90	6.5	138.6	17.4	$8830	12.0	4.3	3	0
2	0.70	0.30	50.6	157.5	8.1	7613	10.9	11.3	1	0
3	0.15	0.52	38.1	144.5	14.0	9306	6.2	7.2	2	0
4	9.35	17.58	14.5	100.6	28.1	10145	14.2	3.3	3	1
5	21.79	3.36	25.9	141.9	12.1	9376	12.4	11.1	3	0
6	27.18	0.65	37.8	169.0	14.3	8570	15.3	10.3	3	1
7	8.93	2.15	16.5	297.4	12.4	9677	11.0	5.0	3	0
8	0.39	0.75	35.4	112.5	26.1	9638	10.4	3.8	1	0
9	5.13	0.52	37.7	171.2	14.7	9049	10.5	12.4	3	0
10	11.19	0.56	55.7	146.0	15.6	9500	10.2	8.5	2	1
11	34.95	2.30	47.1	135.0	11.8	7589	19.1	7.7	3	0
12	4.44	3.00	57.5	143.7	14.5	9802	8.0	11.8	2	0
13	19.38	1.63	93.7	142.6	15.2	10017	10.1	10.7	2	1
14	4.20	1.25	14.8	137.8	12.3	9557	7.9	6.1	2	0
15	0.21	0.73	19.6	128.5	20.8	9604	9.8	4.0	2	0
16	0.42	5.87	18.6	110.3	28.8	9851	11.0	5.1	4	0
17	4.92	29.30	116.1	137.4	15.2	8452	14.5	12.5	4	1
18	1.22	1.66	29.4	140.8	17.4	8604	10.3	7.6	4	0
19	6.15	1.11	25.7	153.7	14.4	9629	8.0	5.8	1	1
20	2.16	0.48	56.7	165.3	9.8	7481	15.4	14.1	3	0
21	21.60	1.83	35.6	136.3	14.1	9271	15.1	5.7	3	1
22	7.46	1.23	60.3	127.9	23.0	9695	10.7	7.6	2	0

(1) Metr. area	(2) % black (1980)	(3) % Spanish (1980)	(4) AFDC recip. per 1000 (1984)	(5) Soc. sec. recip. per 1000 (1982)	(6) % with 16+ yrs school (1980)	(7) Income per cap. (1983)	(8) Poverty rate (1979)	(9) Unemp. rate (1984)	(10) Region	(11) Size
23	0.19	0.38	53.7	155.7	17.3	8994	9.9	6.7	2	0
24	5.58	3.14	58.4	114.3	21.7	9470	10.6	10.1	2	0
25	0.24	0.53	51.0	182.9	10.5	7745	12.6	7.4	1	0
26	9.21	23.97	64.0	123.0	18.4	10608	11.8	7.2	4	1
27	0.50	0.80	15.8	142.7	18.5	10743	7.2	3.0	1	0
28	14.90	23.50	29.5	184.0	16.1	10690	12.7	6.7	3	1
29	29.15	0.92	51.2	137.2	15.3	7857	20.6	9.3	3	0
30	1.01	2.20	51.6	175.8	10.8	8505	10.1	7.1	1	0
31	16.55	1.64	33.7	256.4	9.6	7672	17.9	7.1	3	0
32	3.94	0.49	29.1	160.6	11.0	8519	12.5	9.3	3	0
33	18.18	2.60	66.4	158.1	17.0	10025	11.8	6.8	1	1
34	2.58	2.02	28.0	141.7	19.8	9803	9.0	8.1	4	1
35	26.12	0.83	20.5	123.3	26.6	9922	12.3	3.6	3	1
36	11.76	0.57	30.6	162.5	14.3	9877	11.0	4.6	3	0
37	0.21	0.37	26.9	126.6	14.2	7547	11.5	8.0	2	0
38	3.99	21.17	11.2	150.6	14.7	8987	12.5	4.1	3	0
39	5.17	1.48	9.1	372.5	17.7	11277	9.1	4.3	3	0
40	7.03	1.50	12.8	190.3	12.9	9190	9.7	5.3	3	0
41	6.07	0.56	45.5	164.7	18.3	10635	8.6	7.0	2	0
42	4.82	0.88	39.6	154.1	17.6	9378	10.3	6.0	1	1
43	7.64	3.97	38.9	142.8	19.9	10424	7.8	5.6	2	0
44	6.85	30.44	21.1	116.9	12.0	9398	13.1	6.0	3	0

(Continued)

(1) Metr. area	(2) % black (1980)	(3) % Spanish (1980)	(4) AFDC recip. per 1000 (1984)	(5) Soc. sec. recip. per 1000 (1982)	(6) % with 16+ yrs school (1980)	(7) Income per cap. (1983)	(8) Poverty rate (1979)	(9) Unemp. rate (1984)	(10) Region	(11) Size
45	13.45	4.94	17.1	261.3	17.1	12122	10.1	6.4	3	1
46	1.35	2.20	39.4	170.1	15.4	9476	9.1	5.0	1	1
47	0.93	14.76	64.0	167.2	11.3	8020	15.4	14.5	4	0
48	2.57	0.87	20.3	152.6	11.3	9182	7.2	8.0	1	0
49	10.48	1.27	76.7	169.8	10.5	8574	9.7	12.0	2	1
50	2.74	10.26	102.1	150.0	12.0	7620	13.6	18.2	4	0

Key: Column 1 represents the metropolitan area (1 = Abilene TX; 2 = Altoona PA; 3 = Appleton-Oshkosh-Neenah WI; 4 = Austin TX; 5 = Beaumont-Port Arthur TX; 6 = Birmingham AL; 7 = Bradenton FL; 8 = Burlington VT; 9 = Charleston WV; 10 = Cincinnati-Hamilton OH-KY-IN; 11 = Columbus GA-AL; 12 = Davenport-Rock Island-Moline IA-IL; 13 = Detroit-Ann Arbor MI; 14 = Elkhart-Goshen IN; 15 = Fargo-Moorhead ND-MN; 16 = Fort Collins-Loveland CO; 17 = Fresno CA; 18 = Great Falls MT; 19 = Harrisburg-Lebanon-Carlisle PA; 20 = Huntington-Ashland WV-KY-OH; 21 = Jacksonville FL; 22 = Kalamazoo MI; 23 = La Crosse WI; 24 = Lansing-E. Lansing MI; 25 = Lewiston-Auburn ME; 26 = Los Angeles-Anaheim-Riverside CA; 27 = Manchester-Nashua NH; 28 = Miami-Ft. Lauderdale FL; 29 = Monroe LA; 30 = New Bedford-Fall River-Attleboro MA; 31 = Ocala FL; 32 = Owensboro KY; 33 = Philadelphia-Wilmington-Trenton PA-NJ-DE-MD; 34 = Portland-Vancouver OR-WA; 35 = Raleigh-Durham NC; 36 = Roanoke VA; 37 = St. Cloud MN; 38 = San Angelo TX; 39 = Sarasota FL; 40 = Sherman-Denison TX; 41 = Springfield IL; 42 = Syracuse NY; 43 = Topeka KS; 44 = Victoria TX; 45 = West Palm Beach-Boca Raton-Delray Beach FL; 46 = Worcester-Fitchburg-Leominster MA; 47 = Yakima WA; 48 = York PA; 49 = Youngstown-Warren OH; 50 = Yuba City CA.

Column 2 is the percent of the 1980 population that is black. Column 3 is the percent of the 1980 population that is of Spanish origin. Column 4 is AFDC recipients per 1000 persons in 1984. Column 5 is social security recipients per 1000 population in 1982. Column 6 is the percent of persons age 25+ with 16+ years of school in 1980. Column 7 is income per capita in 1983. Column 8 is the poverty rate (as a percent) in 1979. Column 9 is the unemployment rate (as a percent) in 1984. Column 10 is the region, where 1 represents Northeast, 2 represents Midwest, 3 represents South, and 4 represents West, and we use the Census definitions of these regions. Column 11 is a variable equal to 1 if the metropolitan area has a population of at least 500,000 persons, and 0 if the metropolitan area has a population of less than 500,000 persons, in 1984.

DISCUSSION QUESTIONS FOR CHAPTER 11

1. Explain the reason that if we wish to make a confidence interval for a single mean only half as wide as originally, we must increase the sample size by a factor of 4.

2. Suppose that over the years there have been many attempts at estimating a particular parameter; these estimates have included the construction of confidence intervals. Now, through some technological breakthrough, we know the true value of this parameter. Looking back on the confidence intervals that were constructed, we see that the vast majority of them contain the true parameter, but some of them do not. How do you explain this finding?

3. Suppose that X represents a population that is normally distributed, with mean μ. Explain why a 100% "confidence interval" for μ would be $(-\infty, \infty)$, and why a 0% "confidence interval" would be $(\bar{X}, \bar{X})$—that is, just $\bar{X}$.

Chapter *12*

Hypothesis Testing

In Chapter 11 we introduced the concept of a confidence interval as a way of making an inference (more sophisticated than an estimate) regarding a population parameter. Such an inference uses probability theory to characterize the pattern of variability of our estimator (which is a random variable) so as to exploit the information in our sample more fully. In this chapter we approach the problem of making sophisticated inferences by developing and carrying out hypothesis tests. The hypothesis test approach, though closely related to that for confidence intervals, has a different focus and intention, and therefore it often addresses the problem at hand more directly.

CONFIDENCE INTERVALS VERSUS HYPOTHESIS TESTS

The focus of a confidence interval is on constructing a range of values within which we are quite confident that the parameter of interest lies. (Remember that this parameter is a fixed number and is not randomly determined; we simply do not know the value of this fixed number.) We then, implicitly, conclude that values outside that range, though possible, are not plausible. (By "not plausible" we mean that we are persuaded beyond a reasonable doubt that they are not the case.) In hypothesis tests, we are more concerned with the actual *testing* of some theoretical position in an effort to determine whether it is or is not consistent with our data (taking account of sampling variability). In particular, we essentially ask: Can we say that a certain value of the parameter of interest is implausible given the data that we have observed? Although the distribution theory used is the same as with confidence intervals, the hypothesis test approach is different because of the different focus of the inquiry.[1]

INTUITIVE FOUNDATIONS OF HYPOTHESIS TESTING

In this chapter we shall present hypothesis tests in a context which suggests that they involve a line of reasoning which is quite intuitive and which in fact we would have a natural tendency to use in certain types of situations. We begin by presenting an example in which the underlying rationale for a hypothesis test is reasonably obvious. We then make clear what that rationale is and proceed to use it in the context of other problems for which its applicability is perhaps initially less obvious.

In many situations, there is disagreement or uncertainty as to the "state of the world;" that is, there are competing hypotheses about the characteristics of a situation, and the clearest way to determine which hypothesis is most consistent with the true state of the world is through some sort of empirical investigation. Let us motivate this framework and see how it leads to hypothesis testing through the use of an example.

Suppose that you are invited to participate in the following game:

A die is rolled. If it turns up any number from one to five, you receive $1. If it turns up six, however, you must pay $5.

This situation would be an example of a "fair" game if the die were such that each number occurred with probability one-sixth, where

---

[1] There is another difference, which is of a more technical nature. Formally, the confidence interval approach is strictly equivalent to the performance of a "two-sided" as opposed to a "one-sided" hypothesis test, but for the purpose of interpreting the conclusions that result, this point may often be overlooked without significant harm.

by "fair" we mean that the expected winning is zero. (Can you verify that this is the case?) You suspect, however, that the die is unfair, and in particular, that it is "loaded" in a way such that sixes turn up more than one-sixth of the time (thereby making the game an "unfair" one in which the expected winning is negative). Given that you do not have the knowledge from physics necessary to take the appropriate measurements on the die to determine whether or not it is "loaded," how might you make this determination?

A natural thing to do would be to start rolling the die, and to observe the relative frequency with which a six turns up; after all, you know that if the die were fair that a six would appear about one-sixth of the time.

Suppose that you roll the die 60 times (a relatively large number) and observe that a six appears 11 times. In this case, although the relative frequency of a six is a bit more than one would expect if the die were fair ($11/60 > 1/6$) it is clearly not an extraordinary event for such an outcome to occur. We then would be unable to conclude confidently that the die is loaded. On the other hand, if a six had appeared 30 times, it would be clear that the die is loaded in a way so as to make a six turn up "too often."

But in what sense is it "clear" that the die is loaded if we get 30 sixes? It is important to emphasize that it is *possible* that with a fair die we would get 30 sixes in 60 rolls; in fact, we know how to use the binomial distribution to evaluate this probability. This probability, however, is very small. It is, though possible, not *plausible* that we could get 30 sixes out of 60 rolls of a fair die. Since it is not plausible, we conclude that the die "must" be loaded.

What have we done here regarding these two different possible outcomes of our experiment (11 or 30 sixes)? We have essentially determined in an intuitive and natural way that the probability of 11 (out of 60) sixes occurring (if the die were fair) is reasonably high, but the probability of 30 (out of 60) sixes occurring (if the die were fair) is very low. To clarify further this process, note that we have set up the hypothesis that the die is fair as a type of "straw man" and have investigated the extent to which the data are inconsistent with it. When the data are sufficiently inconsistent with this straw man we reject it (we "knock down" the straw man). In the case where the data are only somewhat inconsistent with the straw man (e.g., the case of getting 11 sixes) we cannot reject it. This statement does not imply that we believe that the straw man is true (after all, 11/60 is still greater than the 1/6 we would expect with a fair die, and this whole investigation was motivated by a suspicion that the die was not fair); it is simply the case that the evidence against it is not sufficient to warrant a confident rejection. That is, although we cannot rule out (beyond a reasonable doubt) the possibility that the die is fair, we do not necessarily believe that it in fact is fair. Notice how the decision to reject or not reject this original hypothesis depends not only on what happens in our sample but also on our implicit use of probability concepts to determine the likelihood of what we observed happening under various states of the world (e.g., a fair die or a loaded die).

The above logic is exactly what we use in constructing and performing hypothesis tests, although it can at times be obscured by the technical details we must necessarily concern ourselves with (such as the determination of the probability of an event occurring under our original hypothesis). We now turn to the issues of more formally stating the problem and our method of investigation, so that we might ultimately be able to make decisions in less clear-cut cases (e.g., we observe 15 sixes out of our 60 rolls of the die), and of using the logic developed above in situations where its applicability is less apparent.

HYPOTHESES AND PROB-VALUES

Our goal is to determine the extent to which our data are consistent or inconsistent with a particular hypothesis; in particular (as in the above example), we attempt to determine whether the data are *sufficiently* inconsistent with the hypothesis that we may conclude that it is implausible. In general, in a hypothesis test we set up this hypothesis as a straw man and try to show that the data are sufficiently inconsistent with it as to warrant its rejection; in the preceding example, we set up the hypothesis of a fair die as a straw man and then attempted to show that what we observed could not plausibly have occurred if this hypothesis were true.[2]

Our initial hypothesis (the straw man) is typically referred to as the *null hypothesis* and is represented as H_0; we do not necessarily believe this hypothesis but are rather trying to determine whether

---

[2] It is important to recall that the parameter of interest is not a random variable, but is rather a constant (but unknown) number. Our uncertainty is as to what the value of this parameter *is,* and not as to the value it will take on as the result of some random process.

the data are sufficiently inconsistent with it to warrant its rejection. In the loaded die example above, the null hypothesis is

$$H_0: \text{die is fair} \quad \left[\text{i.e., } \Pr(\text{six}) = \frac{1}{6} \right]$$

We also specify an *alternative hypothesis* (H_1) to take the place of the null hypothesis if the latter is rejected; in the current example, the alternative hypothesis is

$$H_1: \text{die is loaded} \quad \left[\text{i.e., } \Pr(\text{six}) > \frac{1}{6} \right]$$

As presented above, we observed a sample from the population (the 60 rolls of the die) and inquired as to how likely the resulting outcome would be *if the die were fair;* in the case where that likelihood was small, we concluded that the die was loaded. In the case, however, where that likelihood was not unreasonably small, our decision was that the evidence was insufficient to conclude that the die was loaded.

The key to carrying out a hypothesis test more generally is the calculation of the *prob-value,* which we define as follows:[3]

Definition of a Prob-Value   The prob-value is the answer to the following question: What is the probability of getting a value of our estimator as extreme as in our sample if the null hypothesis is in fact true?

In the above example, we might think of the estimator as the sample proportion (the number of sixes in our sample divided by 60); this estimates the population proportion (which is $1/6$ if the die is fair). When we inquired as to the likelihood of getting the number of

---

[3] We shall present hypothesis testing in the context of the prob-value approach. The more traditional approach to hypothesis testing, based upon the comparison of a test statistic to a critical value of a probability distribution, is discussed in Appendix 1 to this chapter. The prob-value approach, though yielding the same result as the traditional approach, is more consistent with the intuition we have here attempted to develop.

sixes we actually observed in our sample, we were essentially making a determination regarding the prob-value. In the example where we observed 30 sixes, we concluded in essence that the prob-value was very small (and therefore H_0 was untenable), whereas in the example of 11 sixes we concluded that the prob-value was not unreasonably small (and therefore that H_0 was not untenable). In order to pursue this matter further, let us consider a more familiar example in which we consider a sample mean as a basis for inferring the characteristics of a population mean.

Clearly, a key part of our determination of a prob-value is the use of the appropriate piece of distribution theory necessary in order to answer the prob-value question. In the loaded-die example, the answers were really so obvious that formal distribution theory was not necessary. In general, of course, such is not the case, and we must therefore appeal more directly and specifically to the appropriate distribution theory. In the case of the sample mean as the source of inference regarding a population mean, for example, we are in the context of a normal or t distribution. Let us now consider a concrete example involving a mean so as to begin to develop the specifics of the prob-value approach.

Suppose that we are interested in the mean SAT score of members of the freshman class at a state university. In particular, to satisfy the state legislature that we are achieving a particular level of quality in the student body we investigate whether the mean (combined verbal and math) SAT score of freshmen is above 1000. For the purpose of this study, the total freshman class is too large to observe, so we take a simple random sample of 30 freshmen and observe their combined SAT scores. We find that $\bar{X} = 1020$. For simplicity, we assume that the population is normally distributed.

Clearly, we have *some* evidence to suggest that the *population* mean SAT (μ) is over 1000, inasmuch as our estimate of μ is 1020 (>1000). It is, of course, possible that μ is not above 1000 and by chance we picked a "better-than-average" group of students for our sample. So in order to investigate whether this second possibility is plausible, we perform a hypothesis test.

The null hypothesis is

$$H_0: \mu = 1000$$

and we are trying to find persuasive evidence against this hypothesis.

The alternative hypothesis is then

$$H_1: \mu > 1000$$

and if we reject H_0 we wind up concluding that H_1 must be true.[4]

The prob-value is always the answer to the prob-value question: What is the probability of getting a value of our estimator as extreme as in our sample if the null hypothesis is in fact true. In the problem at hand, since our estimator is $\bar{X}$ we say

$$\text{prob-value} = \Pr(\bar{X} \geq 1020) \quad \text{if } \mu = 1000 \quad (12.1)$$

The estimator $\bar{X}$ is as extreme as in our sample if it is 1020 *or more,* since by extreme we mean as compared to the value of μ in H_0 [and clearly numbers larger than 1020 are as extreme (in fact, are more extreme) when compared to H_0 as 1020 is]. And in any case, the prob-value must be for $(\bar{X} \geq 1020)$ as opposed to $(\bar{X} = 1020)$ since the latter probability is *always* zero (why?).

Determining the probability given in Eq. (12.1) involves using the distribution of a random variable $(\bar{X})$ *conditional* on the mean of the parent population (and therefore $\bar{X}$'s mean) being a particular value (1000). If this probability is sufficiently small, then we conclude that it would be quite unlikely for us to have observed our $\bar{X}$ if $\mu = 1000$; this leads to the conclusion that it is implausible that $\mu = 1000$. But how do we actually evaluate this probability? We exploit our knowledge concerning the distribution of $\bar{X}$!

Since the parent population was assumed to be normal, we know that $\bar{X}$ is normally distributed with mean μ and variance σ^2/n, where σ^2 is the variance of the parent population X (in this case, it is the variance of SAT scores among the population of all freshmen). If σ^2 were known, we could evaluate the probability in Eq. (12.1) based upon the fact that

$$\frac{\bar{X} - \mu}{\sigma/\sqrt{n}} \sim N(0, 1)$$

---

[4] A word is perhaps in order about why the null hypothesis is $\mu = 1000$ and ignores the possibility that $\mu < 1000$. In actuality, it makes little difference for our purposes. If the evidence is sufficiently contradictory to the hypothesis $\mu = 1000$, then it will also be contradictory to the hypothesis $\mu < 1000$. And if we do not reject the null hypothesis, we are not really saying that we believe it (after all, $\bar{X}$ was still over 1000). The more interesting problem in this regard is as to why $\mu < 1000$ has been ruled out as a possibility for the *alternative* hypothesis; this issue will be discussed in more depth below.

where the value of μ is as hypothesized in H_0. But, of course, it will generally be the case that σ^2 is unknown. Following the strategy we used in the case of confidence intervals, we can evaluate the prob-value based upon the alternative distribution theory

$$\frac{\bar{X} - \mu}{s/\sqrt{n}} \sim t^{(n-1)}$$

where s^2 is the usual sample variance. (As discussed earlier in Chapter 11, if the degrees of freedom $(n - 1)$ exceeds 30 we can approximate the t distribution with the standard normal.) Suppose that in the above example we found $s^2 = 1600$ (i.e., $s = 40$). To determine the prob-value, we would then proceed as follows:

$$\Pr(\bar{X} \geq 1020) \qquad (\text{if } \mu = 1000)$$

$$= \Pr\left(\frac{\bar{X} - \mu}{s/\sqrt{n}} \geq \frac{1020 - 1000}{40/\sqrt{30}}\right)$$

$$= \Pr(t^{(29)} \geq 2.74)$$

This is our prob-value. How can we determine the value of this probability using the t table that we have available?

In Figure 12.1 we show the t distribution with 29 degrees of freedom; we have included here the $t_{.010}$ and $t_{.005}$ values and the geometric interpretation that the area under the curve to the right of the $t_{.010}$ value is .01 and the area under the curve to the right of

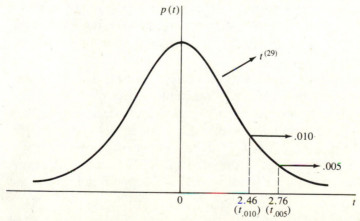

Figure 12.1 Calculating a prob-value with the t distribution.

the $t_{.005}$ value is .005. Since 2.74 is between 2.46 and 2.76, the area under the curve to the right of 2.74 must be between .010 and .005. That is,

$$\Pr(t^{(29)} \geq 2.46) > \Pr(t^{(29)} \geq 2.74) > \Pr(t^{(29)} \geq 2.76)$$

or

$$.010 > \Pr(t^{(29)} \geq 2.74) > .005$$

We then say

$$.010 > \text{prob-value} > .005$$

Although we could attempt to interpolate in the t table to determine a more precise value of the prob-value, this procedure is generally not necessary. We conclude that if μ were really 1000, then somewhere between 1% and .5% of the time we would observe an $\bar{X}$ of 1020 or above. Obviously, this occurrence is fairly unlikely.

In a *hypothesis test* we compare the prob-value to some predetermined cutoff point representing a probability sufficiently low that the null hypothesis is no longer plausible or believable (although it is still possible). This cutoff point is referred to as the *level of significance of the test* (or just the *level of the test*) and is often set at .05. (The determination of the level of the test is similar to the determination of the level of a confidence interval and is in the same sense arbitrary. Although .05 is a common choice, .10 and .01 are other possibilities. The importance of the level of the test shall be elaborated upon below in our discussion of type I and type II errors.) If the prob-value is less than (or equal to) the level of the test (call it α), then we say that we *reject H_0 in favor of H_1 at the α level.*[5] On the other hand, if the prob-value is greater than the level of the test, we say that we *cannot reject H_0 in favor of H_1 at the α level.* In the former case, we say that we are confident that H_0 is false and therefore that H_1 is true. In the latter case, however, our knowledge is not as fully developed. We say "we cannot reject H_0" as opposed to "we accept H_0" because, after all, we do not necessarily believe that H_0 is true (remember, we still have some evidence against it). In the case of "not rejecting H_0" we do not have persuasive evidence in favor of one hypothesis or the other; there is then a clear asymmetry in the type of knowledge we acquire depending on how the test turns out.

---

[5] If the prob-value is equal to the level of the test, we shall adopt the convention that we reject H_0.

Let us suppose that in the above example the level of the test was the standard .05. Then, since the prob-value is less than the level of the test, we reject H_0 in favor of H_1 at the .05 level, and we conclude that the mean SAT score is in fact over 1000.[6]

The arbitrariness of the level of the test is in some ways remedied if we report the prob-value so that readers of our work can determine for themselves the conclusion. By reporting the prob-value, we are reporting in good faith the appropriate information as well as the potential limitations of our conclusions.

Let us stop, before moving on, to emphasize that what has happened in the above example is just what we talked about in the context of the loaded-die problem considered earlier in this chapter. We showed that it was, though possible, very unlikely that we could have a sample mean SAT of 1020 (or more) if the population mean were really only 1000, and we therefore concluded that the population mean must (in all likelihood) be above 1000.

In the above example, "extremeness" was in terms of being "too big"; that is, it was in the positive direction. What happens when we are going in the opposite direction? Let us address this situation within the context of an example.

Suppose that you are interested in the possibility that the mean family income of the residents of a particular neighborhood is less than the "poverty level of income." We shall, for simplicity, confine ourselves to families of four persons; the poverty level of income for families of 4 in 1989 was \$12,675.[7] Due to the large size of the neighborhood you cannot observe the entire population. But in a simple random sample of 25 families (of size 4 each) you find $\bar{X}$ equals \$11,500 with $s = $ \$5000. Can you conclude beyond a reasonable doubt that the mean family has an income less than the poverty level?

Remember the prob-value question: What is the probability of getting a value of our estimator as extreme as in our sample if

---

[6] Strictly speaking, $(\bar{X} - \mu)/(s/\sqrt{n})$ has the $t^{(n-1)}$ distribution only if the parent population is normal. But the t distribution is often used as an approximation when such is not the case, and we shall adopt this practice.

[7] U.S. Bureau of the Census, Current Population Reports, Series P-60, No. 168, *Money Income and Poverty Status in the United States: 1989. Advance Data from the March 1990 Current Population Survey* (Washington, DC: U.S. Government Printing Office, 1990), p. 86. The poverty level of income as determined by the federal government is a measure of the minimum amount of income necessary for a family of the given size to achieve a minimally acceptable standard of living.

the null hypothesis is in fact true? The only difference between the current situation and that considered previously is that now "as extreme" means in the "less than" as opposed to the "greater than" direction.

Formally, the problem is to find the prob-value for and test the hypothesis

$$H_0: \mu = 12,675$$

$$H_1: \mu < 12,675$$

at the (say) .05 level. The prob-value is just the probability of getting a value of $\bar{X}$ *less than* or equal to 11,500 if μ really equals 12,675:

$$\text{prob-value} = \Pr(\bar{X} \le 11,500) \qquad (\text{if } \mu = 12,675)$$

and we determine this probability using the distribution theory and procedure developed above:

$$\Pr(\bar{X} \le 11,500) = \Pr\left(\frac{\bar{X} - \mu}{s/\sqrt{n}} \le \frac{11,500 - 12,675}{5000/\sqrt{25}} \right)$$

$$= \Pr(t^{(24)} \le -1.18)$$

$$= \Pr(t^{(24)} \ge 1.18)$$

(since the t distribution is symmetric about zero)

Therefore

$$.25 > \text{prob-value} > .10$$

(That is, if the population mean really were \$12,675 we could expect to get a sample mean as small as \$11,500 somewhere between 10 and 25 percent of the time.) To perform the hypothesis test, we compare the prob-value to the level of the test (.05); since the prob-value exceeds the level of the test, we cannot reject H_0 in favor of H_1 at the .05 level. That is, our data are not sufficiently contradictory to the hypothesis that the average family is *not* below the poverty level to warrant a rejection of that possibility. We say that "we cannot reject H_0" as opposed to "we accept H_0" inasmuch as we still have evidence against H_0; this evidence, however, is not strong enough to meet our decision-making criterion.

EXAMPLE ■

This example relies on the data contained in Appendix 2 to Chapter 11. For the unemployment rate (in 1984), we find a sample mean $\bar{X} = 7.7\%$ with a sample standard deviation $s = 3.25\%$. Suppose that the question of interest is as to whether the unemployment rate in the (population) mean metropolitan area exceeds 7%. Our problem is to find the prob-value for and test the hypothesis

$$H_0: \mu = 7$$

$$H_1: \mu > 7$$

at the .05 level. The prob-value is

$$\Pr(\bar{X} \geq 7.7) \qquad (\text{if } \mu = 7)$$

$$= \Pr\left(\frac{\bar{X} - \mu}{s/\sqrt{n}} \geq \frac{7.7 - 7}{3.25/\sqrt{50}}\right)$$

$$= \Pr(t^{(49)} \geq 1.52)$$

Since the degrees of freedom exceeds 30, we follow our usual practice of approximating the t with the standard normal:

$$\Pr(Z \geq 1.52) = .0643 = \text{prob-value}$$

Since the prob-value exceeds .05, we cannot reject H_0 in favor of H_1 at the .05 level. We cannot rule out the possibility that the unemployment rate in the mean metropolitan area was as low as 7% and that $\bar{X}$ was 7.7% due simply to random variability. ∎

PROBLEMS

12.1 As part of a demographic study, you are interested in whether the mean number of children for families in a given area is 2. In a simple random sample of 25 families, you find the sample mean number of children $\bar{X} = 2.2$, with the sample variance $s^2 = .64$. Find the prob-value for and test the hypothesis

$$H_0: \mu = 2$$

$$H_1: \mu > 2$$

at the .05 level.

12.2 In a simple random sample of 20 males, you find the sample mean weight $\bar{X} = 170$ (pounds) and $s^2 = 250$. Find the prob-value for and test the hypothesis

$$H_0: \mu = 180$$

$$H_1: \mu < 180$$

at the .05 level.

12.3 In a simple random sample of 25 law school graduates, you find the sample mean starting salary $\bar{X} = 31,500$ (dollars) and the sample standard deviation $s = 3000$. Find the prob-value for and test the hypothesis

$$H_0: \mu = 30,000$$

$$H_1: \mu > 30,000$$

at the .05 level.

12.4 Suppose that the starting salary of college graduates is a random variable with an unknown mean but with a known population standard deviation of $2000. In a simple random sample of 16 college graduates you find the sample mean $\bar{X} = \$20,200$. Find the prob-value for and test the hypothesis

$$H_0: \mu = 21,000$$

$$H_1: \mu < 21,000$$

at the .05 level.

12.5 In a simple random sample of 36 families in a certain country, you find the sample mean income $\bar{X} = \$9500$ and the sample standard deviation $s = \$2500$. Find the prob-value for and test the hypothesis

$$H_0: \mu = 9000$$

$$H_1: \mu > 9000$$

at the .05 level.

12.6 Using the data from problem 11.9, find the prob-value for and test the hypothesis

$$H_0: \mu = 500$$

$$H_1: \mu > 500$$

at the .05 level, where μ is union membership (in thousands) for the mean state.

12.7 Using the data from problem 11.11, find the prob-value for and test the hypothesis

$$H_0: \mu = 55(\%)$$

$$H_1: \mu < 55(\%)$$

at the .05 level, where μ is the percent of bachelor's degrees conferred upon women in the mean state in 1981–82.

12.8 (Requires computer.) Using the data from Appendix 2 to Chapter 11, find the prob-value for and test the hypothesis

$$H_0: \mu = 10,000$$

$$H_1: \mu < 10,000$$

at the .05 level, where μ is income per capita (in 1983) in the mean U.S. metropolitan area.

12.9 (Requires computer.) Using the data from Appendix 2 to Chapter 11, find the prob-value for and test the hypothesis

$$H_0: \mu = 13.0$$

$$H_1: \mu > 13.0$$

at the .05 level, where μ is the percent of persons with 16+ years of schooling (in 1980) in the mean U.S. metropolitan area.

HYPOTHESIS TESTS FOR DIFFERENCES IN MEANS

In Chapter 11 we discussed the difference in means and how many interesting questions can be investigated by inferences regarding this parameter. Confidence intervals were one way of undertaking sophisticated inference in this context; hypothesis tests for differences in means take many of the same issues and principles and utilize them in a slightly different context (i.e., the hypothesis testing context as opposed to the confidence interval context). We shall therefore now begin to extend the concept of a hypothesis test to situations regarding differences in means; what will be apparent is that although the mechanical aspects of the procedure differ (due to the different distribution theory) from what we saw with hypothesis tests for single means, the conceptual framework and the underlying strategy are the same. Let us illustrate this point within the context of an example.

It is generally presumed that individuals with higher amounts of education earn, on average, more than those with lower amounts of education. Although the evidence in support of this assumption is overwhelming, trying to test this hypothesis and noting the methods by which we might attempt to do so is still an interesting exercise and one that illustrates nicely the application of hypothesis tests to problems involving differences in means.

Suppose that, in independent simple random samples of individuals who are (1) high school graduates (with no college) and (2) college graduates, we find that the mean high school graduate earns \$20,000 per year (call this $\bar{X}_h$) and the mean college graduate earns \$23,000 per year (call this $\bar{X}_c$). Suppose that the sample sizes are $n_h = 15$ and $n_c = 10$. The theory suggests that $\mu_c > \mu_h$ (or $\mu_c - \mu_h > 0$) where the μ's are the respective population means. One way to see if this theory is consistent with the data is to find the prob-value for and test the hypothesis

$$H_0: \mu_c - \mu_h = 0$$

$$H_1: \mu_c - \mu_h > 0$$

at the .05 level. If we reject H_0 then we have persuasive evidence to suggest that the underlying theory is valid empirically. How do we calculate the prob-value? *By asking the same question (and interpreting its answer in the same way) as before.* Remember the prob-value question: What is the probability of getting a value of our estimator as extreme as in our sample if the null hypothesis is in fact true? In this case, our estimator is $\bar{X}_c - \bar{X}_h$; in our sample it takes on a value of $23,000 - 20,000$. The prob-value is then

prob-value $= \Pr(\bar{X}_c - \bar{X}_h \geq 23,000 - 20,000)$　　　(if $\mu_c - \mu_h = 0$)

How do we determine this probability? In our discussion in Chapter 11 regarding confidence intervals for differences in means, we based the determination of the confidence interval on the distribution theory that

$$\frac{\bar{X}_c - \bar{X}_h - (\mu_c - \mu_h)}{s_p \sqrt{\dfrac{1}{n_c} + \dfrac{1}{n_h}}} \sim t^{(n_c + n_h - 2)}$$

when the samples are independent (using here the c and h subscripts to denote the independent samples) and the population variances though unknown are assumed equal (and are estimated with the pooled variance s_p^2).[8] Let us suppose that in the problem at hand $s_p = 4000$. The prob-value is then

$$\Pr(\bar{X}_c - \bar{X}_h \geq 23{,}000 - 20{,}000)$$

$$= \Pr\left(\frac{\bar{X}_c - \bar{X}_h - (\mu_c - \mu_h)}{s_p \sqrt{\dfrac{1}{n_c} + \dfrac{1}{n_h}}} \geq \frac{23{,}000 - 20{,}000 - 0}{4000 \sqrt{\dfrac{1}{10} + \dfrac{1}{15}}} \right)$$

$$= \Pr(t^{(23)} \geq 1.84)$$

Thus, we can say that

$$.05 > \text{prob-value} > .025$$

Since the prob-value is less than .05 (the level of the test), we reject the null hypothesis in favor of the alternative hypothesis at the .05 level. That is, we conclude that the data are sufficiently contradictory to the hypothesis that the population means are the same that we may conclude with confidence that the mean earnings for college graduates exceeds the mean earnings for high school graduates. We often say that the difference in the sample means is then "statistically significant"—in other words, that it is unlikely to be due to random variation and therefore is representative of some underlying fundamental difference between the two populations.[9]

Of course, the interpretation of this finding involves all the complications discussed in Chapter 11 regarding the interpretation of confidence intervals. That is, our conclusion simply tells us that the population means are different; it does not tell us why they are different. In particular, determining the "why" part must involve an appeal to the theory, including perhaps a consideration of whether

---

[8] Strictly speaking, this expression has exactly a t distribution only if the parent population is normal. But it is often used as an approximation even when such is not the case, and we shall do so here.

[9] When we say that a difference is statistically significant we are saying that the difference in sample statistics is unlikely to result from an element of chance; we are therefore confident that this difference reflects a difference in the underlying population parameters. This conclusion does not imply, however, that the difference is "interesting" in an application framework; for example, if the (population) mean college graduate makes $5 per year more than the (population) mean high school graduate, this difference is hardly relevant for most policy issues.

the study was observational or experimental (in the problem at hand, of course, it is observational). In particular, the fact that college graduates earn on average more than high school graduates does not imply that the additional education is the reason behind the difference! (See problem 12.10 below for further thought regarding this issue.)

EXAMPLE ■

This example uses the data from Appendix 2 to Chapter 11. Suppose that we are interested in the relationship between income per capita (in 1983) in the mean "large" metropolitan area (at least 500,000 persons) versus the income per capita in the mean "small" metropolitan area (less than 500,000 persons). We find the following sample statistics:

$$\bar{X}_L = 9761 \qquad s_L = 915 \qquad n_L = 16$$

$$\bar{X}_S = 9039 \qquad s_S = 1018 \qquad n_S = 34$$

(where the L and S subscripts indicate large and small, respectively). Our problem is to find the prob-value for and test the hypothesis

$$H_0: \mu_L - \mu_S = 0$$

$$H_1: \mu_L - \mu_S > 0$$

at the .05 level.

The prob-value is (note that from the data above we can calculate $s_p = 987$)

$$\Pr(\bar{X}_L - \bar{X}_S \geq 9761 - 9039) \qquad (\text{if } \mu_L - \mu_S = 0)$$

$$= \Pr\left(\frac{\bar{X}_L - \bar{X}_S - (\mu_L - \mu_S)}{s_p \sqrt{\dfrac{1}{n_L} + \dfrac{1}{n_S}}} \geq \frac{9761 - 9039 - 0}{987 \sqrt{\dfrac{1}{16} + \dfrac{1}{34}}} \right)$$

$$= \Pr(t^{(48)} \geq 2.41)$$

Since the degrees of freedom exceeds 30 we may approximate the t with the standard normal:

$$\Pr(Z \geq 2.41) = .0080 = \text{prob-value}$$

Since the prob-value is below .05, we reject H_0 in favor of H_1 at the .05 level, and conclude that per capita income in the mean "large" U.S. metropolitan area exceeds per capita income in the mean "small" U.S. metropolitan area. ■

PROBLEMS

12.10 Can you construct a theory by which the difference in mean earnings between high school and college graduates uncovered in the preceding discussion could be due to a process other than one in which educational differences are the *cause* of the earnings differences?

12.11 Suppose that you are investigating whether male professors make on average more than their female counterparts. To control for rank, you take simple random samples of ten males and ten females who are assistant professors. You find the following sample statistics:

$$\bar{X}_m = \$30,000 \qquad s_m^2 = 1,000,000$$

$$\bar{X}_f = \$28,000 \qquad s_f^2 = 250,000$$

Assuming that the population variances are equal, find the prob-value for and test the hypothesis

$$H_0: \mu_m - \mu_f = 0$$

$$H_1: \mu_m - \mu_f > 0$$

at the .05 level. (*Hint:* Do you remember how to calculate s_p^2?)

12.12 As a development economist, you are interested in whether individuals in country A have a lower mean income than those in country B. You have available independent simple random samples of size 50 each from the two populations. In these samples,

$$\bar{X}_A = \$500 \qquad s_A = 200$$

$$\bar{X}_B = \$600 \qquad s_B = 300$$

Assuming that the population variances are equal, find the prob-value for and test the hypothesis

$$H_0: \mu_B - \mu_A = 0$$

$$H_1: \mu_B - \mu_A > 0$$

at the .05 level.

12.13 Suppose that you are investigating whether the graduates of a private university earn on average more than the graduates of a public university. In a simple random sample of 20 graduates from the private university you find the sample mean $\bar{X}_{priv} = \$22,000$ with a sample standard deviation $s_{priv} = \$1000$. In a simple random sample of 15 graduates from the public university you find the sample mean $\bar{X}_{pub} = \$21,000$ with a sample standard deviation $s_{pub} = \$2000$. Assuming that the population variances are equal, find the prob-value for and test the hypothesis

$$H_0: \mu_{priv} - \mu_{pub} = 0$$

$$H_1: \mu_{priv} - \mu_{pub} > 0$$

at the .05 level.

12.14 To investigate the respective mean family sizes in two different regions (A and B) you take independent simple random samples of size 15 each. You find

$$\bar{X}_A = 4.8 \text{ persons} \qquad s_A^2 = .8$$

$$\bar{X}_B = 4.4 \text{ persons} \qquad s_B^2 = .7$$

Find the prob-value for and test the hypothesis

$$H_0: \mu_A - \mu_B = 0$$

$$H_1: \mu_A - \mu_B > 0$$

at the .05 level, assuming that the population variances are equal.

12.15 Using the data from problem 11.20, find the prob-value for and test the hypothesis

$$H_0: \mu_w - \mu_s = 0$$

$$H_1: \mu_w - \mu_s > 0$$

at the .05 level.

12.16 Using the data from problem 11.21, find the prob-value for and test the hypothesis

$$H_0: \mu_s - \mu_w = 0$$

$$H_1: \mu_s - \mu_w > 0$$

at the .05 level.

12.17 (Requires computer.) Using the data from Appendix 2 to Chapter 11, find the prob-value for and test the hypothesis

$$H_0: \mu_s - \mu_{mw} = 0$$

$$H_1: \mu_s - \mu_{mw} > 0$$

at the .05 level, where μ_s is the percent of the population (in 1980) that is black in the (population) mean southern metropolitan area, and μ_{mw} is the corresponding value for the (population) mean midwestern metropolitan area. (Assume equal population variances.)

TWO-SIDED HYPOTHESIS TESTS

In the examples considered above it has always been the case that the null hypothesis was that the parameter of interest was equal to a particular value (e.g., $\mu = a$) and the alternative hypothesis was that this parameter was either greater than or less than that value (e.g., $\mu > a$), depending on the particulars of the situation. That is, we always had what is referred to as a "one-sided" H_1 because for some reason we were willing to rule out the "other side" (e.g., $\mu < a$) before the fact. This approach is often the case when we are considering hypothesis tests for means or differences in means; for example, in the earnings problem considered above we could fairly easily justify not considering the possibility that high school graduates earn more on average than college graduates.

However, such a one-sided hypothesis test is not always appropriate, particularly in economics regarding the parameters of interest in regression analysis, which we shall begin to consider in the next chapter. For now, we shall merely lay the foundation for a "two-sided" hypothesis test so that it can be easily extended to the case of regression.

A two-sided hypothesis test is appropriate if the theory (or previous knowledge) is not sufficiently precise so as to be able to rule out one direction of possibility. As an example, consider the effects of monetary expansion on interest rates. Monetary expansion causes a loosening of credit markets, and we then might expect the interest rate to fall. But on the other hand, monetary expansion may create inflationary expectations, causing the (nominal) interest rate to rise (so as to preserve the real interest rate).

Suppose that we know that the mean interest rate over a long time period is 6 percent. We then observe 16 months in which monetary expansion is higher than normal, and the interest rate in each such period. In this case the theory allows for both possibilities ($\mu < 6$ or $\mu > 6$, where μ is the mean interest rate during a period of rapid monetary expansion).

Suppose that we find $\bar{X} = 8$ and $s^2 = 9$. To construct our hypothesis test, we find the prob-value for and test the hypothesis

$$H_0: \mu = 6$$

$$H_1: \mu \neq 6$$

at the (say) .05 level.

To evaluate the prob-value, we now build into the prob-value question the possibility that μ can be below 6 as well as above 6. If we had a one-sided H_1, the prob-value would be $\Pr(\bar{X} \geq 8)$ (if $\mu = 6$). But $\bar{X} = 4$ is as extreme (relative to 6) as 8 is; why not consider then numbers less than or equal to 4? In the case of a one-sided test, the reason is clear; we are willing to rule out these possibilities before the fact. In the case of a two-sided test, however, we must consider these possibilities (since no such ruling out has occurred). That is, in the case of a two-sided test, the prob-value would be $\Pr(\bar{X} \geq 8$ or $\bar{X} \leq 4)$ (if $\mu = 6$). Notice that the logic behind the prob-value question is the same as in the one-sided case, we just must be a bit more careful in terms of how we go about specifying it. [Our discussion of two-sided hypothesis tests will be confined to cases based on the t (or standard normal) distributions. Additional complications arise when the relevant probability distribution is asymmetric, in particular, in terms of what "extremeness" in the two directions means.]

Note that $\bar{X}$ is above 8 or below 4 if the difference between $\bar{X}$ and 6 is greater than 2 (in absolute value). That is, this event occurs if $|\bar{X} - \mu| \geq |8 - 6|$, where the absolute value signs make all distances positive, as is the usual convention. In the problem at hand,

$$\text{prob-value} = \Pr(|\bar{X} - \mu| \geq |8 - 6|) \quad (\text{if } \mu = 6)$$

How do we evaluate such a probability? As before, the key is to transform the left-hand side; the only difference is that now the left-hand side will be the absolute value of a t-distributed random variable:

$$\text{prob-value} = \Pr\left(\left|\frac{\bar{X} - \mu}{s/\sqrt{n}}\right| \geq \left|\frac{8 - 6}{3/\sqrt{16}}\right|\right)$$

$$= \Pr(|t^{(15)}| \geq 2.67)$$

Since the t distribution is symmetric about zero, we know that this equals

$$2\Pr(t^{(15)} \geq 2.67)$$

Since $.010 > \Pr(t^{(15)} \geq 2.67) > .005$, we know that

$$.020 > \text{prob-value} > .010$$

(we just doubled the two probabilities that were the upper and lower bounds). Since the prob-value is less than the level of the test (.05), we reject the null hypothesis in favor of the alternative hypothesis at the .05 level. We conclude that the interest rate during periods of rapid monetary expansion is different than it is normally.

The prob-value associated with a two-sided test will be double that of a one-sided test due to the symmetry of the t distribution. But given the general arbitrariness involved in selecting the level of the hypothesis test in the first place, it is easy to exaggerate the importance in practice of the distinction between a one-sided and two-sided test.

For a two-sided test involving a difference in means, we would proceed with the same general strategy. For example, if $\bar{X}_1 = 3$ and $\bar{X}_2 = 2$ and the null hypothesis is that the population means are equal, the prob-value would be

$$\Pr(|\bar{X}_1 - \bar{X}_2 - (\mu_1 - \mu_2)| \geq |3 - 2 - 0|)$$

(since, under H_0, $\mu_1 - \mu_2 = 0$) and we would ultimately get to a question involving the absolute value of a t-distributed random variable with $n_1 + n_2 - 2$ degrees of freedom.

We now turn to some further interpretive issues regarding the results of hypothesis tests.

PROBLEMS

12.18 We know from years of experience that the mean life of light bulbs produced by a particular company is 1000 hours. To investigate the production situation in a new factory, we took a simple random sample of 30 light bulbs and found a sample mean length of life of 1020 hours with a sample standard deviation of 100 hours. Find the prob-value for and test (at the .05 level)

$$H_0: \mu = 1000$$

$$H_1: \mu \neq 1000$$

where μ is the mean life of light bulbs from the new factory. Clearly interpret your findings.

12.19 A particular course was optional for many years, and from that experience we know that the mean grade for the course was 82 (out of 100). This course has recently become required, however, and the first class after this change (which had 40 students) had a mean grade $\bar{X} = 75$ (with $s = 12$). Find the prob-value for and test the hypothesis (at the .05 level)

$$H_0: \mu = 82$$

$$H_1: \mu \neq 82$$

where μ is the mean score for all students once the course is required.

12.20 A simple random sample of 20 students at a high school yields a sample mean (combined) SAT score of 960 with a sample standard deviation of 100. Find the prob-value for and test the hypothesis

$$H_0: \mu = 1000$$

$$H_1: \mu \neq 1000$$

at the .05 level.

12.21 Suppose that you are investigating the relative wages of male and female workers in a given industry. In a simple random sample of 12 males you find $\bar{X}_m = \$500$ (per week) and $s_m = \$100$. In a simple random sample of 9 females you find $\bar{X}_f = \$350$ and $s_f = \$140$. Find the prob-value for and test the hypothesis

$$H_0: \mu_m - \mu_f = 0$$

$$H_1: \mu_m - \mu_f \neq 0$$

at the .05 level. (Assume that the population variances are equal.)

12.22 An automobile producer is investigating the relative fuel efficiencies of two new engine designs. In a simple random sample of seven cars with design A engines, the sample mean miles per gallon is 23.4 with a sample variance of 4. In a simple random sample of four cars with design B engines, the sample mean miles per gallon is 24.2 with a sample variance of 3. Find the prob-value for and test the hypothesis that the population means are the same, versus the alternative that they are different, at the .05 level. (Assume that the population variances are equal.)

12.23 (Requires computer.) Using the data from Appendix 2 to Chapter 11, find the prob-value for and test the hypothesis

$$H_0: \mu_L - \mu_S = 0$$

$$H_1: \mu_L - \mu_S \neq 0$$

at the .05 level, where μ_L is AFDC recipients per 1000 persons (in 1984) in the mean U.S. metropolitan area that is "large" (i.e., population of at least 500,000 persons) and μ_S is the comparable value for the mean U.S. metropolitan area that is "small" (i.e., population below 500,000 persons). (Assume that the population variances are equal.)

TYPE I VERSUS TYPE II ERRORS

Many issues regarding the content, as well as the limitations, of hypothesis tests can be seen through a discussion of type I and type II errors. We shall discuss these issues here in the context of a one-sided hypothesis test for a single population mean. Let us suppose that our problem is to test the hypothesis

$$H_0: \mu = \mu_0$$

$$H_1: \mu > \mu_0$$

at the .05 level. There are two possible decisions for us to make: we can decide to reject H_0, or we can decide to not reject H_0. If we reject H_0, our decision is "correct" if H_0 really is false, but our decision is wrong if H_0 is in fact true. If we reject a null hypothesis that is in fact true, we commit a *type I* error. On the other hand, if we do not reject H_0 our decision is correct if H_0 really is true, but it is wrong if H_0 is in fact false. If we do not reject a null hypothesis that is in fact false, we commit a *type II* error. Understanding these two possible mistakes and the relationships between the probabilities with which they occur is very useful for broadening our general understanding of the hypothesis testing framework.

Let us begin by considering type I errors. In general, we set up our framework and methodology so that the probability of a type I error is small; that is, the probability of rejecting H_0 when H_0 is true is small. This approach suggests that if we reject H_0 we can be very confident regarding our conclusion. To determine the probability of a type I error, consider Figure 12.2. Under the null hypothesis $\mu = \mu_0$, the distribution of $\bar{X}$ is $p(\bar{X}|H_0)$—that is, the distribution of $\bar{X}$ given that H_0 is true. Calculation of the prob-value involves observing the value of $\bar{X}$ (say $\bar{X}_0$) and calculating the area under the curve to the right of this point; this area gives us the prob-value as shown. We reject H_0 when this area is less than (say) .05; to investigate type I errors, let us assume that this is in fact what happens; after all, we can make a type I error only when we reject H_0.

The quantity $\bar{X}_c$ is that value of $\bar{X}$ such that

$$\Pr((\bar{X} \geq \bar{X}_c)|H_0) = .05$$

that is, it is that value of $\bar{X}$ such that the area under $p(\bar{X}|H_0)$ to the right of it is .05. We then reject H_0 whenever $\bar{X} \geq \bar{X}_c$, since in this case prob-value $\leq .05$. A type I error occurs whenever we reject a true H_0; a type I error then occurs if $\bar{X} \geq \bar{X}_c$ when H_0 is true. What is the probability of this occurring? .05! Notice that this probability is exactly the level of the test! Since the probability of a type I error

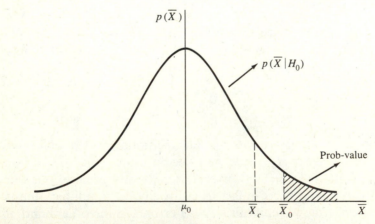

Figure 12.2 Illustrating the probability of a type I error.

is the level of the test, by making the level of the test small we have implicitly made the probability of a type I error small. The implication is that if we in fact reject the null hypothesis we can be very confident in doing so; rejection of the null hypothesis is then a very powerful conclusion.

Consider Figure 12.3, which builds upon Figure 12.2. In Figure 12.3 we have added a dashed probability distribution that lies to the right of the original. If $\bar{X} \geq \bar{X}_c$, we then conclude that it is implausible that $\mu = \mu_0$ [i.e., that $p(\bar{X}|H_0)$ is correct] and that the true distribution of $\bar{X}$ "must" lie somewhere further to the right (although we are not saying exactly where). The dashed distribution then becomes a possibility. Figure 12.3 is helpful for a discussion of the probability of making a type II error. Suppose that the null hypothesis is false and that the true distribution of $\bar{X}$ is indeed the dashed one. A type II error occurs if we do not reject H_0 when H_0 is false. But, we do not reject H_0 if $\bar{X} < \bar{X}_c$ (regardless of whether H_0 is true or false, this is our decision-making rule). Clearly, there is a potentially substantial probability that $\bar{X} < \bar{X}_c$ even when H_0 is false (i.e., when the dashed distribution is the case); this condition is especially true when the true mean μ is only somewhat above μ_0. That is, there will often be a fairly substantial probability of making a type II error.

Let us note two things about the relationship between the probabilities of type I and type II errors. First, note that it will generally *not* be the case that the two probabilities sum to one (it is tempting to think that they do since in one case we reject the null hypothesis but in the other case we do not). The simplest way to

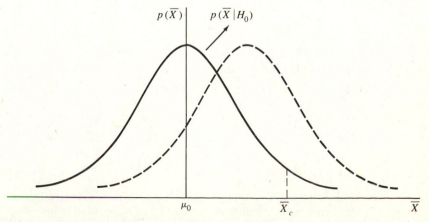

Figure 12.3 Illustrating the probability of a type II error.

see this situation is to note that these probabilities are essentially conditional probabilities, but they are based upon *different* conditions: the probability of a type I error is $\Pr(\bar{X} \geq \bar{X}_c)$ *given* that H_0 is true, whereas the probability of a type II error is $\Pr(\bar{X} < \bar{X}_c)$ *given* that H_0 is false.

Second, and more important, there is a trade-off between these two probabilities; that is, reducing the probability of a type I error necessarily involves increasing the probability of a type II error (and vice versa). Notice that (in Figure 12.2) in order to reduce the probability of a type I error we would have to increase the value of $\bar{X}_c$ so as to make the area under the curve to the right of this value smaller (that is, we are reducing the level of the test, to .01 perhaps). In so doing, however, we make it *more* likely that $\bar{X} < \bar{X}_c$ when H_0 is false (i.e., with a distribution such as the dashed one in Figure 12.3). That is, by reducing the probability of a type I error we have increased the probability of a type II error. Similarly, in order to reduce the probability of a type II error we would have to decrease the value of $\bar{X}_c$, necessarily increasing the probability of a type I error. It then is not surprising that the fairly low probabilities of type I errors are very often associated with fairly high probabilities of type II errors. Since we can only make a type II error when we do not reject the null hypothesis, we then arrive at a situation of being very confident of our result when we reject H_0 but not so confident about it when we do not.[10]

We can further elaborate on this issue in the context of an analogy to the U.S. criminal justice system. In a criminal trial, we essentially have

H_0: defendant is innocent

H_1: defendant is guilty

We reject H_0 only when the evidence is overwhelmingly against it— that is, when it is implausible that our observed evidence could have occurred if H_0 were in fact really true. This situation makes the probability of a type I error (convicting someone who is innocent) small, which is what we want given the concept of only convicting those who are "guilty beyond a reasonable doubt." But a side effect of this low probability of a type I error may be a relatively high probability of a type II error (a verdict of "not guilty" for a guilty

---

[10] To decrease the probability of a type II error without increasing the probability of a type I error would require the acquisition of more information—that is, a larger sample size.

defendant); that is, since we make it hard to reject H_0 (convict the defendant) we make it more likely that we will fail to reject H_0 even when it is false. When we fail to reject H_0 we do not necessarily believe that the defendant is innocent (although we may); what we are really concluding is that he or she is not "guilty beyond a reasonable doubt." Clearly, we are generally much more confident in our conclusion when the verdict is "guilty" than when it is "not guilty."

How might we reduce the probability of a type II error? That is, how might we lessen the number of guilty defendants who go free? By making the evidentiary criteria for conviction less severe. But we must then expect the side effect that there will be an increase in the fraction of innocent defendants who are convicted (i.e., the probability of a type I error will increase). Both probabilities could be decreased by acquiring more information—a more detailed investigation, more witnesses, and the like.

SUMMARY

In this chapter we have developed the concept of hypothesis testing and have illustrated its use in some relatively straightforward situations. Hypothesis testing becomes an important way of making inferences in the framework of regression analysis; its extensions into that area differ from what we have discussed here *only* in terms of the specifics of the procedure. The conceptual framework and strategy behind the tests are identical to what we have already presented.

We now have the background in statistical and probability theory necessary for us to develop and interpret regression analysis in its varied forms. We now turn to this problem.

REVIEW PROBLEMS FOR CHAPTER 12

12.24 A clinic is conducting an experimental study of the effects of a particular drug on a disease. A patient, upon first diagnosis of the disease, is given either the drug or a placebo; which one the patient gets is determined by a coin flip. The following survival times (in months) for patients in the

study, organized by whether they received the drug or
the placebo, are as follows:

Drug	Placebo
8 months	6 months
6	5
9	13
11	9
24	7
14	5
	4
	15

The drug is not believed to have any negative side effects.
What can you conclude about the effectiveness of the drug?
Fully explain your answer.

12.25 Suppose that you are considering the relative effectiveness
of two different SAT preparation courses. You observe a
group of students and their SAT scores before any course;
then, some students take preparation course A and some
take preparation course B, and we note the change in the
score for each student. We find

$$\bar{X}_A = +30 \text{ points} \qquad \bar{X}_B = +22 \text{ points}$$

where $\bar{X}_A$ and $\bar{X}_B$ are the sample mean changes in SAT
scores of those students taking preparation courses A and
B, respectively. The sample standard deviations are $s_A = 5$
and $s_B = 4$. The sample sizes were 20 students for course
A and 15 students for course B.

(a) Find the prob-value for and test the hypothesis

$$H_0: \mu_A - \mu_B = 0$$

$$H_1: \mu_A - \mu_B > 0$$

at the .05 level, where the μ's are the respective population means.

(b) Suppose that course A is more expensive, and you have
judged that it is "better" than B only if the mean improvement for A students exceeds that for B students
by at least 5 points. Is A "better"?

12.26 As a restaurant manager, you have purchased 100 cases of
frozen lobsters from a wholesaler. You need 1000 lobsters
for a large function you are hosting. The cases do not, how-

ever, all contain the same number of lobsters. You open four cases (selected at random from the 100 you have purchased) and find 8, 9, 9, and 10 lobsters in the four cases. Can you present a convincing argument that you will not have enough lobsters for the function?

12.27 We observe, in the population as a whole, two types of persons: those who own an expensive sports car and those who do not. In a simple random sample of 12 persons owning sports cars, we find a sample mean pulse rate of 75 (beats per minute) with a sample standard deviation of 10 (beats per minute). In a simple random sample of 10 persons who do not own such sports cars, we find a sample mean pulse of 65 with a sample standard deviation of 8.

(a) Do those owning sports cars have a higher pulse on average than those who do not? Fully explore this question.

(b) Does this result imply that sports cars make the pulse rates of their owners increase? Explain.

12.28 We have a sample of firms and have recorded their levels of sales (in thousands of dollars) and their profit rates (in percent) for a given year:

Firm number	Sales	Profit rate
1	150	3
2	160	4
3	190	5
4	70	2
5	70	3
6	40	6
7	75	3
8	170	5
9	210	3
10	60	2
11	65	4
12	200	4
13	220	4
14	180	4
15	70	3
16	80	3

You want to address the question as to whether small firms have different profit rates (on average) than large firms do. Define a small firm as one that has sales less than $100,000, and define a large firm as one that has sales over $100,000. Let μ_L be the mean profit rate for large firms, and let μ_S be the mean profit rate for small firms. Assuming the population variances to be equal, find the prob-value for and test the hypothesis

$$H_0: \mu_L - \mu_S = 0$$

$$H_1: \mu_L - \mu_S \neq 0$$

at the .05 level.

*12.29 Suppose that X represents a population that is normally distributed with unknown mean μ but known variance $\sigma^2 = 25$. The only other information you have on X is that another researcher has calculated an estimate of μ called $\hat{\mu}$, where

$$\hat{\mu} = \frac{1}{6} X_1 + \frac{1}{6} X_2 + \frac{1}{3} X_3 + \frac{1}{3} X_4$$

($X_1, \ldots, X_4$ are a simple random sample), and found $\hat{\mu} = 10$. Find the prob-value for and test the hypothesis

$$H_0: \mu = 3$$

$$H_1: \mu \neq 3$$

at the .05 level. [*Hints:* (1) $\hat{\mu} \neq \bar{X}$; (2) what is the distribution of $\hat{\mu}$?; (3) what is the general strategy behind carrying out a hypothesis test?]

*12.30 It is known that 60% of all persons in a particular country live in a house that they own. In a sample of ten residents of a particular city within that country we find that nine of them live in a house that they own. Find the prob-value for and test the hypothesis

$$H_0: \pi = .6$$

$$H_1: \pi > .6$$

at the .05 level, where π is the fraction of all persons *in this city* who live in a house that they own. Carefully interpret your result. (*Hint:* The key here is *not* the t distribution.)

APPENDIX 1 TO CHAPTER 12

In the text to this chapter, we have presented hypothesis testing in the context of the prob-value approach. This somewhat modern approach on the surface is quite different from the traditional approach to hypothesis testing. In this appendix we shall (1) describe the traditional approach to hypothesis testing and (2) show how it is in fact fundamentally the same as the prob-value approach discussed and used in the text.

In the traditional approach we specify a null hypothesis and an alternative hypothesis as described in our discussion of the prob-value approach. Carrying out the test, however, involves the calculation of a *test statistic* and a comparison of it to a *critical value* of the appropriate probability distribution.

Let us illustrate this procedure for a one-sided hypothesis test for a single mean. The null hypothesis is

$$H_0: \mu = \mu_0$$

and the alternative hypothesis is

$$H_1: \mu > \mu_0$$

The test is at the .05 level. We assume that the parent population is normally distributed.

In a traditional hypothesis test we calculate a test statistic with a known distribution, in this case

$$T = \frac{\bar{X}_0 - \mu_0}{s/\sqrt{n}} \qquad (1A12.1)$$

where $\bar{X}_0$ is the value of $\bar{X}$ in our sample. Note that in this case $T \sim t^{(n-1)}$; that is, T has a t distribution with $(n-1)$ degrees of freedom. In a traditional hypothesis test, if $T \geq t_{.05}^{(n-1)}$, then we would reject H_0 in favor of H_1 at the .05 level. On the other hand, if $T < t_{.05}^{(n-1)}$ then we would be unable to reject H_0 in favor of H_1 at the .05 level.

Let us visualize this situation graphically; see Figure 1A12.1. The $t_{.05}$ value is that value of the t distribution such that $\Pr(t \geq t_{.05}) = .05$; that is, the area under the curve to its right is .05. In a classical hypothesis test, we reject H_0 if $T \geq t_{.05}$ (e.g., T_1), and we fail to reject H_0 if $T < t_{.05}$ (e.g., T_2). Now let us compare this result to that of the prob-value approach.

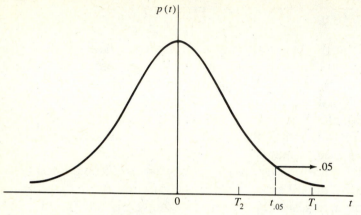

Figure 1A12.1 The traditional approach to hypothesis testing.

In this case the prob-value is

$$\Pr(\bar{X} \geq \bar{X}_0) \qquad \text{if } \mu = \mu_0 \qquad (H_0 \text{ is true})$$

To evaluate this probability we change it into a statement regarding a t-distributed random variable:

$$\Pr(\bar{X} \geq \bar{X}_0) = \Pr\left(\frac{\bar{X} - \mu}{s/\sqrt{n}} \geq \frac{\bar{X}_0 - \mu_0}{s/\sqrt{n}}\right)$$

$$= \Pr(t^{(n-1)} \geq T)$$

since $T = (\bar{X}_0 - \mu_0)/(s/\sqrt{n})$ by Eq. (1A12.1). The prob-value will be less than or equal to .05 if $T \geq t_{.05}^{(n-1)}$, in which case we would reject H_0. But this criterion is exactly the one for rejecting H_0 in the traditional approach to hypothesis testing (see, for example, T_1 in Figure 1A12.1). And the prob-value will be above .05 if $T < t_{.05}^{(n-1)}$, in which case we would not reject H_0. And this is exactly the criterion for not rejecting H_0 in the traditional approach (e.g., T_2 in Figure 1A12.1).

To summarize, if we are testing

$$H_0: \mu = \mu_0$$

$$H_1: \mu > \mu_0$$

we form the test statistic

$$T = \frac{\bar{X}_0 - \mu_0}{s/\sqrt{n}}$$

(where $\bar{X}_0$ is the value taken on by $\bar{X}$ in our sample) and reject H_0

if $T \geq t_{.05}^{(n-1)}$. If we had an H_1 in the opposite direction ($\mu < \mu_0$ and, presumably $\bar{X}_0 < \mu_0$), then we would reject H_0 if $T \leq -t_{.05}^{(n-1)}$, since $T < 0$.

EXAMPLE ■

Suppose that we are to test

$$H_0: \mu = 5$$

$$H_1: \mu > 5$$

at the .05 level. In a simple random sample of size 25, we find $\bar{X} = 7$ and $s = 3$. We form the test statistic

$$T = \frac{\bar{X}_0 - \mu_0}{s/\sqrt{n}} = \frac{7 - 5}{3/\sqrt{25}} = 3.33$$

Since $t_{.05}^{(24)} = 1.71$, we have $T \geq t_{.05}^{(24)}$. We therefore would reject H_0 in favor of H_1 at the .05 level.   ■

For a two-sided H_1, that is,

$$H_0: \mu = \mu_0$$

$$H_1: \mu \neq \mu_0$$

we form T as before, Now, we reject H_0 at the .05 level if $|T| \geq t_{.025}^{(n-1)}$ (the absolute value signs are necessary because T could be negative). Note that the two-sided nature of the test requires there to be an area of .025 in each tail (instead of an area of .05 in one). (Can you see how this is equivalent to the prob-value approach in this situation?)

EXAMPLE ■

Suppose that we are to test, at the .05 level,

$$H_0: \mu = 10$$

$$H_1: \mu \neq 10$$

In a simple random sample of size 16 we find $\bar{X} = 9$ and $s = 4$. Our test statistic is

$$T = \frac{\bar{X}_0 - \mu_0}{s/\sqrt{n}} = \frac{9 - 10}{4/\sqrt{16}} = -1$$

Since $t_{.025}^{(15)} = 2.13$, we know that $|T| < t_{.025}^{(15)}$. We therefore cannot reject H_0 in favor of H_1 at the .05 level.   ■

We can generalize traditional hypothesis tests for a single mean as follows. We form the test statistic T as above. If the test is at the α level (which may differ from .05), then in a one-sided test we reject H_0 if $T \geq t_\alpha^{(n-1)}$ or $T \leq -t_\alpha^{(n-1)}$, the former if H_1 contains a ">" sign and the latter if H_1 contains a "<" sign. In a two–sided test we reject H_0 if $|T| \geq t_{\alpha/2}^{(n-1)}$.

For a hypothesis test on a difference in means (in the case of independent samples from normal populations when the unknown population variances are equal), our test statistic is

$$T = \frac{\bar{X}_1 - \bar{X}_2 - (\mu_1 - \mu_2)}{s_p \sqrt{\dfrac{1}{n_1} + \dfrac{1}{n_2}}} \qquad (1A12.2)$$

Note that $T \sim t^{(n_1+n_2-2)}$. We would reject H_0: $\mu_1 - \mu_2 = 0$ in favor of H_1: $\mu_1 - \mu_2 > 0$ (at the .05 level) if $T \geq t_{.05}^{(n_1+n_2-2)}$. Two-sided tests are structured analogously to the single-mean situation.

To generalize hypothesis tests for differences in means (in the case of independent samples from normal populations when the unknown population variances are equal) at the α level, in a one-sided test we reject H_0 if $T \geq t_\alpha^{(n_1+n_2-2)}$ or $T \leq t_\alpha^{(n_1+n_2-2)}$, the former if H_1 contains a ">" sign and the latter if H_1 contains a "<" sign. In a two-sided test we reject H_0 if $|T| \geq t_{\alpha/2}^{(n_1+n_2-2)}$.

EXAMPLE ■

Suppose that we test the following hypothesis at the *10%* level:

$$H_0: \mu_1 - \mu_2 = 0$$

$$H_1: \mu_1 - \mu_2 < 0$$

In independent samples of size 8 each from the two populations (which we assume have equal variances) we find $\bar{X}_1 = 3$ and $\bar{X}_2 = 6$. The pooled variance estimate is $s_p^2 = 4$. The test statistic, from Eq. (1A12.2), is

$$T = \frac{\bar{X}_1 - \bar{X}_2 - (\mu_1 - \mu_2)}{s_p \sqrt{\dfrac{1}{n_1} + \dfrac{1}{n_2}}} = \frac{3 - 6 - 0}{2 \sqrt{\dfrac{1}{8} + \dfrac{1}{8}}} = -3$$

Since $t_{.10}^{(14)} = 1.34$, we know that $T \leq -t_{.10}^{(14)}$. We can therefore reject H_0 in favor of H_1 at the .10 level.   ■

APPENDIX 2 TO CHAPTER 12

We based our discussion of hypothesis tests for differences in means upon the assumption that the population variances, though unknown, were equal. But can we test this assumption's validity? The purpose of this appendix is to illustrate how this test can be accomplished under certain assumptions. In developing this procedure, we shall introduce a new probability distribution—the F distribution—which has important future uses in applied regression analysis.

Let's begin by introducing the F distribution. [For a further discussion, see, for example, John E. Freund and Ronald E. Walpole, *Mathematical Statistics,* 4th ed. (Englewood Cliffs, NJ: Prentice-Hall, 1987), pp. 292–295.]

Definition of the *F* distribution   Suppose that W_1 is a random variable that has a chi-square distribution with k_1 degrees of freedom. Suppose that W_2 is an independent random variable that has a chi-square distribution with k_2 degrees of freedom. Then the quotient

$$V = \frac{W_1/k_1}{W_2/k_2} \tag{2A12.1}$$

is said to have an F distribution with k_1 and k_2 degrees of freedom (i.e., $V \sim F^{(k_1,k_2)}$).

(For a discussion of the chi-square distribution see the appendix to Chapter 6.)

In Figure 2A12.1 we depict the probability density function for the random variable V defined above as an example of an F distribution. Given its definition we know that V is a continuous random variable (since it is a function of continuous chi-squares) and that $V \geq 0$ (since W_1, W_2, k_1, and k_2 are all nonnegative). Note that the F distribution has the same general shape as the chi-square distribution; in particular, it is asymmetric and skewed to the right.

The F distribution is characterized by two parameters: the "degrees of freedom in the numerator" (k_1 in the above example) and the "degrees of freedom in the denominator" (k_2 in the above example). Table V depicts the $F_{.05}$ values for various pairs of values of these parameters, where the $F_{.05}$ value is defined such that the area in the right-hand tail is .05; that is, $\Pr(F \geq F_{.05}) = .05$.

Let us now apply this new distribution to our problem of testing

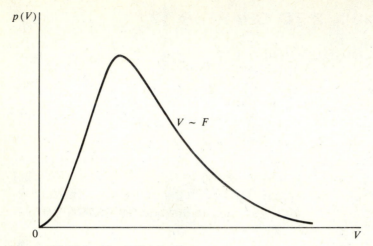

Figure 2A12.1 The *F* distribution.

the validity of the assumption of equal variances. We shall confine ourselves to the situation of independent simple random samples from populations that are normally distributed.

Suppose that X_1 represents a normal population with variance σ_1^2 and that we have a simple random sample of size n_1 from this population. Suppose that X_2 represents a normal population with variance σ_2^2 and that we have a simple random sample of size n_2 from this population. Suppose finally that these simple random samples are independent. The null hypothesis is

$$H_0: \sigma_1^2 = \sigma_2^2$$

and the alternative hypothesis is

$$H_1: \sigma_1^2 > \sigma_2^2$$

where we have defined variables so that "1" represents the population with the larger sample variance. (We shall confine ourselves to the case of a one-sided H_1 because two-sided tests are complicated by the asymmetric nature of the *F* distribution.) It is useful to restate the hypotheses as

$$H_0: \frac{\sigma_1^2}{\sigma_2^2} = 1$$

$$H_1: \frac{\sigma_1^2}{\sigma_2^2} > 1$$

thereby putting the hypothesis test into the context of a single parameter (equal to σ_1^2/σ_2^2).

In the prob-value approach, we estimate the parameter of interest (σ_1^2/σ_2^2) and then ask: What is the probability of getting a value of our estimator as extreme as in our sample if the null hypothesis is in fact true? To estimate σ_1^2/σ_2^2 we form s_1^2/s_2^2 (the ratio of sample variances). The prob-value is then

$$\Pr\left[\frac{s_1^2}{s_2^2} \geq \left(\frac{s_1^2}{s_2^2}\right)_0\right] \quad \text{if } \frac{\sigma_1^2}{\sigma_2^2} = 1 \quad \text{(i.e., if } H_0 \text{ is true)} \quad (2A12.2)$$

where $(s_1^2/s_2^2)_0$ is the value of s_1^2/s_2^2 calculated from our samples. How do we evaluate this probability?

In this situation it turns out that

$$\frac{\dfrac{(n_1-1)s_1^2}{\sigma_1^2} \bigg/ (n_1-1)}{\dfrac{(n_2-1)s_2^2}{\sigma_2^2} \bigg/ (n_2-1)} \sim F^{(n_1-1,n_2-1)} \qquad (2A12.3)$$

Why? Refer to Eq. (2A12.1). The term $(n_1-1)s_1^2/\sigma_1^2$ has a chi-square distribution with (n_1-1) degrees of freedom (since we are assuming that the parent population is normal; see the appendix to Chapter 9) and is therefore like W_1 in Eq. (2A12.1); (n_1-1) is then like k_1. Similarly, $(n_2-1)s_2^2/\sigma_2^2$ has a chi-square distribution with (n_2-1) degrees of freedom and is therefore like W_2 in Eq. (2A12.1); (n_2-1) is then like k_2. [Since the assumption of independent samples tells us that $(n_1-1)s_1^2/\sigma_1^2$ and $(n_2-1)s_2^2/\sigma_2^2$ are independent, Eq. (2A12.3) is like Eq. (2A12.1); it therefore has an F distribution.] But

$$\frac{\dfrac{(n_1-1)s_1^2}{\sigma_1^2} \bigg/ (n_1-1)}{\dfrac{(n_2-1)s_2^2}{\sigma_2^2} \bigg/ (n_2-1)} = \frac{\left(\dfrac{s_1^2}{\sigma_1^2}\right)}{\left(\dfrac{s_2^2}{\sigma_2^2}\right)} = \left(\frac{s_1^2}{s_2^2}\right)\left(\frac{\sigma_2^2}{\sigma_1^2}\right) = \frac{s_1^2}{s_2^2}$$

if H_0 is true (in which case $\sigma_2^2/\sigma_1^2 = 1$). That is, under H_0,

$$\frac{s_1^2}{s_2^2} \sim F^{(n_1-1,n_2-1)}$$

The prob-value from Eq. (2A12.2) can then easily be determined:

$$\Pr\left[\frac{s_1^2}{s_2^2} \geq \left(\frac{s_1^2}{s_2^2}\right)_0\right] = \Pr\left[F^{(n_1-1, n_2-1)} \geq \left(\frac{s_1^2}{s_2^2}\right)_0\right]$$

Comparing the prob-value to the level of the test then completes the hypothesis test.

EXAMPLE ■

Suppose that X_1 and X_2 represent normal populations. In a simple random sample of size 10 from X_1 we find $s_1^2 = 20$. In an independent simple random sample of size 12 from X_2 we find $s_2^2 = 15$. Our problem is to find the prob-value for and test the hypothesis

$$H_0: \frac{\sigma_1^2}{\sigma_2^2} = 1$$

$$H_1: \frac{\sigma_1^2}{\sigma_2^2} > 1$$

at the .05 level.

The prob-value, from Eq. (2A12.2), is

$$\Pr\left(\frac{s_1^2}{s_2^2} \geq \frac{20}{15}\right) = \Pr(F^{(9,11)} \geq 1.33)$$

Since $F_{.05}^{(9,11)} = 2.90$ (see Table V, page 392) we know that $\Pr(F^{(9,11)} \geq 1.33) > .05$. Since the prob-value exceeds .05, we cannot reject H_0 in favor of H_1 at the .05 level. We do not have evidence that contradicts beyond a reasonable doubt the assumption of equal population variances.   ■

DISCUSSION QUESTIONS FOR CHAPTER 12

1. Suppose that you undertook 100 studies of a particular hypothesis (the null hypothesis) and found that 5 times the evidence was persuasively contradictory to it, but the other 95 times it was not. How might you interpret this series of events?

2. Suppose that you undertook 100 studies of a particular hypothesis (the null hypothesis) and found that 50 times the evidence was persuasively contradictory to it, but the other 50 times it was not. How might you interpret this series of events?

REGRESSION ANALYSIS

Chapter 13

The Two-Variable Regression Model

In this chapter we shall begin to consider regression analysis and its role in economic research. In particular, we shall develop the foundations of regression analysis in the context of the two-variable model. This framework will be extended to more complicated situations involving more than two variables in Chapter 14.

Many economic issues that we investigate empirically can be expressed in terms of two (or more) variables and their underlying relationships. Regression analysis, which builds upon the probability theory and statistical concepts that we have developed throughout this book, provides a comprehensive methodology for undertaking these investigations.

GENERAL PURPOSES

There are two broadly constructed purposes for undertaking a regression analysis in economics. The first is to *quantify* economic relationships. For example, microeconomic theory suggests that there is an inverse relationship between price and quantity demanded (holding other things constant); we might express this relationship as

$$q^d = \alpha + \beta p$$

where q^d is the quantity demanded of the commodity, p is its price, α and β are parameters, and we have for simplicity expressed the demand curve in linear form. Economic theory tells us that there is an inverse relationship between q^d and p, and therefore the parameter β is a negative number.[1] This statement is a *qualitative* one about the nature of the relationship between the two variables. Economic theory does not, however, tell us the values of α and β; we do not know, for example, whether β is a small negative number (suggesting a relatively inelastic relationship) or a large negative number (suggesting a relatively elastic relationship). Determining the value of β (and α) would allow us to put the theory into a *quantitative* context, thereby clearly enhancing our understanding of the underlying market fundamentals characterizing this commodity. Regression analysis ultimately allows us to make inferences regarding the values of α and β and therefore to quantify the underlying theory.

The second broadly defined purpose of regression analysis is as a tool to verify or refute economic theories or to test competing theories. Economic theories can often be expressed in terms of a value or range of values of a parameter, and the empirical determination of the value of that parameter can then be seen as a way of evaluating the quality of the theory in terms of its consistency with reality. In the demand function example, we might attempt to test whether in fact β is less than zero; we might estimate β and then do a hypothesis test based on that estimate in order to carry out this procedure. Obviously, if we do not conclude that β is negative we might want to reconsider the underlying theory.

Regression analysis can also be used to test competing theories; as an example, suppose that there is debate as to whether a good is a normal or an inferior good. We might specify the quantity demanded (q^d) of the good as a function of both price (p) and income (i) and write

$$q^d = \alpha + \beta p + \gamma i$$

If $\gamma > 0$, we have a normal good; but if $\gamma < 0$, we have an inferior good. By estimating the value of γ and performing an appropriate hypothesis test, in essence we test the competing theories in an effort

---

[1] The possible exception, of course, is the case of a Giffen good; however, given the rarity of the occurrence of Giffen goods (if they exist at all), we ignore this complication.

to see which is more consistent with the data. (This procedure would actually involve a use of the multiple regression framework inasmuch as there are more than two variables; we shall discuss multiple regression in Chapter 14.)

Let us now begin to develop the framework for using regression analysis to pursue the types of questions just suggested. In this chapter, we shall confine ourselves to situations where there is only one variable on the right-hand side of the equation (that is, a two-variable regression). In Chapter 14, we shall extend this analysis to the case of more than one right-hand-side variable (that is, a multiple regression).

THE BASIC REGRESSION FRAMEWORK

Suppose there are two variables X and Y whose relationship we are interested in investigating. Using the material we have studied up to this point, we could: (1) investigate whether or not the variables are independent (although this tells us nothing about the specific nature of any relationship that might be present), or (2) determine the correlation coefficient (to see whether the variables are positively or negatively related in a linear context and to find the strength of that relationship). Even with the correlation coefficient, however, we have little information about the specifics of the ways in which changes in the two variables are related. We do not know, for example, whether a change in X is associated with a large or a small change in $Y;$ rather, we just know how closely these changes are related. With regression analysis we seek to explore more completely the nature of the relationship between the two variables so that we can learn more about the specifics of the ways in which their behaviors are linked. Our goal at this point is to develop the conceptual foundations of a simple regression analysis.

Suppose, for simplicity, that X and Y are such that changes in X will *cause* changes in $Y,$ but the causality does not go the other way. In general, the nature of the causal structure of the problem must come from the theoretical framework in which we are operating. For example, imagine that the Federal Reserve each month pursues policies which cause the money supply to rise at a specified rate; X_i is the rate of growth of the money supply during month i. Monetary expansion typically is associated (at least in part) with inflation; Y_i is the inflation rate during month i. Economic theory

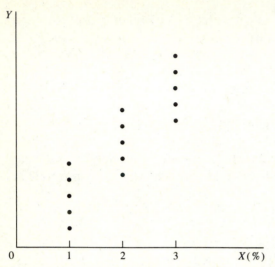

Figure 13.1   Rate of growth of the money supply and inflation.

suggests that increases in X should be associated with increases in Y. (We ignore here the fact that these changes are undoubtedly not instantaneous—that is, that there is a lag between monetary expansion and the resulting inflation.) Presumably, the causality goes from X to Y but not the other way, since the value of X is the result of a policy decision on the part of the Federal Reserve. Though this is a simple example, it is very useful for beginning to lay out the conceptual framework of regression analysis. We generally refer to Y as the *dependent* variable and X as the *independent* variable.

Let us suppose that we consider three different rates of monetary expansion (X is 1%, 2%, or 3%—at an annual rate, say) that are implemented over and over again by the Federal Reserve. We observe the inflation rate (Y) for each month, for the given rate of increase of the money supply. In Figure 13.1 we plot the pairs of values of X and Y that we observe.

Notice that the value of X does not completely determine the value of Y in that for each value of X there are different possible values of $Y;$ sometimes monetary expansion has a bigger effect than it does at other times as the result of differences in the other factors that influence price behavior. The fact that the value of X does not completely determine the value of Y is a key observation that plays a crucial role in the underlying specification of the regression framework. Clearly, X and Y are not perfectly correlated; nevertheless, we can say that there appears to be a positive relationship *on average*

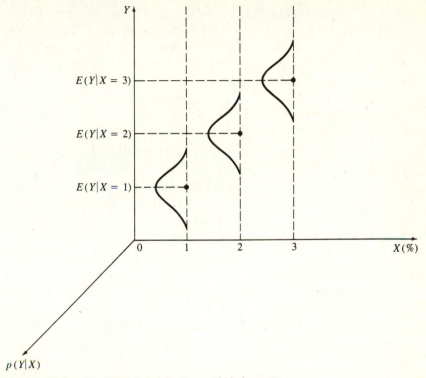

Figure 13.2 Conditional distributions of Y given X.

between the two variables; that is, as the value of X rises, there is a *tendency* for the value of Y also to rise. In other words, on average, increases in the rate of monetary expansion are associated with increases in the rate of inflation.

   We can view the observed values of Y as draws from a probability distribution conditional on the associated value of X. The observations over $X = 1$, for example, are all draws from the conditional distribution $p(Y \mid X = 1)$—that is, the conditional probability distribution of the inflation rate (Y) given that the rate of growth of the money supply (X) is 1%. Once the value of X has been determined, we know (at least in principle) the distribution characterizing the behavior of Y as a random variable. These conditional distributions and their characteristics are a key element of the conceptual framework of regression analysis.

   What do the conditional distributions of Y given X look like? In Figure 13.2 we have expanded upon the previous figure in order to create a three-dimensional perspective in which the third axis is

$p(Y \mid X)$; this axis is perpendicular to each of the other two axes. We show $p(Y \mid X)$ for each of the three values of X; these curves should be viewed as "coming off the paper," where how far off the paper they rise represents the value of the probability density function for the associated value of Y.[2] (We've omitted the actual observed values of Y from Figure 13.1 for simplicity.)

Probability density functions have means. The mean of $p(Y \mid X = 1)$ is $E(Y \mid X = 1)$—that is, what we expect the value of Y to be given that $X = 1$. We can similarly define $E(Y \mid X = 2)$ and $E(Y \mid X = 3)$. These conditional means are also depicted in Figure 13.2. Notice that as X rises, $E(Y \mid X)$ rises. This result means that as the rate of growth of the money supply (X) rises, the value of the associated inflation rate that we expect to occur rises as well. This trend is exactly what we referred to above when we said that there appears to be *on average* a positive relationship between X and Y; all we have done here is to put this intuitively pleasing observation into more formal language in a more completely developed framework. That is, the regression model takes a conclusion we would naturally make and formalizes (and elaborates upon) it. We cannot say that when X rises, Y will rise, because of the randomness involved in the determination of the value of Y. But we can say that when X rises, we *expect* Y to rise.

Figure 13.2 and the associated analysis suggest a very helpful way for investigating the relationship between X and Y. If we knew the relationship between X and $E(Y \mid X)$, we would know what to expect concerning the change in Y resulting from a change in X. But there is a problem: We don't know $E(Y \mid X)$. Why?

In order to know $E(Y \mid X)$ we must know $p(Y \mid X)$, which characterizes an entire population. That is, the conditional distribution of Y given X characterizes all possible values of Y associated with the given value of X. (In the above example, this value would include all months in which the money supply was changing at a particular rate.) But we do not observe all such possible values of Y; we merely observe a sample from the population, and this sample is composed of the individual observations depicted in Figure 13.1. Since we do not observe the whole population, we do not know its mean [i.e., $E(Y \mid X)$]. This conclusion is an extension of the insight

---

[2] We have drawn these distributions as having the normal shape for convenience; sometimes this assumption is a reasonable approximation of the conditional behavior of Y and sometimes it is not.

discussed earlier in this book that we do not know a mean unless we know the entire population.

The problem of investigating the relationship between X and Y becomes that of estimating several unknown population means. In the problem at hand, we could take the sample mean of Y when $X = 1$, the sample mean of Y when $X = 2$, and the sample mean of Y when $X = 3$ and use them as our estimates. We are interested, however, in the relationship between the two variables for the entire range of possible values of X and not just those we have actually observed. In addition, our problem is to infer not only the individual expected values but also how they are related to each other—that is, how they change as X changes. (We are interested in the behavior of the inflation rate not only for rates of growth of the money supply of 1, 2, or 3 percent but for other rates of growth as well; in addition, we are interested in how the inflation rate is affected by a change in the rate of monetary expansion.)

The typical technique is to assume some functional form describing the relationship between the conditional expected values of Y and the values of X and then attempt to quantify the nature of that relationship. In *linear regression analysis* we assume that the conditional expected values are collinear:

$$E(Y \mid X) = \alpha + \beta X \qquad (13.1)$$

where α and β are unknown parameters. This linearity assumption represents a simplifying approach that we take in order to make the problem more manageable. Notice that, as a result, the problem becomes one of making inferences about the two parameters α and β that characterize the relationship on average between X and Y.[3] However, before we can discuss the process of inferring the values of these parameters, we must elaborate further on the underlying structure of the relationship between X and Y.

Equation (13.1) suggests that if we were to plot the conditional mean of Y for each possible value of X, then these points would form a line. In Figure 13.3 we plot this line in conjunction with the actual observations that form our sample. The fact that most of the

---

[3] It is not necessary to assume a linear relationship between X and $E(Y \mid X)$, although this assumption is most commonly made. One could, for example, assume the quadratic relationship $E(Y \mid X) = \alpha + \beta X + \gamma X^2$—an example of a nonlinear regression framework. We can often get some indication as to the reasonableness of the linearity assumption by plotting the data and observing whether the response of Y to X seems constant or variable as X changes.

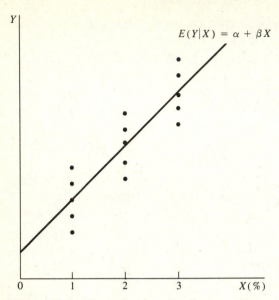

Figure 13.3   The true regression line.

observed points are off the line indicates that, in general, $Y_i \neq E(Y_i \mid X_i)$, where the i subscript denotes a particular observation; this key result will now be specifically incorporated into the model. [In the context of the money/inflation example, the actual inflation rate is not typically equal to the expected (average) inflation rate *given* the rate of increase of the money supply with which it is associated.]

   Equation (13.1) describes the relationship between X and Y in terms of what we expect Y to be (given X) as opposed to the actual value of Y. We know, however, that for each observation i in our sample

$$Y_i = E(Y_i \mid X_i) + [Y_i - E(Y_i \mid X_i)] \qquad (13.2)$$

It is clear that Eq. (13.2) is true algebraically because the right-hand side equals Y_i. This equation, however, has important content; it suggests that the observed value Y_i can be decomposed into (1) what we expect Y_i to be given the value of X_i [i.e., $E(Y_i \mid X_i)$] and (2) the deviation of the actual observation from that expectation. If we define

$$\varepsilon_i = Y_i - E(Y_i \mid X_i)$$

then Eq. (13.2) may be rewritten as

$$Y_i = \alpha + \beta X_i + \varepsilon_i \qquad (13.3)$$

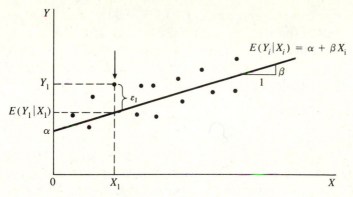

Figure 13.4 The error term ε_i.

[given that $E(Y_i \mid X_i) = \alpha + \beta X_i$; see Eq. (13.1)]. This equation represents an alternative way of stating the model, and it focuses on the determination of the actual value of Y_i.

Let us pursue this discussion graphically in the context of a general problem. In particular, let us presume that there are many different values of X for which we have associated values of Y; perhaps we have the results of an observational study where the values of X are not fixed by experimental design. Figure 13.4 demonstrates such a situation.

In this figure we have plotted the actual observations (the pairs of values of X_i and Y_i that comprise our sample). We have also plotted the line connecting the expected values of Y conditional on X; this line, referred to as the *true regression line*, describes the relationship on average between X and Y. To understand further the relationship between our actual observations and this line, let us consider the point we have picked out by the arrow in the graph; the coordinates of this point are (X_1, Y_1). The true regression line tells us that given the value X_1 we expect Y to equal $E(Y_1 \mid X_1)$, which is the point on the true regression line associated with this value of X. The vertical distance between our observed point and the true regression line is $\varepsilon_1 = Y_1 - E(Y_1 \mid X_1)$. This equation gives us an interpretation of ε as the distance between the observed point and the true regression line. In this case, $\varepsilon > 0$; if our observation were below the true regression line, then ε would be negative.

The value ε is often referred to as the *error term* (or sometimes the *disturbance term*); it represents the error one would make by predicting Y with $E(Y \mid X)$.

From an interpretive point of view, the keys to understanding the relationship between X and Y are the values of the parameters

α and β. The parameter α represents $E(Y \mid X = 0)$; that is, it is the vertical intercept of the true regression line.[4] The parameter β represents the change in $E(Y \mid X)$ for a one unit change in X; it is then the slope of the true regression line. These geometric interpretations of α and β are incorporated into Figure 13.4.

Unfortunately, the true regression line is unknown. Each point on it represents a population mean; since we do not observe the entire populations, however, these means are unknown. Our problem then becomes one of estimating them. To express this concept differently, the parameters α and β that characterize the true regression line are unknown and must be estimated; in so doing we implicitly form estimates of $E(Y \mid X)$ for each possible value of X. We now begin to explore this issue.

TRUE VERSUS ESTIMATED REGRESSION LINES

In Figure 13.5 we illustrate our situation incorporating the concept of an *estimated regression line*. Let us refer to our estimate of α as $\hat{\alpha}$ and our estimate of β as $\hat{\beta}$, without saying how these estimates are determined (it is a method for their determination that we are pursuing). Since it will generally be the case that $\hat{\alpha} \neq \alpha$ and $\hat{\beta} \neq \beta$ (why?), the estimated regression line will in general be different from the true regression line. They have therefore been drawn differently in Figure 13.5; in reality, of course, we would not know their relative positions since the true regression line would be unknown. We refer to the estimated regression line as

$$\hat{Y}_i = \hat{\alpha} + \hat{\beta} X_i$$

where the "$\hat{}$'s" indicate to us that we are referring to the estimated (as opposed to the true) regression line. $\hat{Y}_i$ is, in effect, our estimate of $E(Y_i \mid X_i)$ (the expected value of Y_i conditional on the associated

---

[4] This interpretation of α is actually an oversimplification. For example, suppose that X represents rainfall and Y represents the yield of our cornfield (bushels per acre). It is likely that $\alpha \neq 0$ given the placement and slope of the true regression line in terms of the range of X values we observe. But, of course, $E(Y \mid X = 0)$ may very well be 0 (with no rain we get no corn). The linearity assumption may be a good approximation over the range of X's we observe even if over the entire range of possible X's it is not (i.e., there may be global nonlinearities). In such a case, α simply represents the intercept needed for the proper placement of the regression line.

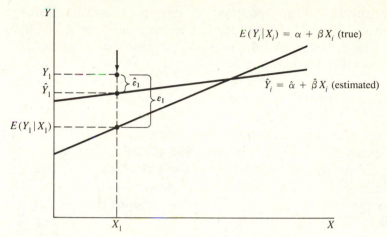

Figure 13.5   True versus estimated regression lines.

value of X_i). This equation, once estimated, is often referred to as the *regression of Y on X*.

Consider the point picked out by the arrow in Figure 13.5. For $X = X_1$ the mean value of Y is $E(Y_1 \mid X_1)$, which is unknown. The point on the estimated regression line associated with X_1 is $\hat{Y}_1$, which is our estimate of $E(Y_1 \mid X_1)$. We presume that the true regression line goes "through" the cluster of observations we have plotted; after all, it is just as likely for an observed Y to exceed its expected value as it is to be less than it (that is, the observations are just as likely to be above the true regression line as below it).[5] We therefore, in the process of constructing our estimated regression line, want to draw a line that more or less goes through the cluster of observations—or, as we often say, "fits the data well." Note that for the point we have picked out with the arrow the deviation from the estimated regression line, $Y_1 - \hat{Y}_1$, is different from ε_1; such will be the case everywhere except where X equals the value at which the estimated and true regression lines cross.

We often express the difference between the observed point and the estimated regression line $(Y_i - \hat{Y}_i)$ as $\hat{\varepsilon}_i$ and refer to it as a *residual;* $\hat{\varepsilon}_i$ in some sense estimates ε_i (i.e., the deviation from the estimated regression line estimates the deviation from the true regression line). The distinction between $\hat{\varepsilon}_i$ and ε_i is illustrated in Figure 13.5.

[5] Although this statement might not be true in a technical sense, if the values of Y are not symmetrically distributed about the true regression line, there is little harm at the present time in operating as if it were true.

Intuitively, we want to draw our estimated regression line so that the $\hat{\varepsilon}_i$'s are small—that is, so that the estimated regression line fits the data reasonably well. What procedure might we follow?

ESTIMATING THE REGRESSION LINE

Intuitively, we want to construct an estimate of the true regression line that fits the data well, meaning that the deviations of the observed points from the estimated regression line are small (the $\hat{\varepsilon}_i$'s are small). Since there are many observations, it is suggested that some index created by some method of aggregating these deviations should be made as small as possible. One way of aggregating the residuals ($\hat{\varepsilon}_i$'s) would be simply to add them:

$$\sum_{i=1}^{n} \hat{\varepsilon}_i = \sum_{i=1}^{n} (Y_i - \hat{Y}_i)$$

The problem with this approach is that since some residuals are positive and some are negative, there will be some canceling out. In reality, we are not so much concerned with whether the residual is positive or negative but rather with its absolute magnitude. It is thus suggested that we might want to add the absolute values of the residuals, or the squares of the residuals, for the purposes of creating the index. Typically, we follow the latter suggestion and construct the estimated regression line so as to minimize the sum of squared residuals. This approach, commonly referred to as the *least squares* criterion, has many desirable characteristics which we shall develop below. To be more specific, we choose the estimated regression line so as to

$$\text{minimize} \sum_{i=1}^{n} \hat{\varepsilon}_i^2$$

over the observations in our sample. Since $\hat{\varepsilon}_i = Y_i - \hat{Y}_i$, the least squares criterion is to

$$\text{minimize} \sum_{i=1}^{n} (Y_i - \hat{Y}_i)^2$$

Since

$$\hat{Y}_i = \hat{\alpha} + \hat{\beta} X_i$$

we can restate our problem as

$$\text{minimize} \sum_{i=1}^{n} (Y_i - \hat{\alpha} - \hat{\beta} X_i)^2 \qquad (13.4)$$

That is, we must choose the values of $\hat{\alpha}$ and $\hat{\beta}$ (and therefore the estimated regression line) so as to make the sum in Eq. (13.4) as small as possible.

For the reader with a background in calculus, this task is a fairly standard minimization problem in two unknowns. In Appendix 1 to this chapter we provide the derivation of the solutions. If you do not have background in calculus, all you need to know is that there are some fairly standard techniques for determining the values of $\hat{\alpha}$ and $\hat{\beta}$ that minimize the value of Eq. (13.4). These values of $\hat{\alpha}$ and $\hat{\beta}$ then become the least squares estimates of α and β.

It turns out that the values of $\hat{\alpha}$ and $\hat{\beta}$ that minimize Eq. (13.4) can be determined as follows:

$$\hat{\beta} = \frac{\sum_{i=1}^{n} (X_i - \bar{X}) Y_i}{\sum_{i=1}^{n} (X_i - \bar{X})^2}$$

$$\hat{\alpha} = \bar{Y} - \hat{\beta} \bar{X}$$

Let us now consider an example to illustrate the use of these formulas.

Suppose that we have five observations on X and Y, as listed in Table 13.1. Here we carry out the calculation of the least squares estimators by hand by building a type of manual spreadsheet. To calculate $\hat{\beta}$, we need to determine the value of $(X_i - \bar{X}) Y_i$ for each

Table 13.1 CALCULATING AN ESTIMATED REGRESSION LINE

Y_i	X_i	$X_i - \bar{X}$	$(X_i - \bar{X}) Y_i$	$(X_i - \bar{X})^2$
4	2	−4	−16	16
5	4	−2	−10	4
7	6	0	0	0
10	8	2	20	4
9	10	4	36	16
$\bar{Y} = 7$	$\bar{X} = 6$		$\Sigma = 30$	$\Sigma = 40$

observation. We first calculate $\bar{X} = 6$. In the column headed by $(X_i - \bar{X})$ we determine the value of this expression for each observation; in the following column we calculate $(X_i - \bar{X})Y_i$. Summing this column then gives us the numerator in the formula for $\hat{\beta}$. To get the denominator in the formula for $\hat{\beta}$, we (in the final column) form $(X_i - \bar{X})^2$ and sum the values therein. The result is as follows:

$$\hat{\beta} = \frac{\sum\limits_{i=1}^{n}(X_i - \bar{X})Y_i}{\sum\limits_{i=1}^{n}(X_i - \bar{X})^2} = \frac{30}{40} = .75$$

To calculate $\hat{\alpha}$, the only other thing we need is the value of $\bar{Y}$, which is shown in the table. Therefore

$$\hat{\alpha} = \bar{Y} - \hat{\beta}\bar{X} = 7 - (.75)6 = 2.5$$

The estimated regression line is then

$$\hat{Y}_i = 2.5 + .75X_i$$

We plot this estimated regression line in Figure 13.6.

Our estimate of α is 2.5, which of course is just the vertical intercept. We also estimate that the expected change in Y for a one-unit change in X is .75 units; this is the slope of the estimated regression line. That is, we estimate that the relationship between X and Y is such that a one-unit change in X is associated with, on average, a .75-unit change in Y.

We shall further discuss interpretation issues later in this chapter. Notice, however, that by forming the estimated regression line,

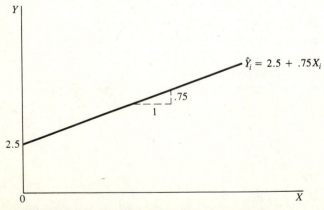

Figure 13.6   Plotting the estimated regression line.

we have made an inference as to the *quantitative* nature of the relationship between X and Y. Of course, $\hat{\alpha}$ and $\hat{\beta}$ are just estimates of the true (but unknown) parameters α and β. What can we say about the quality of these estimates? And can we accomplish more sophisticated types of inference such as confidence intervals or hypothesis tests? In order to pursue these questions, we must be more careful in specifying the assumptions that underlie the analysis we have already undertaken. We now turn to the exposition of a set of assumptions that are an elaboration of our theoretical framework and under which we shall eventually discuss the properties of the least squares estimators.

PROBLEMS

13.1 To investigate the relationship between educational attainment and earnings, you have information on years of schooling completed (X) and annual earnings (Y) for a simple random sample of five individuals:

Y	X
14,000	11
21,000	14
38,000	16
25,000	16
30,000	18

Estimate the regression of Y on X; that is, calculate the estimated regression line $\hat{Y}_i = \hat{\alpha} + \hat{\beta} X_i$.

13.2 Suppose that you are interested in investigating the relationship between mortgage rates and purchases of residential housing in a particular community. X represents the average interest rate (as a percent) for a 30-year mortgage at the lending institutions in the community during a given month, and Y represents the total volume of sales of residential housing (in millions of dollars) for that month. You have 5 months of observations:

Y	X
10	9.2
9	9.4
8	10.1
8	10.1
5	11.2

Estimate the regression of Y on X; that is, calculate the estimated regression line $\hat{Y}_i = \hat{\alpha} + \hat{\beta}X_i$.

13.3 As a state policymaker you are investigating the relationship between property tax rates and the level of corporate activity across different communities. You suspect that corporations have a tendency to locate their operations in communities with relatively low taxes, and you therefore expect there to be an inverse relationship across communities between property tax rates and corporate activity. To investigate your hypothesis, you take a simple random sample of six communities and observe the property tax rate (as a percent of appraised value) X and the level of corporate output (in millions of dollars) Y. You have the following observations:

Y	X
3.1	0.5
2.8	1.2
2.6	0.8
2.7	0.5
2.9	0.9
0.9	0.9

Estimate the regression of Y on X; that is, calculate the estimated regression line $\hat{Y}_i = \hat{\alpha} + \hat{\beta}X_i$.

THE CLASSICAL LINEAR REGRESSION MODEL

The properties of the estimators we have developed, typically referred to as the Ordinary Least Squares (OLS) estimators (so as to distinguish them from other types of least squares estimators used in more complicated situations) depend on the underlying characteristics of the situation under investigation. In this section we shall present a set of six assumptions that together comprise a version of the Classical Linear Regression Model (CLRM), which is the simplest framework for regression analysis and is one in which the OLS estimators have very desirable properties. We shall now proceed to list these six assumptions and to discuss their implications.

Assumption 1: $Y_i = \alpha + \beta X_i + \varepsilon_i$

This assumption is a restatement of the model and of the linearity assumption underlying the analysis.

Assumption 2: X_i's are fixed numbers and not random variables.

This assumption is potentially the most confusing of the six. It suggests that the X's we observe, though *variables,* are not *random variables.* That is, the process by which X varies is one that contains no element of chance. This assumption suggests, in general, an experimental study in which the values of X are fixed by the researcher. The researcher performs the experiment and observes the values of Y that result. In principle, the researcher could then repeat the experiment with the *exact* same X's, although of course a different set of Y's would undoubtedly result (since the Y's depend in part on the ε's). In the example presented earlier in this chapter where we observed the inflation rate resulting from money supply increases of 1, 2, or 3 percent, we could view ourselves as in the general context described by assumption 2; the Federal Reserve could, in principle, repeatedly increase the money supply by 1, 2, or 3 percent and observe the resulting inflation rates.

This assumption makes the analysis far simpler, for it suggests that the sole source of *random* variation in Y_i is the error term ε_i. (The remaining four assumptions of the CLRM concern ε_i as a random variable.) The simplicity gained by the fact that the X's are not *random* variables will become quite apparent when we determine the properties of the OLS estimators under these assumptions. The problem, however, is that at least in economics it is rarely the case that assumption 2 is true! Why?

In economics we most commonly work with data obtained in the context of observational studies. The researcher does not here control the values of X, and they are generally determined by a process that is, at least in part, random. (In practice, for example, the Federal Reserve does not have complete control over the money supply because this quantity is determined in part by, among other things, the behavior of commercial banks.) So why do we make assumption 2 when it is typically wrong in the types of problems economists consider?

It turns out that there is an alternative version of assumption 2 that is more relevant to observational studies and leads to essentially the same set of conclusions (regarding the properties of the OLS estimators) as those derived below; these conclusions are, however, significantly more difficult to derive. This alternative to assumption 2 is that the X's and ε's are independent random variables.

Much can be learned about the nature of the model by employing our original assumption 2; we just need to keep in mind that for economic problems we are really depending upon the alternative version just described. We do not pursue further the alternative assumption 2 because the intuition and general understanding of the regression framework are best uncovered with our original assumption 2.[6]

Notice that assumption 2 implies that the error term ε_i is the key to determining the *random* behavior of Y_i; the rest of our assumptions then focus on the behavior of ε_i as a random variable.

Assumption 3: $E(\varepsilon_i) = 0$ for each i.

This assumption implies that the mean deviation (over the population) of observed values from the true regression line is zero; that is, it implies that the expected value of Y_i (given X_i) really is $\alpha + \beta X_i$.

Assumption 4:   var (ε_i) is the same for each i (we shall call this common variance σ^2).

This assumption suggests that the extent to which individual observations have a tendency to deviate from the true regression line is constant across observations. In particular, it suggests that this variance does not change as the value of X gets larger or smaller.

Assumption 5: ε_i's are a set of mutually independent random variables.

In the money supply–inflation example, this assumption would suggest that a higher-than-expected inflation rate (given money supply growth) in period 1 (i.e., $\varepsilon_1 > 0$) is unassociated with the behavior of the inflation rate in period 2 (or any other period). In general, this assumption suggests that the process by which Y is determined (given X) is unrelated across observations, and there is no systematic compounding or canceling of errors (ε's) from one observation to another. (Note that this result may not be very likely in the context of the money supply–inflation example; if inflation is higher than expected in period 1, it is more likely than not that it will be higher than expected in period 2 because of the cyclical nature of the vari-

---

[6] See Jan Kmenta, *Elements of Econometrics*, 2d ed. (New York: Macmillan, 1986), pp. 334–337 for a further discussion of this issue.

able. It is then suggested that assumption 5 is one that might often be violated in certain types of economic problems; in such cases we say that the error term is characterized by *serial correlation*. We will return to this issue later in this chapter.)

Assumption 6: ε_i's are normally distributed.

This assumption suggests a symmetric distribution of errors from the true regression line, which is often a reasonable assumption. It also plays a key role in the development of the appropriate distribution theory for confidence intervals and hypothesis tests.

These six assumptions are not the only way of expressing the framework for the CLRM, but they represent the set of assumptions from which the properties of the estimators come most directly and easily. Given our goal of providing the intuitive foundation for the concepts, we proceed with this version of the problem and ignore many technical subtleties.[7] We now turn to a discussion of the properties of OLS estimators in the CLRM framework.

PROPERTIES OF OLS ESTIMATORS IN THE CLRM

We now derive the properties of OLS estimators under the assumptions of the CLRM. In particular, these estimators have very desirable properties in terms of unbiasedness and efficiency.

We begin by considering $\hat{\beta}$. We know from our discussion above that

$$\hat{\beta} = \frac{\Sigma(X_i - \bar{X})Y_i}{\Sigma(X_i - \bar{X})^2}$$

Notice that since X_i is not a random variable (assumption 2), $\bar{X}$ and $\Sigma(X_i - \bar{X})^2$ are not random variables. This assumption implies that the only source of random variation on the right-hand side of the equation for $\hat{\beta}$ is Y_i. To determine the properties of $\hat{\beta}$ as a random variable we need then to explore the randomness in Y_i more completely. To isolate Y_i, let us define

$$k = \Sigma(X_i - \bar{X})^2$$

---

[7] For example, strictly speaking, assumption 5 could be that the individual ε_i's are uncorrelated, for a set of uncorrelated normals are necessarily independent (this is a special characteristic of the normal distribution).

and rewrite $\hat{\beta}$ as

$$\hat{\beta} = \Sigma \left(\frac{X_i - \bar{X}}{k} \right) Y_i$$

Let us notice that $\hat{\beta}$ is then of the form

$$\hat{\beta} = \Sigma c_i Y_i \qquad\qquad (13.5)$$

where

$$c_i = \frac{X_i - \bar{X}}{k}$$

is a fixed number (because of the nonrandom nature of X_i).

Determining the properties of $\hat{\beta}$ as an estimator involves determining the distribution of $\hat{\beta}$ as a random variable. The randomness in $\hat{\beta}$ is due to the Y's, but the randomness in the Y's is due to the ε's. So to learn more about $\hat{\beta}$ let us rewrite Eq. (13.5) in terms of ε_i (which we know a lot about given the assumptions of the CLRM). How do we proceed?

Since

$$Y_i = \alpha + \beta X_i + \varepsilon_i \qquad \text{for each } i \text{ (assumption 1)}$$

we can rewrite Eq. (13.5) as

$$\hat{\beta} = \Sigma c_i (\alpha + \beta X_i + \varepsilon_i)$$

We now have $\hat{\beta}$ in terms of ε_i and the nonrandom values c_i, α, β, and X_i. We can determine the properties of $\hat{\beta}$ based on this function and the properties of ε_i.

Note that $\hat{\beta}$ may be rewritten as

$$\hat{\beta} = \alpha \Sigma c_i + \beta \Sigma c_i X_i + \Sigma c_i \varepsilon_i \qquad\qquad (13.6)$$

which is accomplished by exploiting the linearity property of Σ and the fact that multiplicative constants (α and β) can be equivalently written inside or outside the Σ sign. This step is very helpful because it turns out that it will always be the case that

$$\Sigma c_i = 0$$

and

$$\Sigma c_i X_i = 1$$

Substituting[8] these into Eq. (13.6) yields

$$\hat{\beta} = \beta + \Sigma c_i \varepsilon_i$$

$$= \beta + c_1 \varepsilon_1 + c_2 \varepsilon_2 + \cdots + c_n \varepsilon_n \qquad (13.7)$$

Equation (13.7) is the key to understanding the distribution of $\hat{\beta}$ as a random variable and therefore its properties as an estimator. In particular, note that since the c's are a set of fixed numbers, $\hat{\beta}$ is simply a linear function of n random variables (the ε's). The properties of $\hat{\beta}$ can then be determined by exploiting what we know about linear functions in combination with the six assumptions of the CLRM. To find the expected value of $\hat{\beta}$, we apply the expected value operator to each side of Eq. (13.7):

$$E(\hat{\beta}) = \beta + c_1 E(\varepsilon_1) + c_2 E(\varepsilon_2) + \cdots + c_n E(\varepsilon_n)$$

$$= \beta \qquad [\text{since } E(\varepsilon_i) = 0 \text{ for each } i \text{ by assumption 3}]$$

Thus, $\hat{\beta}$ is an unbiased estimator of β.

The variance of $\hat{\beta}$ plays a role in efficiency discussions as well as in procedures for confidence intervals and hypothesis tests. Equation (13.7) tells us that $\hat{\beta}$ is a linear function of n random variables; assumption 5 tells us that these random variables are mutually independent, and assumption 4 tells us that they have a common variance.

---

[8] To prove the first proposition, note that

$$\Sigma c_i = \frac{\Sigma(X_i - \bar{X})}{k} = \frac{1}{k}\Sigma(X_i - \bar{X}) = \frac{1}{k}(\Sigma X_i - \Sigma\bar{X}) = \frac{1}{k}(n\bar{X} - n\bar{X}) = 0$$

To prove the second, note that

$$\Sigma c_i X_i = \frac{1}{k}\Sigma(X_i - \bar{X})X_i = 1 \qquad \text{if } \Sigma(X_i - \bar{X})X_i = k$$

To show that $\Sigma(X_i - \bar{X})X_i = k$, note that

$$k = \Sigma(X_i - \bar{X})^2 = \Sigma(X_i - \bar{X})(X_i - \bar{X})$$

$$= \Sigma[(X_i - \bar{X})X_i - (X_i - \bar{X})\bar{X}]$$

$$= \Sigma(X_i - \bar{X})X_i - \Sigma(X_i - \bar{X})\bar{X}$$

$$= \Sigma(X_i - \bar{X})X_i - \bar{X}\Sigma(X_i - \bar{X})$$

$$= \Sigma(X_i - \bar{X})X_i$$

since $\Sigma(X_i - \bar{X}) = 0$, as shown in the $\Sigma c_i = 0$ proof. Since $k = \Sigma(X_i - \bar{X})X_i$, we know that $\Sigma c_i X_i = 1$.

In Chapter 8 we learned how to determine the variance of a linear function of many random variables that are mutually independent:

$$\text{var}(\hat{\beta}) = c_1^2 \text{var}(\varepsilon_1) + c_2^2 \text{var}(\varepsilon_2) + \cdots + c_n^2 \text{var}(\varepsilon_n)$$

$$= c_1^2 \sigma^2 + c_2^2 \sigma^2 + \cdots + c_n^2 \sigma^2$$

$$= \sigma^2(c_1^2 + c_2^2 + \cdots + c_n^2)$$

$$= \sigma^2 \Sigma c_i^2$$

[recall that $\text{var}(\varepsilon_i) = \sigma^2$ for all i by assumption 4]. Now

$$\Sigma c_i^2 = \Sigma \left(\frac{X_i - \bar{X}}{k} \right)^2$$

$$= \frac{1}{k^2} \Sigma (X_i - \bar{X})^2$$

$$= \frac{1}{\Sigma(X_i - \bar{X})^2} \qquad \text{since } k = \Sigma(X_i - \bar{X})^2$$

and therefore[9]

$$\text{var}(\hat{\beta}) = \frac{\sigma^2}{\Sigma(X_i - \bar{X})^2}$$

Finally, note that due to assumption 6, $\hat{\beta}$ is a linear function of n independent random variables, all of which are normally distributed. Since linear functions of independent normals are normal (see Chapter 9), we then conclude by saying that

$$\hat{\beta} \sim N\left(\beta, \frac{\sigma^2}{\Sigma(X_i - \bar{X})^2} \right) \qquad (13.8)$$

(where the first term in parentheses is the mean and the second is the variance). This equation then fully characterizes the distribution of $\hat{\beta}$.

It can be shown that, under the assumptions of the CLRM, $\hat{\beta}$ has the smallest variance of any unbiased estimator of β. This

---

[9] Note that $\text{var}(\hat{\beta})$ falls as $\Sigma(X_i - \bar{X})^2$ rises. This finding suggests that $\hat{\beta}$ is more reliable [var$(\hat{\beta})$ is smaller] the greater the degree of dispersion in the X's we observe. It also suggests that $\hat{\beta}$ is typically more reliable [var$(\hat{\beta})$ is smaller] as the number of observations increases because more observations typically cause $\Sigma(X_i - \bar{X})^2$ to rise.

result becomes a primary justification for the use of the OLS estimator. We do not pursue this efficiency question, however, because it requires techniques beyond the scope of this text.[10]

In economics, we are generally much more interested in slope coefficients than intercept terms, because we generally focus on changes in variables as opposed to their levels; that is, β (or $\hat{\beta}$) is generally of greater interest than α (or $\hat{\alpha}$) is. This observation, in conjunction with the fact that most of what we learn by going through the complicated derivations has already been achieved in our discussion of $\hat{\beta}$, leads us to present here simply the distribution of $\hat{\alpha}$ (it will be formally derived in Appendix 2 to this chapter). It turns out that

$$\hat{\alpha} \sim N\left\{\alpha,\ \sigma^2\left[\frac{1}{n} + \frac{\bar{X}^2}{\Sigma(X_i - \bar{X})^2}\right]\right\} \qquad (13.9)$$

This relation suggests that $\hat{\alpha}$ is an unbiased estimator; it also turns out that it has the same efficiency properties as $\hat{\beta}$ does under the assumptions of the CLRM.

We do not at this point discuss the properties of the OLS estimators if any of the assumptions of the CLRM are not valid; we shall briefly discuss this problem later in this chapter. For the most part, however, such a discussion is the subject of econometrics courses, which build upon the statistical foundations discussed in this book.

To summarize, we now know that the OLS estimators $\hat{\alpha}$ and $\hat{\beta}$ are (under the CLRM assumptions) unbiased and have desirable efficiency properties. As is the case with any estimator, however, they are likely to be wrong! That is, even though $E(\hat{\beta}) = \beta$, since $\hat{\beta}$ is a continuous random variable, $\Pr(\hat{\beta} = \beta) = 0$. This finding suggests that we need to consider confidence intervals and hypothesis tests as more sophisticated forms of inference. We now turn to this problem.

---

[10] For a discussion of the efficiency question see, for example, Kmenta (see note 6), pp. 223–224. It is important to note that if the ε's are not normally distributed, then generally it will not be the case that the OLS estimators have the smallest variance of any unbiased estimator; they would, however, have the smallest variance of any *linear* unbiased estimator (i.e., of any unbiased estimator that is a linear function of the Y_i's).

PROBLEMS

13.4 Trace through the derivation of Eq. (13.8) and summarize the role played by each of the six assumptions of the CLRM in achieving that result.

13.5 Assumption 1 of the CLRM is based on there being a linear relationship between X and Y. Comment on the likely appropriateness of this assumption in the context of the money supply–inflation example considered earlier in this chapter.

13.6 From Eq. (13.8) we know that

$$\text{var}\,(\hat{\beta}) = \frac{\sigma^2}{\Sigma(X_i - \bar{X})^2}$$

which suggests that $\hat{\beta}$ is a more reliable estimator of β the greater the degree of dispersion in the X's in our sample. Provide an intuitive explanation for this result.

SOPHISTICATED INFERENCE IN THE TWO-VARIABLE MODEL

If σ^2 were known, then

$$\frac{\hat{\beta} - \beta}{\sigma/\sqrt{\Sigma(X_i - \bar{X})^2}} \sim N(0, 1)$$

could be the basis for the construction of a 95% confidence interval for β, following the procedures used in our discussion of confidence intervals in Chapter 11. This confidence interval would be

$$\hat{\beta} - Z_{.025}\,\frac{\sigma}{\sqrt{\Sigma(X_i - \bar{X})^2}} \quad , \quad \hat{\beta} + Z_{.025}\,\frac{\sigma}{\sqrt{\Sigma(X_i - \bar{X})^2}} \quad (13.10)$$

that is,

$$\hat{\beta} \pm Z_{.025}\,\frac{\sigma}{\sqrt{\Sigma(X_i - \bar{X})^2}}$$

Our strategy for operationalizing Eq. (13.10) is the same as that employed in Chapter 11; we first estimate σ^2 (and implicitly, σ) and then use this estimate in place of σ in Eq. (13.10). This procedure will change the relevant distribution from the standard normal to something else; if we choose this estimator carefully, however, this new distribution can be modeled with the t distribution. As before, we are substituting an estimate of the variance for its true value and then using the t distribution to account for the additional variability.

How do we estimate σ^2? If the population were known we would know σ^2 since

$$\sigma^2 = \text{var}(\varepsilon) = \frac{1}{m} \sum_{i=1}^{m} [\varepsilon_i - E(\varepsilon_i)]^2 = \frac{1}{m} \sum_{i=1}^{m} \varepsilon_i^2$$

where m is the size of the population (and the Σ is over this entire population). To estimate σ^2 we might then consider forming

$$\frac{1}{\text{scale factor}} \sum_{i=1}^{n} \varepsilon_i^2$$

where the sum is over our sample of size n and the scale factor is related to that sample size. Unfortunately, even this technique is not operational. Our problem is not simply that we do not observe the entire population of ε's; we in fact do not observe *any* of the ε's (not even the ones for the observations in our sample). Why?

Remember that ε_i represents the deviation of the observed value Y_i from the true regression line. Since the true regression line is unknown, we do not know ε_i for any of our observations. We do, however, know the residual $\hat{\varepsilon}_i$ (the deviation of our observation from the estimated regression line). Since the estimated regression line estimates the true regression line, the deviation from the estimated regression line in effect estimates the deviation from the true regression line (i.e., $\hat{\varepsilon}_i$ estimates ε_i). This conclusion suggests an estimate of σ^2 equal to

$$\frac{1}{\text{scale factor}} \sum_{i=1}^{n} \hat{\varepsilon}_i^2$$

What is the appropriate scale factor? This question is to a large extent a technical one, but it turns out that if we form

$$s_\varepsilon^2 = \frac{1}{n-2} \sum_{i=1}^{n} \hat{\varepsilon}_i^2$$

then s_ε^2 is an unbiased estimator of σ^2. (Note that s_ε^2 differs from the sample variance s^2 presented in Chapter 4, and thus the ε subscript.)[11]

It turns out that

$$\frac{\hat{\beta} - \beta}{s_\varepsilon / \sqrt{\Sigma(X_i - \bar{X})^2}} \sim t^{(n-2)} \tag{13.11}$$

which is operational for the formation of a confidence interval. Intuitively, we have substituted an estimate of σ^2 for its true value and then used the t distribution, with $(n - 2)$ degrees of freedom,[12] to account for the additional variability.

From Eq. (13.11) we may then derive the formula for a 95% confidence interval for β:

$$\hat{\beta} \pm t_{.025}^{(n-2)} \frac{s_\varepsilon}{\sqrt{\Sigma(X_i - \bar{X})^2}} \tag{13.12}$$

For the intercept α, we go through a similar procedure. We know from Eq. (13.9) that

$$\frac{\hat{\alpha} - \alpha}{\sigma \sqrt{\dfrac{1}{n} + \dfrac{\bar{X}^2}{\Sigma(X_i - \bar{X})^2}}} \sim N(0, 1)$$

It turns out that

$$\frac{\hat{\alpha} - \alpha}{s_\varepsilon \sqrt{\dfrac{1}{n} + \dfrac{\bar{X}^2}{\Sigma(X_i - \bar{X})^2}}} \sim t^{(n-2)} \tag{13.13}$$

[11] We divide by $(n - 2)$ because it provides us with an unbiased estimator and ultimately leads to a t distribution framework for confidence intervals and hypothesis tests. Notice that it is only to the extent that we have more than two observations that s_ε^2 will differ from zero, since with only two observations both points would be on the estimated regression line (they would in fact form it) and therefore $\Sigma \hat{\varepsilon}_i^2 = 0$. It is then the case that there are only $(n - 2)$ observations contributing to variation in s_ε^2. This situation is analogous to the intuition for why we divide by $(n - 1)$ when forming the sample variance s^2.

[12] Under the CLRM assumptions, $(n - 2)s_\varepsilon^2 / \sigma^2 \sim \chi^2_{(n-2)}$, which is analogous to the result presented in the appendix to Chapter 9 for the sample variance s^2. The determination of the distribution of Eq. (13.11) as a t distribution is similar to that discussed in the appendix to Chapter 11. The degrees of freedom is $(n - 2)$ because there are only $(n - 2)$ sources of variation in the construction of s_ε^2, inasmuch as the first two observations are needed to form the regression line.

A 95% confidence interval for the parameter α may then be formed as

$$\hat{\alpha} \pm t_{.025}^{(n-2)} s_\varepsilon \sqrt{\frac{1}{n} + \frac{\bar{X}^2}{\Sigma(X_i - \bar{X})^2}} \qquad (13.14)$$

Let us apply these results in the context of the example contained above in Table 13.1.

The first step is to calculate the $\hat{\varepsilon}_i$'s. Remember that $\hat{\varepsilon}_i = Y_i - \hat{Y}_i$; we therefore in Table 13.2 build upon the start provided in Table 13.1 by adding columns for $\hat{Y}_i$ $(= \hat{\alpha} + \hat{\beta}X_i = 2.5 + .75X_i)$, $\hat{\varepsilon}_i$ $(= Y_i - \hat{Y}_i)$, and $\hat{\varepsilon}_i^2$. We find $\Sigma\hat{\varepsilon}_i^2 = 3.5$, suggesting that

$$s_\varepsilon^2 = \frac{1}{n-2} \Sigma\hat{\varepsilon}_i^2 = \frac{1}{3}(3.5) = 1.17$$

We can now build 95% confidence intervals for α and β. For β we would get

$$\hat{\beta} \pm t_{.025}^{(3)} \frac{s_\varepsilon}{\sqrt{\Sigma(X_i - \bar{X})^2}} = .75 \pm (3.18)\frac{\sqrt{1.17}}{\sqrt{40}}$$

That is, a 95% confidence interval for β is $(.21, 1.29)$. For the intercept α we would obtain

$$\hat{\alpha} \pm t_{.025}^{(3)} s_\varepsilon \sqrt{\frac{1}{n} + \frac{\bar{X}^2}{\Sigma(X_i - \bar{X})^2}} = 2.5 \pm (3.18)\sqrt{1.17}\sqrt{\frac{1}{5} + \frac{36}{40}}$$

That is, a 95% confidence interval for α is $(-1.11, 6.11)$.

The confidence intervals are subject to the same interpretations as the intervals developed in Chapter 11. In particular, if we were to take many samples of size 5 from this population and calculate for each of them a 95% confidence interval for $\beta(\alpha)$, just about 95% of those confidence intervals would contain $\beta(\alpha)$. Since 95% of all such created confidence intervals contain the true parameter, we are 95% confident that the one associated with our sample does.

Table 13.2 CALCULATING s_ε^2

$\hat{\varepsilon}_i^2$	$\hat{\varepsilon}_i = Y_i - \hat{Y}_i$	$\hat{Y}_i = 2.5 + .75X_i$	Y_i	X_i	$X_i - \bar{X}$	$(X_i - \bar{X})Y_i$	$(X_i - \bar{X})^2$
0	0	4	4	2	-4	-16	16
.25	$-.5$	5.5	5	4	-2	-10	4
0	0	7	7	6	0	0	0
2.25	1.5	8.5	10	8	2	20	4
1	-1	10	9	10	4	36	16
$\Sigma = 3.5$			$\bar{Y} = 7$	$\bar{X} = 6$		$\Sigma = 30$	$\Sigma = 40$

The distribution theory used above also suggests what we need in order to carry out hypothesis tests within the two-variable regression model. It is important to emphasize that the conceptual framework, general strategy, and interpretive context of hypothesis tests in the two-variable regression model are the same as in the case of means or differences in means; what differs is the specific distribution theory involved. In particular, we set up a null hypothesis, an alternative hypothesis, and ask the prob-value question: What is the probability of getting a value of our estimator as extreme as in our sample if H_0 is in fact true? By comparing the prob-value to the prespecified level of the test we then carry out the hypothesis test.

Let us illustrate this procedure in the context of this example. Suppose that we consider the problem of finding the prob-value for and testing the hypothesis

$$H_0: \beta = 0$$

$$H_1: \beta \neq 0$$

at the .05 level. As an aside, this test is often very interesting because in it we are asking whether it can be determined from our data that there is a statistically significant link between changes in X and changes in Y (on average). Often in exploratory research this result is exactly what we are interested in determining, for it helps us to find out whether or not the variables are related.

The prob-value is a two-sided one:

$$\text{prob-value} = \Pr(|\hat{\beta} - \beta| \geq |.75 - 0|)$$

and we evaluate it by changing the left-hand side of the inequality into (the absolute value of) a t distribution. If we transform this into

$$\text{prob-value} = \Pr\left(\left|\frac{\hat{\beta} - \beta}{s_e/\sqrt{\Sigma(X_i - \bar{X})^2}}\right| \geq \left|\frac{.75 - 0}{\sqrt{1.17}/\sqrt{40}}\right|\right)$$

then the left-hand side of the inequality is the absolute value of a t distribution with 3 degrees of freedom; see Eq. (13.11). The prob-value is then

$$\Pr(|t^{(3)}| \geq 4.39)$$

Since $\Pr(t^{(3)} \geq 4.39)$ is between .025 and .010, we know that

$$.05 > \text{prob-value} > .02$$

In any case, since the prob-value is less than .05, we reject H_0 in favor of H_1 at the .05 level. We conclude that there is persuasive evidence that changes in X are associated on average with changes in Y.

For completeness, let us illustrate a hypothesis test for α, although α values are generally less interesting in economics (why?). Let us find the prob-value for and test

$$H_0: \alpha = 0$$

$$H_1: \alpha > 0$$

at the .05 level. Equation (13.13) gives us the distribution theory needed to determine the prob-value:

prob-value $= \text{Pr}(\hat{\alpha} \geq 2.5)$ (if $\alpha = 0$)

$$= \text{Pr}\left(\frac{\hat{\alpha} - \alpha}{s_\epsilon \sqrt{\dfrac{1}{n} + \dfrac{\bar{X}^2}{\Sigma(X_i - \bar{X})^2}}} \geq \frac{2.5 - 0}{\sqrt{1.17}\sqrt{\dfrac{1}{5} + \dfrac{36}{40}}}\right)$$

$$= \text{Pr}(t^{(3)} \geq 2.20)$$

Thus

$$.10 > \text{prob-value} > .05$$

and since the prob-value exceeds .05, we cannot reject H_0 in favor of H_1 at the .05 level.

By calculating the estimated regression line and forming confidence intervals and hypothesis tests we obtain much more complete knowledge concerning the quantitative nature of the relationship between X and Y than we would through, say, the correlation coefficient. In particular, we learn about the change in Y that we would expect to occur as a result of a change in X. What we do not get directly through a regression, however, is a measure of the closeness (or strength) of the relationship between the two variables (and we do get this from the correlation coefficient). Knowledge concerning the degree of correlation between the two variables can, however, be obtained from a regression, and in so doing we obtain a superior interpretation of the concept of correlation than we have heretofore obtained. We now turn to an investigation of this issue.

PROBLEMS

13.7 Using the data of problem 13.1:

(a) Build 95% confidence intervals for α and β.

(b) Find the prob-value for and test the hypothesis

$$H_0: \beta = 0$$

$$H_1: \beta > 0$$

at the .05 level.

(c) Find the prob-value for and test the hypothesis

$$H_0: \alpha = 0$$

$$H_1: \alpha \neq 0$$

at the .05 level.

13.8 Using the data of problem 13.2:
   (a) Build 95% confidence intervals for α and β.
   (b) Find the prob-value for and test the hypothesis

$$H_0: \beta = 0$$

$$H_1: \beta < 0$$

at the .05 level.
   (c) Find the prob-value for and test the hypothesis

$$H_0: \alpha = 20$$

$$H_1: \alpha > 20$$

at the .05 level.

13.9 Using the data of problem 13.3:
   (a) Build 95% confidence intervals for α and β.
   (b) Find the prob-value for and test the hypothesis

$$H_0: \beta = 0$$

$$H_1: \beta \neq 0$$

at the .05 level.
   (c) Find the prob-value for and test the hypothesis

$$H_0: \alpha = 0$$

$$H_1: \alpha > 0$$

at the .05 level.

GOODNESS OF FIT

Consider Figure 13.7 and the two different situations depicted therein. In situation (a), there is clearly a closer relationship between X and Y than there is in situation (b); operationally, this statement

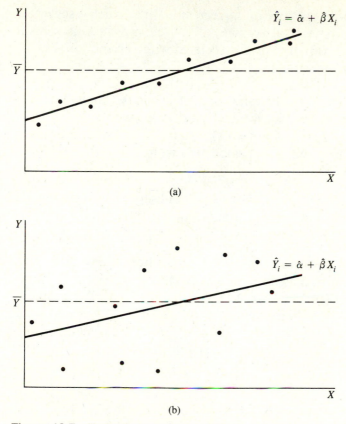

Figure 13.7   Illustrating goodness of fit.

is evidenced by the fact that the observed points are on average closer to the estimated regression line in (a) than they are in (b). Clearly, this closeness is an important characteristic of the relationship between X and Y that a simple estimation of the regression line does not bring out. However, we can slightly extend the regression framework to consider not only the relationship between X and Y *on average* but also how the actual Y's compare to what we expect them to be (given X). That is, we can extend the regression framework so as to consider how closely the variations in X and Y are linked.

The total amount of variation in Y in our sample can be measured by

$$\Sigma(Y_i - \bar{Y})^2$$

Some of this variation is associated with variation in X; this portion is associated with the slope of the regression line, which is suggestive of the responsiveness (on average) of Y to changes in X. In general, that part of the variation in Y associated with variation in X is

$$\Sigma(\hat{Y}_i - \bar{Y})^2$$

This expression is what the variation in Y would be (about its mean) if knowing the value of X implied that we knew exactly the value of Y (that is, if $Y_i = \hat{Y}_i$ for each i). On the other hand, the part of the variation in Y that is not associated with X is

$$\Sigma(Y_i - \hat{Y}_i)^2$$

which focuses on the deviations of the observed points from the estimated regression line. It turns out that

$$\Sigma(Y_i - \bar{Y})^2 = \Sigma(\hat{Y}_i - \bar{Y})^2 + \Sigma(Y_i - \hat{Y}_i)^2 \qquad (13.15)$$

Equation (13.15) is essentially a decomposition of the variation in Y into (1) that part associated with X (i.e., $\Sigma(\hat{Y}_i - \bar{Y})^2$) and (2) that part not associated with X (i.e., $\Sigma(Y_i - \hat{Y}_i)^2$).[13] A measure of how well the regression line fits the data is

---

[13] Equation (13.15) is derived by noting that

$$Y_i - \bar{Y} = (\hat{Y}_i - \bar{Y}) + (Y_i - \hat{Y}_i)$$

and therefore

$$(Y_i - \bar{Y})^2 = (\hat{Y}_i - \bar{Y})^2 + (Y_i - \hat{Y}_i)^2 + 2(\hat{Y}_i - \bar{Y})(Y_i - \hat{Y}_i)$$

and

$$\Sigma(Y_i - \bar{Y})^2 = \Sigma(\hat{Y}_i - \bar{Y})^2 + \Sigma(Y_i - \hat{Y}_i)^2 + 2\Sigma(\hat{Y}_i - \bar{Y})(Y_i - \hat{Y}_i)$$

Since the last term is always 0, Eq. (13.15) then follows. To show that the last term is always zero, note that

$$\hat{Y}_i - \bar{Y} = \hat{\alpha} + \hat{\beta}X_i - \hat{\alpha} - \hat{\beta}\bar{X} = \hat{\beta}(X_i - \bar{X})$$

(since $\hat{\alpha} = \bar{Y} - \hat{\beta}\bar{X}$ we know that $\bar{Y} = \hat{\alpha} + \hat{\beta}\bar{X}$). Thus

$$
\begin{aligned}
\Sigma(\hat{Y}_i - \bar{Y})(Y_i - \hat{Y}_i) &= \hat{\beta}\Sigma(X_i - \bar{X})(Y_i - \hat{Y}_i) \\
&= \hat{\beta}\Sigma(X_i - \bar{X})(Y_i - \hat{\alpha} - \hat{\beta}X_i) \\
&= \hat{\beta}[\Sigma(X_i - \bar{X})Y_i - \hat{\alpha}\Sigma(X_i - \bar{X}) - \hat{\beta}\Sigma(X_i - \bar{X})X_i] \\
&= \hat{\beta}[\Sigma(X_i - \bar{X})Y_i - \hat{\beta}\Sigma(X_i - \bar{X})^2]
\end{aligned}
$$

$$(\text{since } \Sigma(X_i - \bar{X}) = 0 \text{ and } \Sigma(X_i - \bar{X})X_i = \Sigma(X_i - \bar{X})^2)$$

$$= \hat{\beta}[\Sigma(X_i - \bar{X})Y_i - \Sigma(X_i - \bar{X})Y_i]$$

$$\left(\text{since } \hat{\beta} = \frac{\Sigma(X_i - \bar{X})Y_i}{\Sigma(X_i - \bar{X})^2}\right)$$

$$= 0$$

$$R^2 = \frac{\Sigma(\hat{Y}_i - \bar{Y})^2}{\Sigma(Y_i - \bar{Y})^2} \qquad (13.16)$$

that is, that fraction of the total variation in Y associated with the variation in X. R^2 is often referred to as the *coefficient of determination*. Obviously, $0 \le R^2 \le 1$, and the closer to 1 the better the fit of the regression line. $R^2 = 1$ would be a perfectly fitting regression line (i.e., $\hat{\varepsilon}_i = 0$ for each i).

Note that we may rewrite R^2 as

$$R^2 = 1 - \frac{\Sigma(Y_i - \hat{Y}_i)^2}{\Sigma(Y_i - \bar{Y})^2} = 1 - \frac{\Sigma\hat{\varepsilon}_i^2}{\Sigma(Y_i - \bar{Y})^2} \qquad (13.17)$$

which comes from dividing Eq. (13.15) through by $\Sigma(Y_i - \bar{Y})^2$ and then rearranging terms. Equation (13.17) is the form from which R^2 typically is calculated.

Using the data from Table 13.2, we find $\Sigma(Y_i - \bar{Y})^2 = 26$. Remember that $\Sigma\hat{\varepsilon}_i^2 = 3.5$. Thus

$$R^2 = 1 - \frac{3.5}{26} = .865$$

suggesting that 86.5% of the variation in Y is associated with variation in X.[14]

How does the R^2 measure of goodness of fit compare to the correlation coefficient? It turns out that the R^2 from a two-variable regression always equals the square of the sample correlation coefficient discussed in Chapter 8.[15] This finding suggests that R^2 represents (1) an old concept reappearing and (2) a good interpretation of the correlation coefficient. Note also that this result suggests the

---

[14] We say "associated with" as opposed to "explained by," inasmuch as the latter implies causality; causality, however, is not suggested by the data but can rather only be suggested by the underlying theory.

[15] To prove this result, note that

$$\Sigma(\hat{Y}_i - \bar{Y})^2 = \Sigma[\hat{\alpha} + \hat{\beta}X_i - (\hat{\alpha} + \hat{\beta}\bar{X})]^2 = \hat{\beta}^2\Sigma(X_i - \bar{X})^2$$

$$= \frac{[\Sigma(X_i - \bar{X})(Y_i - \bar{Y})]^2}{\Sigma(X_i - \bar{X})^2} = \frac{[(n-1)\widehat{\text{cov}(X, Y)}]^2}{(n-1)s_x^2}$$

since $\Sigma(X_i - \bar{X})Y_i = \Sigma(X_i - \bar{X})(Y_i - \bar{Y})$ (why?). Therefore,

$$R^2 = \left(\frac{[(n-1)\widehat{\text{cov}(X, Y)}]^2}{(n-1)s_x^2}\right) \bigg/ \Sigma(Y_i - \bar{Y})^2$$

$$= \frac{[(n-1)\widehat{\text{cov}(X, Y)}]^2}{[(n-1)s_x s_y]^2} = \left(\frac{\widehat{\text{cov}(X, Y)}}{s_x s_y}\right)^2 = \hat{\rho}^2$$

link between the correlation coefficient and *linear* relationships between X and Y, which is consistent with our interpretation of the covariance (and thus the correlation coefficient) and its relationship to independence as discussed in Chapter 8.[16]

We now move on to a consideration of examples that apply regression analysis to economic problems.

PROBLEMS

13.10 Calculate R^2 for the data contained in each of problems 13.1, 13.2, and 13.3.

13.11 Prove that $\Sigma \hat{\varepsilon}_i$ is always 0, where the $\hat{\varepsilon}_i$'s are the residuals from an OLS regression.

13.12 We know that, for each observation,

$$Y_i - \bar{Y} = (\hat{Y}_i - \bar{Y}) + (Y_i - \hat{Y}_i)$$

which is a decomposition of the deviation of Y_i from its mean into (1) that part associated with X_i (i.e., $\hat{Y}_i - \bar{Y}$) and (2) that part not associated with X_i (i.e., the residual $Y_i - \hat{Y}_i$). Clearly,

$$\Sigma(Y_i - \bar{Y}) = \Sigma(\hat{Y}_i - \bar{Y}) + \Sigma(Y_i - \hat{Y}_i)$$

and we might then be led to suggest the following measure of goodness of fit:

$$\frac{\Sigma(\hat{Y}_i - \bar{Y})}{\Sigma(Y_i - \bar{Y})}$$

The numerator is an aggregation of behavior associated with X_i, and the denominator is an aggregation of the total behavior of Y_i.

(a) Prove that $\Sigma(Y_i - \bar{Y})$ always equals zero.
(b) Prove that $\Sigma(\hat{Y}_i - \bar{Y})$ always equals zero.
(c) Comment then on the quality of this alternative measure of goodness of fit.

---

[16] Recall that if two variables are uncorrelated, there is the absence of a relationship that can be expressed in the form of a linear function.

EXAMPLE 1: A CONSUMPTION FUNCTION

In this section we present an example of an application of the two-variable regression model to a fairly common economic problem: a standard Keynesian consumption function as developed and used in introductory courses. In the simple version of the Keynesian model, we say that

$$C = \alpha + \beta YD \qquad (13.18)$$

where C is aggregate consumption, YD is disposable income, and both are measured in constant dollars. This specification of the model, if interpreted literally, suggests that if we were to know the value of YD, then we would also know the value of C, but this conclusion is clearly unrealistic. To make the theory more realistic, we should make the specification *stochastic* (that is, we should incorporate a random element), for example,

$$C_i = \alpha + \beta YD_i + \varepsilon_i \qquad (13.19)$$

where ε_i is an error term similar to those we have been discussing.

The theory tells us certain characteristics of the parameters (e.g., $0 \le \beta \le 1$, since β is the marginal propensity to consume) but does not tell us their precise values. The theory then needs to be quantified, and we can do so by estimating the parameters α and β using the OLS estimation technique. To achieve this result we took data on the U.S. economy from 1959 to 1987 for C_i and YD_i (both measured in constant 1982 dollars), and used these data as the observations on our Y and X variables respectively.[17] (This example is, by the way, that of a *time series* data set, wherein the observations are on a given entity observed at different points in time. This type of data is in contrast to a *cross-sectional* data set, wherein the observations are a set of different entities all observed at the same point in time; for a consumption function, an example might be a set of countries all observed for the same year.)

To estimate this result we want to use a statistical program on a computer, because the computations become quite burdensome with this number of observations. There are many such programs available for personal computers or mainframes, and we presume that the reader has access to such a program.

---

[17] These data are listed in Appendix 3 to this chapter (along with some additional data we shall use in Chapter 14).

Using such a program, we achieved the following OLS estimates of α and β in the consumption function example:

$$\hat{\alpha} = -32.6$$

$$\hat{\beta} = .926$$

Our estimated regression equation is then

$$\hat{C}_i = -32.6 + .926YD_i$$

In particular, our estimate of the parameter β (the marginal propensity to consume) is .926, which indicates that on average about 93 cents out of each additional dollar of disposable income goes to consumption. The R^2 for this regression is .997, a result that suggests a very close relationship in our sample between consumption and disposable income.

Confidence intervals for α and β can be built so as to get a picture of the likely margin of error involved in the estimation of the parameters. Equation (13.12) suggests that a 95% confidence interval for β can be formed as

$$\hat{\beta} \pm t_{.025}^{(n-2)} \frac{s_\varepsilon}{\sqrt{\Sigma(X_i - \bar{X})^2}}$$

Most statistical programs provide us with the value of $s_\varepsilon/\sqrt{\Sigma(X_i - \bar{X})^2}$, which is often referred to as $s_{\hat{\beta}}$ (i.e., the estimated standard deviation of the estimator $\hat{\beta}$). In the current example, $s_{\hat{\beta}} = .011$. Since $t_{.025}^{(27)} = 2.05$, our 95% confidence interval for β is then

$$.926 \pm (2.05)(.011) = (.903, .949)$$

To build a 95% confidence interval for α, Eq. (13.14) tells us to form

$$\hat{\alpha} \pm t_{.025}^{(n-2)} s_\varepsilon \sqrt{\frac{1}{n} + \frac{\bar{X}^2}{\Sigma(X_i - \bar{X})^2}}$$

Most programs provide us with a value of

$$s_\varepsilon \sqrt{\frac{1}{n} + \frac{\bar{X}^2}{\Sigma(X_i - \bar{X})^2}}$$

which is often referred to as $s_{\hat{\alpha}}$. In this example, $s_{\hat{\alpha}} = 20.0$, and so a 95% confidence interval for α is

$$-32.6 \pm (2.05)20.0 = (-73.6, 8.4)$$

To verify the theory of the consumption function, we might find the prob-value for and test the hypothesis

$$H_0: \beta = 0$$

$$H_1: \beta > 0$$

at the .05 level. (We use a one-sided alternative hypothesis inasmuch as the theory clearly rules out the possibility $\beta < 0$.) Given the definition of $s_{\hat{\beta}}$ above, our hypothesis test is based upon the fact that

$$\frac{\hat{\beta} - \beta}{s_{\hat{\beta}}} \sim t^{(n-2)}$$

The prob-value is

$$\text{prob-value} = \Pr(\hat{\beta} \geq .926) \qquad \text{if } \beta = 0$$

$$= \Pr\left(\frac{\hat{\beta} - \beta}{s_{\hat{\beta}}} \geq \frac{.926 - 0}{.011}\right)$$

$$= \Pr(t^{(27)} \geq 84.2)$$

Clearly,

$$\text{prob-value} < .001$$

Since the prob-value is less than .05 (the level of the test), we can reject H_0 in favor of H_1 at the .05 level. We then conclude that increases in disposable income are on average associated with increases in consumption.

Similarly, we might find the prob-value for and test the hypothesis

$$H_0: \alpha = 0$$

$$H_1: \alpha \neq 0$$

(at the .05 level) to see whether our intercept is significantly different from zero. (Perhaps the negative value of $\hat{\alpha}$ is due simply to random variability.) The prob-value is

$$\text{prob-value} = \Pr(|\hat{\alpha} - \alpha| \geq |-32.6 - 0|)$$

$$= \Pr\left(\left|\frac{\hat{\alpha} - \alpha}{s_{\hat{\alpha}}}\right| \geq \left|\frac{-32.6 - 0}{20.0}\right|\right)$$

$$= \Pr(|t^{(27)}| \geq 1.63)$$

Thus

$$.20 > \text{prob-value} > .10$$

and we cannot reject H_0 in favor of H_1 at the .05 level. It is plausible that the intercept α of the consumption function is zero and that $\hat{\alpha} < 0$ because of random variability.

EXAMPLE 2: STATEWIDE VARIATION IN POVERTY RATES

In the example above, the data were of a time series nature; that is, the observations were on the same entity (the United States) at different periods in time. Our second example is based on cross-sectional data, in which the units of observation are different entities observed at the same point in time.[18] The issue under consideration here is the relationship between the poverty rate (i.e., the percent of individuals living in families with incomes below the poverty level) and the unemployment rate. There is a fairly straightforward suggestion that increases in the unemployment rate should be expected to be associated with increases in the poverty rate—but what happens if we investigate this assumption empirically?

In Appendix 4 to this chapter we present data on the 51 United States (including the District of Columbia) for the poverty rate, unemployment rate, and other variables. (These other variables will be used for an example in Chapter 14 of a multiple regression.) To illustrate a two-variable regression in a cross-sectional framework, we regressed the poverty rate (Y) on the unemployment rate (X) and achieved the following estimated regression line:

$$\hat{Y}_i = 10.2 + .389X_i$$
$$(2.12)\ (.329)$$

where the numbers in parentheses under the estimated coefficients

[18] Whether the data are a time series or a cross section implies different things regarding the likelihood with which certain assumptions of the CLRM are violated. We will address this issue below, specifically in the context of time series data.

are the estimated standard errors of the parameter estimates (i.e., $s_{\hat{\alpha}} = 2.12$ and $s_{\hat{\beta}} = .329$). In this regression, $R^2 = .028$. The fact that $\hat{\beta}$ is positive indicates that we have evidence to suggest that increases in X are on average associated with increases in Y; that is, we have evidence to suggest that, on average, states with higher unemployment rates have higher poverty rates. But is this persuasive evidence? That is, can we rule out the possibility that there is no (population) relationship between unemployment and poverty (i.e., $\beta = 0$) and that $\hat{\beta} > 0$ simply because of random variability?

To test this possibility we find the prob-value for and test the hypothesis

$$H_0: \beta = 0$$

$$H_1: \beta > 0$$

at the .05 level. The prob-value is

$$\Pr(\hat{\beta} \geq .389) = \Pr\left(\frac{\hat{\beta} - \beta}{s_{\hat{\beta}}} \geq \frac{.389 - 0}{.329}\right) = \Pr(t^{(49)} \geq 1.18)$$

Since the degrees of freedom exceeds 30, we can approximate the prob-value as $\Pr(Z \geq 1.18) = .1190$. Since the prob-value exceeds the level of the test, we cannot reject the null hypothesis in favor of the alternative hypothesis at the .05 level. That is, we cannot rule out the possibility that the relationship we observed between unemployment and poverty is due to randomness and not to some underlying structural relationship. (Note that this conclusion is consistent with the very small R^2 reported above.)

PREDICTION IN THE TWO-VARIABLE MODEL

Suppose that, in the context of the consumption function example just considered, we believed that several years in the future disposable income (YD) would be (in billions of 1982 dollars) 3000. Can we make a prediction or forecast concerning the level of consumption at that point?

Our estimated consumption function is

$$\hat{C}_i = -32.6 + .926 YD_i$$

An obvious way to predict the value of consumption for that time is to substitute 3000 in for YD_i in this equation and to determine then the associated value of $\hat{C}_i$; we are here taking our prediction of C as the point on the estimated regression line associated with that value of YD. If we do this, we would get

$$\hat{C}_i = -32.6 + .926(3000) = 2745.4$$

Of course, it is likely that the actual C for that time will be different from $\hat{C}$ because of the random nature of the problem, but nevertheless we have a prediction.

But how good is that prediction? One important factor to consider is whether the value of the X variable upon which we are basing the prediction is in a range for which we have information about the relationship between X and Y. Consider Figure 13.8. We have here drawn the estimated regression line, and have also noted the range of values of disposable income observed in our sample (1067.2 for 1959 is the low value and 2686.3 for 1987 is the high value). We have lots of information regarding the relationship between YD and C in the range of values of YD between the low value of 1067.2 and the high value of 2686.3. But we have no such information regarding the relationship between YD and C for values of YD outside that range. And it is possible that the relationship between YD and C is different outside that range; for example, it is possible that once YD gets to a particular level the marginal propensity to consume (β) declines and hence the true regression relationship flattens out. We do not know. If the relationship on average between YD and C tapers off after 2686.3 as suggested in the figure, then the prediction made above is likely to overstate the actual consumption figure for that period in time. Obviously, the

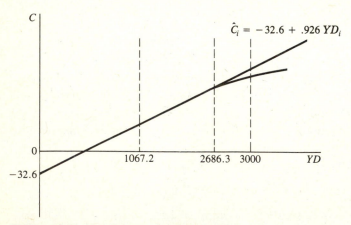

Figure 13.8 The danger of extrapolation.

more the value of X (upon which the prediction is conditioned) is outside the range of X's for which we have observations, the greater the possible extent of this problem.

We must confront this situation as a possibility in many varied settings. Adding an additional bag of fertilizer might increase my corn crop yield by 1 bushel, but we should not necessarily expect that increasing the amount of fertilizer by 10 bags will have ten times this effect; such predictions are often subject to great inaccuracies. We refer to this general issue as the *danger of extrapolation* and emphasize that a prediction within a range for which we have information is a very different result from one based on possibilities outside that range. The problem of course is that the assumption of linearity, though perhaps reasonable for the range of values in our sample, might become unreasonable as we move to different domains.

ISSUES REGARDING THE VALIDITY OF THE ASSUMPTIONS OF THE CLRM

The unbiasedness and efficiency results, as well as the probability distributions, that characterize the OLS estimators are based upon the six assumptions of the CLRM we have specified.[19] One must always wonder whether these assumptions are in fact valid in the problem at hand, and what the properties of the estimators become should any of these assumptions not be valid.[20]

To elaborate, let us refer back to the consumption function example. Assumption 5 of the CLRM states that the error term ε_i is independent across observations. In this problem, this assumption then implies that if (say) consumption in year 1 is higher than expected (given disposable income), this fact has no effect on whether consumption in year 2 will be higher or lower than expected (given disposable income). But does it not seem likely that if consumption in year 1 is higher than expected (given disposable income), that

---

[19] As stated above, the six assumptions used here are not the least restrictive assumptions for the CLRM, but they are the assumptions from which the results come most easily.

[20] In fact, this problem is a big part of the subject matter of an econometrics course, wherein we spend considerable time discussing the violation of assumptions of the CLRM, the impact of this on the properties of the OLS estimators, and how and if the estimation procedure should therefore be changed.

is, $\varepsilon_1 > 0$, then consumption in year 2 will also be higher than expected (given disposable income), that is, $\varepsilon_2 > 0$? After all, this result would just be a manifestation of the cyclical nature of many macroeconomic variables over time.

In time series data such a pattern often occurs; that is, the error terms are often correlated over time. Assumption 5 of the CLRM would be violated in such a case; this problem is referred to as one of *serial correlation.* It is therefore implied that when we have time series data we must address questions such as:

1. How can we tell whether the error terms are correlated over time?
2. If they are correlated, how are the properties of OLS estimators affected?
3. How, if at all, should we alter our estimation technique to address this problem?

Also of interest in the problem at hand is assumption 2 of the CLRM, in which we assume that the X's are fixed numbers and not random variables. Clearly such is not the case inasmuch as disposable income (the X variable) is observed and undoubtedly determined in part by random factors. But remember that there is an alternative to assumption 2, which is that the X's are independent of the ε's. Is this situation likely in the problem at hand?

Economic theory tells us not only that changes in disposable income cause changes in consumption, but also that changes in consumption potentially cause changes in disposable income. This latter statement comes from a full development of the Keynesian model of national income determination; an autonomous increase in consumption represents an increase in desired spending in the economy as a whole (graphically, an upward shift in the $C + I + G$ function) and such increases are held to have a stimulating effect on output and income. So we might expect that above-normal consumption ($\varepsilon > 0$) will cause disposable income (our X variable) to rise; if such is the case, however, then the behavior of ε and X are related (i.e., they are not independent).

What we have just uncovered is the general problem of two-way causality (that is, changes in X cause changes in Y but there is also a context in which changes in Y cause changes in X), which when present generally causes assumption 2 to be violated. If two-way causality leads to undesirable properties of the OLS estimators (and it does), how should we proceed? We would need to build a more sophisticated model that captures the relationship not only as it goes from disposable income to consumption but also as it goes from consumption to disposable income, and then consider the issue of how to estimate appropriately the parameters of the model.

We do not at this point develop mechanisms for dealing with these problems; our purpose has simply been to illustrate in the context of the consumption function example how we must consider whether the assumptions of the CLRM are valid. In other examples, our focus might be on the validity of different assumptions. In a course in econometrics, a great deal of attention would be given to just these kinds of questions.

SUMMARY

In this chapter we have laid the foundations of regression analysis and have developed in detail the two-variable model. But many relationships can be more accurately modeled with more than one variable on the right-hand side; this case involves the *multiple regression* framework, wherein, for example, quantity demanded is a function of both price and income. Most of the foundational aspects of our approach carry over into these more complicated situations, although there are some new issues that need to be addressed. Given that situations with more than one right-hand-side variable are common, we now turn to them and an exposition of the statistical issues that arise.

REVIEW PROBLEMS FOR CHAPTER 13

13.13 Using the data in problem 13.1, form predictions of Y for $X = 13$ and $X = 8$. Comment on the likely quality of these two predictions.

13.14 Accompanying this problem are data from the 1980 Census for the 51 states (including the District of Columbia). Calculate the regression of MIM on PMHS and carefully and fully interpret your results. (The remaining data will be used in a problem in Chapter 14.)

> MIM = mean income of males (having income) who are 18 years of age or older

PMHS = percent of males 18+ who are high school
graduates
PURBAN = percent of total population living in an
urban area
MAGE = median age of males

STATE	MIM ($)	PMHS (%)	PURBAN (%)	MAGE (years)
ME	12112	69.1	47.5	29.2
NH	14505	73.0	52.2	29.2
VT	12711	71.6	33.8	28.4
MA	15362	74.0	83.8	29.6
RI	13911	65.1	87.0	30.1
CT	17938	71.8	78.8	30.6
NY	15879	68.9	84.6	30.3
NJ	17639	70.0	89.0	30.7
PA	15225	68.0	69.3	30.4
OH	16164	69.0	73.3	28.6
IN	15793	68.8	64.2	28.0
IL	17551	68.9	83.3	28.6
MI	17137	69.3	70.7	27.8
WI	15417	70.9	64.2	28.3
MN	15878	73.5	66.9	28.3
IA	15249	71.9	58.6	28.7
MO	14743	66.2	68.1	29.3
ND	13835	68.0	48.8	27.5
SD	12406	68.3	46.4	27.9
NE	14873	74.2	62.9	28.6
KS	15504	74.5	66.7	28.7
DE	16081	70.4	70.6	28.7
MD	17321	69.2	80.3	29.2
DC	15861	67.9	100.0	29.9
VA	15506	64.3	66.0	28.6
WV	13998	58.6	36.2	29.1
NC	12529	58.2	48.0	28.1
SC	12660	58.2	54.1	26.7
GA	13966	60.4	62.4	27.3
FL	14651	68.0	84.3	32.9

STATE	MIM ($)	PMHS (%)	PURBAN (%)	MAGE (years)
KY	13328	55.8	50.9	27.8
TN	13349	59.0	60.4	28.7
AL	13301	59.9	60.0	27.8
MS	11968	57.2	47.3	26.1
AR	12274	58.3	51.6	29.2
LA	15365	61.3	68.6	26.2
OK	14818	68.7	67.3	28.6
TX	16135	65.3	79.6	27.1
MT	14256	73.8	52.9	28.4
ID	14297	73.5	54.0	27.0
WY	17615	77.9	62.7	26.7
CO	16672	79.1	80.6	27.9
NM	14057	70.6	72.1	26.6
AZ	15269	73.4	83.8	28.2
UT	15788	80.4	84.4	23.8
NV	16820	76.0	85.3	30.0
WA	17042	77.5	73.5	29.0
OR	15833	75.1	67.9	29.5
CA	17128	74.3	91.3	28.9
AK	21552	81.9	64.3	26.3
HI	15283	76.9	86.5	27.6

Sources: MIM and PMHS are from U.S. Bureau of the Census, *1980 Census of Population, Vol. 1, Characteristics of the Population, Chapter D: Detailed Population Characteristics,* parts 2 through 52 (Washington, DC: U.S. Government Printing Office, 1983), table 237. PURBAN is calculated from U.S. Bureau of the Census, *1980 Census of Population, Vol. 1, Characteristics of the Population, Chapter B: General Population Characteristics,* part 1: U.S. Summary (Washington, DC: U.S. Government Printing Office, 1983), table 62. MAGE is from the same source as PURBAN, table 67.

13.15 The accelerator model of investment posits that spending on additions to the capital stock (i.e., net investment) is a function of the change in the level of sales (since the capital stock is related to the level of sales, the change in the capital stock is related to the change in sales). Below you are given data for the U.S. economy on net fixed nonresidential investment (excluding inventory accumulations) and the change in the level of final sales, both measured in billions of 1982 dollars. Calculate the regression of this measure of investment on the change in the level of final sales, and carefully and fully interpret your results. (The data on the other variable, RINT, will be used in a problem in Chapter 14.)

NETINV = net fixed nonresidential investment (excluding inventory accumulations), in billions of 1982 dollars

CFSAL = change in the level of final sales from the previous year, in billions of 1982 dollars

RINT = real interest rate on AAA rated corporate bonds, calculated as the nominal percentage rate minus the percent change in the Consumer Price Index for all items from the previous year

YEAR	NETINV (bil. $)	CFSAL (bil. $)	RINT (%)
1970	89.3	9.8	2.34
1971	76.1	57.3	2.99
1972	85.3	121.6	4.01
1973	116.5	117.3	1.24
1974	106.9	−8.1	−2.43
1975	60.8	11.8	−0.27
1976	61.8	96.8	2.63
1977	85.2	124.9	1.52
1978	111.6	148.9	1.13
1979	124.3	99.0	−1.67
1980	101.3	16.6	−1.56
1981	105.5	31.0	3.87
1982	65.5	−34.5	7.59
1983	50.4	95.0	8.84
1984	103.3	153.6	8.41
1985	116.1	170.5	7.77
1986	80.5	96.7	7.12
1987	77.7	106.3	5.78

Sources: All data are from the *Economic Report of the President, 1989* (Washington, DC: U.S. Government Printing Office, 1989). Data on NETINV are from table B-17. Data on CFSAL are calculated from table B-2. Data on RINT are calculated as the yield on AAA corporate bonds (table B-71) minus the percent change in the Consumer Price Index for all items (table B-61).

13.16 A farmer has observed the following data on annual rainfall in inches (X) and the yield of her corn crop Y (in bushels per acre) over the past 5 years:

Y	X
300	16
340	20
330	18
400	24
380	22

(a) Estimate the regression of Y on X (i.e., $\hat{Y}_i = \hat{\alpha} + \hat{\beta} X_i$) and discuss the effect of rainfall on yield.

(b) Estimate the regression of X on Y (i.e., $\hat{X}_i = \hat{\gamma} + \hat{\delta} Y_i$) and determine the significance (or lack thereof) of the slope coefficient.

(c) How do you reconcile the results of parts (a) and (b)?

13.17 Consider a linear demand function

$$QD_i = \alpha + \beta P_i + \varepsilon_i$$

where QD_i is quantity demanded in year i and P_i is the commodity's price during that period.

(a) Suppose there is a demand shock, such that QD_i rises for a reason other than a change in P_i. What does this situation suggest regarding the value of ε_i?

(b) What would you expect to happen to P_i as a result of this change?

(c) Comment on the appropriateness of the assumptions of the CLRM in this situation.

13.18 Suppose that X represents the floor space of an apartment (in square feet) and Y represents its annual rent. You estimate the regression line

$$\hat{Y}_i = \hat{\alpha} + \hat{\beta} X_i$$

and find $\hat{\alpha} = 1000$ and $\hat{\beta} = 5$. If your data on Y were in terms of monthly instead of annual rent, what would the values of $\hat{\alpha}$ and $\hat{\beta}$ have been? How would your interpretation of these results compare with that of the original results? Provide both an algebraic and an intuitive explanation for your response. (*Hint:* You can determine the exact numerical values of the new $\hat{\alpha}$ and $\hat{\beta}$.)

*13.19 Testing whether $\hat{\beta}$ is significantly different from 0 is equivalent to testing whether R^2 is significantly greater than 0. Provide an intuitive as well as an algebraic justification for this statement. (*Hint:* Do not worry about the distribution theory relevant to testing the latter hypothesis.)

*13.20 Prove that $\hat{Y}_i = \bar{Y}$ when $X_i = \bar{X}$, where $\hat{Y}_i$ is based on the OLS estimate of the regression line.

*13.21 Suppose that the correct relationship between X and Y is

$$Y_i = \alpha + \beta X_i^2 + \varepsilon_i$$

but you mistakenly form the OLS estimated regression line

$$\hat{Y}_i = \hat{\alpha} + \hat{\beta} X_i$$

(a) Is $\hat{\beta}$ an unbiased estimator of β? Fully explain.

(b) Suggest an unbiased estimator of β, and prove that it is unbiased assuming that the remaining CLRM assumptions are valid.

*13.22 Prove that $\Sigma \hat{\varepsilon}_i X_i$ is always 0, where the $\hat{\varepsilon}_i$'s are the residuals from an OLS regression. (*Hint:* First prove that $\Sigma \hat{\varepsilon}_i (X_i - \bar{X})$ is always 0.)

APPENDIX 1 TO CHAPTER 13

In this appendix we derive the OLS estimators $\hat{\alpha}$ (of α) and $\hat{\beta}$ (of β). This derivation requires multivariate calculus and is therefore only accessible to those with background in that area.

The least squares criterion is to choose the estimated regression line so as to

$$\text{minimize } \Sigma(Y_i - \hat{Y}_i)^2$$

Since

$$\hat{Y}_i = \hat{\alpha} + \hat{\beta} X_i$$

we may restate the criterion as

$$\text{minimize } \Sigma(Y_i - \hat{\alpha} - \hat{\beta} X_i)^2$$

We minimize this result by choosing the values of $\hat{\alpha}$ and $\hat{\beta}$, the two "unknowns."

The procedure involves taking the partial derivatives (with respect to $\hat{\alpha}$ and $\hat{\beta}$) of the function to be minimized, setting these new functions equal to zero (thereby creating the "first-order conditions"), and then solving for $\hat{\alpha}$ and $\hat{\beta}$:

$$\frac{\partial \Sigma(Y_i - \hat{Y}_i)^2}{\partial \hat{\alpha}} = -2\Sigma(Y_i - \hat{\alpha} - \hat{\beta}X_i)$$

$$\frac{\partial \Sigma(Y_i - \hat{Y}_i)^2}{\partial \hat{\beta}} = -2\Sigma(Y_i - \hat{\alpha} - \hat{\beta}X_i)X_i$$

We set these two equations equal to zero, and then solve for $\hat{\alpha}$ and $\hat{\beta}$:

$$\Sigma(Y_i - \hat{\alpha} - \hat{\beta}X_i) = 0 \qquad (1A13.1)$$

$$\Sigma(Y_i - \hat{\alpha} - \hat{\beta}X_i)X_i = 0 \qquad (1A13.2)$$

Equation (1A13.1) can be solved for $\hat{\alpha}$:

$$\Sigma Y_i - n\hat{\alpha} - \hat{\beta}\Sigma X_i = 0$$

$$n\hat{\alpha} = \Sigma Y_i - \hat{\beta}\Sigma X_i$$

$$\hat{\alpha} = \bar{Y} - \hat{\beta}\bar{X}$$

which is the OLS estimator! To derive $\hat{\beta}$, we substitute this expression for $\hat{\alpha}$ into Eq. (1A13.2) and solve for $\hat{\beta}$:

$$\Sigma(Y_i - \bar{Y} + \hat{\beta}\bar{X} - \hat{\beta}X_i)X_i = 0$$

$$\Sigma[(Y_i - \bar{Y}) - \hat{\beta}(X_i - \bar{X})]X_i = 0$$

$$\Sigma[(Y_i - \bar{Y})X_i - \hat{\beta}(X_i - \bar{X})X_i] = 0$$

$$\Sigma(Y_i - \bar{Y})X_i - \hat{\beta}\Sigma(X_i - \bar{X})X_i = 0$$

and thus

$$\hat{\beta} = \frac{\Sigma(Y_i - \bar{Y})X_i}{\Sigma(X_i - \bar{X})X_i}$$

Since $\Sigma(X_i - \bar{X})X_i = \Sigma(X_i - \bar{X})^2$ and $\Sigma(Y_i - \bar{Y})X_i = \Sigma(X_i - \bar{X})Y_i$ (see note 7 for a proof of the former assertion; the latter follows from the same logic), we then have

$$\hat{\beta} = \frac{\Sigma(X_i - \bar{X})Y_i}{\Sigma(X_i - \bar{X})^2}$$

which is the OLS estimator!

To show formally that these values minimize the function of interest, we would need to consider the second-order conditions; we omit that discussion here, merely stating that $\hat{\alpha}$ and $\hat{\beta}$ are the true minimizing values of the function. Intuitively, there is no maximum value of $\Sigma(Y_i - \hat{Y}_i)^2$ because the value of this function can always be increased by moving the estimated regression line farther away from the observations. Thus, the above solution must be a minimum (remember that the function of interest is quadratic and not of a higher order).

APPENDIX 2 TO CHAPTER 13

In this appendix we provide the justification for Eq. (13.9), which summarizes the distribution of $\hat{\alpha}$ under the assumptions of the CLRM.

By definition,

$$\hat{\alpha} = \bar{Y} - \hat{\beta}\bar{X}$$

$$= \frac{1}{n}\Sigma Y_i - \bar{X}(\beta + \Sigma c_i \varepsilon_i)$$

where

$$c_i = \frac{X_i - \bar{X}}{k} = \frac{X_i - \bar{X}}{\Sigma(X_i - \bar{X})^2}$$

as defined earlier in this chapter, and $\hat{\beta} = \beta + \Sigma c_i \varepsilon_i$, as in Eq. (13.7). Since

$$Y_i = \alpha + \beta X_i + \varepsilon_i$$

we can say that

$$\hat{\alpha} = \frac{1}{n}\Sigma(\alpha + \beta X_i + \varepsilon_i) - \bar{X}\beta - \Sigma(\bar{X}c_i)\varepsilon_i$$

$$= \alpha + \beta\bar{X} + \frac{1}{n}\Sigma\varepsilon_i - \bar{X}\beta - \Sigma(\bar{X}c_i)\varepsilon_i$$

$$= \alpha + \Sigma\left(\frac{1}{n}\right)\varepsilon_i - \Sigma(\bar{X}c_i)\varepsilon_i$$

The two sums involving ε_i can be combined to form

$$\hat{\alpha} = \alpha + \Sigma\left[\frac{1}{n} - \bar{X}c_i\right]\varepsilon_i \qquad (2A13.1)$$

This equation suggests that $\hat{\alpha}$ is a linear function of the ε_i's; notice that the $[(1/n) - \bar{X}c_i]$'s are a set of fixed numbers, given assumption 2 of the CLRM.

Since the ε_i's are normal (assumption 6) and independent (assumption 5), and since $\hat{\alpha}$ is a linear function of the ε_i's, we know that $\hat{\alpha}$ has a normal distribution. To find its mean, we say

$$E(\hat{\alpha}) = \alpha + \sum \left[\frac{1}{n} - \bar{X}c_i\right]E(\varepsilon_i) = \alpha$$

since $E(\varepsilon_i) = 0$ for all i (assumption 3).

To find var($\hat{\alpha}$), we note that since the ε_i's are a set of mutually independent random variables (assumption 5), we can derive the variance of $\hat{\alpha}$ using what we learned about linear functions of many random variables in Chapter 8:

$$\text{var}(\hat{\alpha}) = \sum \left[\frac{1}{n} - \bar{X}c_i\right]^2 \text{var}(\varepsilon_i)$$

$$= \sigma^2 \sum \left[\frac{1}{n} - \bar{X}c_i\right]^2$$

since var(ε_i) = σ^2 for all i (assumption 4). Note that

$$\sum \left[\frac{1}{n} - \bar{X}c_i\right]^2 = \sum \left[\frac{1}{n^2} - \left(\frac{2\bar{X}}{n}\right)c_i + \bar{X}^2 c_i^2\right]$$

$$= \frac{1}{n} - \left(\frac{2\bar{X}}{n}\right)\Sigma c_i + \bar{X}^2 \Sigma c_i^2$$

$$= \frac{1}{n} + \frac{\bar{X}^2}{\Sigma(X_i - \bar{X})^2}$$

since $\Sigma c_i = 0$ and $\Sigma c_i^2 = 1/\Sigma(X_i - \bar{X})^2$, as shown earlier in this chapter. Thus

$$\text{var}(\hat{\alpha}) = \sigma^2 \left[\frac{1}{n} + \frac{\bar{X}^2}{\Sigma(X_i - \bar{X})^2}\right]$$

as asserted earlier.

APPENDIX 3 TO CHAPTER 13

DATA FOR THE CONSUMPTION FUNCTION EXAMPLE

Year	Disposable income (billions of 1982 $)	Consumption (billions of 1982 $)	Money supply M_1 (billions of current $)	Consumption price deflator (1982 = 100)
1959	1067.2	979.4	140.0	32.3
1960	1091.1	1005.1	140.7	32.9
1961	1123.2	1025.2	145.2	33.3
1962	1170.2	1069.0	147.9	33.9
1963	1207.3	1108.4	153.4	34.4
1964	1291.0	1170.6	160.4	35.0
1965	1365.7	1236.4	167.9	35.6
1966	1431.3	1298.9	172.1	36.7
1967	1493.2	1337.7	183.3	37.6
1968	1551.3	1405.9	197.5	39.3
1969	1599.8	1456.7	204.0	41.0
1970	1668.1	1492.0	214.5	42.9
1971	1728.4	1538.8	228.4	44.9
1972	1797.4	1621.9	249.4	46.7
1973	1916.3	1689.6	263.0	49.6
1974	1896.6	1674.0	274.4	54.8
1975	1931.7	1711.9	287.6	59.2
1976	2001.0	1803.9	306.5	62.6
1977	2066.6	1883.8	331.4	66.7
1978	2167.4	1961.0	358.7	71.6
1979	2212.6	2004.4	386.1	78.2
1980	2214.3	2000.4	412.2	86.6
1981	2248.6	2024.2	439.1	94.6
1982	2261.5	2050.7	476.4	100.0
1983	2331.9	2146.0	522.1	104.1
1984	2469.8	2249.3	551.9	108.1
1985	2542.8	2354.8	620.1	111.6
1986	2640.9	2455.2	725.4	114.3
1987	2686.3	2521.0	750.8	119.5

Sources: Disposable income (table B-27); consumption (table B-2), M_1 (table B-67); consumption price deflator (table B-3); all from the *Economic Report of the President, 1989* (Washington, DC: U.S. Government Printing Office, 1989).

APPENDIX 4 TO CHAPTER 13

DATA FOR THE POVERTY RATE EXAMPLE

State	White (%)	High school grad. (%)	Unemp. rate (%)	AFDC ($)	Poverty rate (%)
Maine	98.8	70.2	7.5	332	13.0
N. Hamp.	98.9	73.7	4.8	382	8.5
Vermont	99.2	73.4	6.3	524	12.1
Mass.	94.3	74.1	5.0	419	9.6
Rhode Is.	95.5	64.2	7.0	389	10.3
Conn.	91.7	71.9	4.6	446	8.0
New York	81.6	68.2	7.1	476	13.4
New Jersey	85.4	69.2	6.6	386	9.5
Penn.	90.8	67.6	7.4	373	10.5
Ohio	89.8	68.8	8.0	327	10.3
Indiana	92.1	68.2	7.8	275	9.7
Illinois	83.1	68.3	7.1	315	11.0
Michigan	86.5	69.8	10.9	480	10.4
Wisconsin	95.4	71.7	6.6	492	8.7
Minnesota	97.2	74.9	5.4	454	9.5
Iowa	97.8	73.2	5.0	419	10.1
Missouri	89.6	65.7	6.8	270	12.2
N. Dakota	96.8	70.2	5.2	389	12.6
S. Dakota	94.5	70.7	4.8	361	16.9
Nebraska	95.7	75.1	3.6	370	10.7
Kansas	92.7	74.6	3.9	350	10.1
Delaware	84.3	70.7	6.2	287	11.9
Maryland	77.0	69.2	5.6	294	9.8
DC	31.4	70.1	6.6	253	18.6
Virginia	80.6	64.6	4.7	263	11.8
W. Virginia	96.3	58.8	8.4	249	15.0
N. Carolina	78.1	58.3	5.3	210	14.8
S. Carolina	71.9	57.5	5.8	142	16.6
Georgia	74.8	59.1	5.7	170	16.6
Florida	86.6	67.9	5.0	230	13.5
Kentucky	92.7	56.2	8.3	235	17.6

DATA FOR THE POVERTY RATE EXAMPLE

State	White (%)	High school grad. (%)	Unemp. rate (%)	AFDC ($)	Poverty rate (%)
Tennessee	85.1	59.0	7.3	148	16.5
Alabama	76.4	59.3	7.4	148	18.9
Mississippi	68.3	57.6	7.0	120	23.9
Arkansas	85.0	58.0	6.9	188	19.0
Louisiana	72.1	60.8	5.9	187	18.6
Oklahoma	87.6	67.8	4.0	349	13.4
Texas	80.8	64.1	3.9	140	14.7
Montana	95.2	75.4	8.1	331	12.3
Idaho	96.1	74.2	7.9	366	12.6
Wyoming	95.6	78.1	4.1	340	7.9
Colorado	90.3	78.7	4.9	327	10.1
N. Mexico	78.0	70.0	6.9	242	17.6
Arizona	85.5	73.0	6.1	240	13.2
Utah	95.0	80.0	5.4	389	10.3
Nevada	89.1	75.6	5.8	297	8.7
Washington	92.4	77.7	7.2	483	9.8
Oregon	95.3	75.8	8.2	456	10.7
California	78.7	73.7	6.4	487	11.4
Alaska	79.5	81.8	8.7	450	10.7
Hawaii	34.4	75.9	4.2	546	9.9

Sources: Percent white (for persons 18 years +) is from U.S. Bureau of the Census, *1980 Census of Population, Vol. 1: Characteristics of the Population, Chapter B: General Population Characteristics,* Part 1: U.S. Summary (Washington, DC: U.S. Government Printing Office, 1983), table 67. Percent (of persons 18 years +) who are high school graduates is from U.S. Bureau of the Census, *1980 Census of Population, Vol. 1: Characteristics of the Population, Chapter D: Detailed Population Characteristics,* parts 2 through 52 (Washington, DC: U.S. Government Printing Office, 1983), table 237. Unemployment rate (persons 16 years +) is from U.S. Bureau of the Census, *1980 Census of Population, Vol. 1: Characteristics of the Population, Chapter C: General Social and Economic Characteristics,* part 1: U.S. Summary (Washington, DC: U.S. Government Printing Office, 1983), table 240. The AFDC variable is the monthly payment to a family with one adult and three children and no countable income as of 9/30/79 and is from U.S. Department of Health and Human Services, *Characteristics of State Plans for Aid to Families with Dependent Children,* 1980 ed. (Washington, DC: U.S. Government Printing Office, 1980), pp. 235–236. The poverty rate (for 1979) is from U.S. Bureau of the Census, *State and Metropolitan Area Data Book,* 1986 (Washington, DC: U.S. Government Printing Office, 1986), p. 561.

DISCUSSION QUESTIONS FOR CHAPTER 13

1. Suppose that you are investigating the relationship between the labor-hours employed by a firm and the output of the firm. Comment on the appropriateness of the CLRM for investigating and attempting to quantify that relationship. (Assume that we are in a short-run situation wherein labor is the only variable factor of production.)

2. Is two-way causality more likely to be a problem in an experimental study or in an observational study?

Chapter 14

Multiple Regression

In Chapter 13 we discussed regression analyses involving two variables, one of which was dependent (Y) and one of which was independent (X). In this chapter we extend this discussion to the case of one dependent variable and two or more independent variables ($X_1, X_2, \ldots, X_k$); this case is referred to as the *multiple regression* framework. Many of the insights from the two-variable model carry over into this multiple-variable framework, although some new interpretive issues arise.

Our approach here is not to investigate all the technical issues involved in multiple regression, inasmuch as this presentation often requires mathematics (especially matrix algebra) beyond the scope of this book. We will rather focus on issues involved in carrying out and interpreting the results of multiple regressions. Appendixes 2 and 3 contain further developments of the general regression framework.

MOTIVATION

In the two-variable regression framework we are, of course, restricted in our modeling of the relationship under investigation. For example, in modeling the demand of person i for a particular commodity we could say

$$q_i^d = \beta_0 + \beta_1 p_i$$

where q_i^d is the quantity demanded of the commodity in question by person i, p_i is the price faced by person i, and β_0 and β_1 are parameters. (We switch from α and β notation to β_0 and β_1 and we shall eventually add β_2, β_3, . . . as part of the multiple regression.) However, other factors that affect demand should perhaps be incorporated; income is an example. A more complete model of consumer demand is

$$q_i^d = \beta_0 + \beta_1 p_i + \beta_2 I_i$$

where we have added one additional variable (I = income). Of course, even once p_i and I_i are known, there will be random variation in q_i^d. That is, there is still variation in q_i^d even when we restrict ourselves to a group of individuals facing the same price and having the same income. To incorporate this variation we add the error term ε_i:

$$q_i^d = \beta_0 + \beta_1 p_i + \beta_2 I_i + \varepsilon_i$$

We would expect $\beta_1 < 0$ and $\beta_2 > 0$ (as long as the commodity is not an inferior good).

Many of the issues involved in interpreting this framework are the same as in the two-variable context; there is, however, a new interpretive issue that is of key importance, and we shall now pursue it.

Suppose that p_i rises by one unit. What change would we then expect in q_i^d? In order to answer this question, we must know what if any change in I_i occurs; if p and I were related, the change in p_i would be associated with a change in I_i which would further affect q_i^d.

If, however, I_i were constrained to remain constant, the expected change in q_i^d (for this one unit increase in p_i) would be β_1 units. We then say that β_1 is the expected change in q^d for a one-unit change in p, *holding I constant.* Similarly, β_2 is the expected change in q^d for a one-unit change in I, *holding p constant.*

What we have here in essence is a way of separating the marginal effects of the two variables; β_1 represents the independent effect of p on q^d, and β_2 represents the independent effect of I on q^d. Note

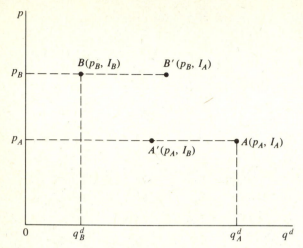

Figure 14.1   Quantity demanded, price, and income.

that β_1 and β_2 have the interpretation as the effects on the dependent variable of *ceteris paribus* changes in p and I (respectively).[1] But we typically view a demand function exactly in this way. Multiple regression is a natural technique for economic problems because so much of economic theory is in the context of *ceteris paribus* changes, which are what a multiple regression is designed for investigating. With a multiple regression we then investigate the effects on the dependent variable of a change in an X variable holding all the other X variables constant. That is, we investigate the relationship between Y and an X variable, controlling for the effects of the other X variables included in the model.

   To illustrate the link between the multiple regression framework and economic theory, consider Figure 14.1. Individual A faces price p_A and has income I_A and sits at point A consuming q_A^d of the good. Person B faces price p_B and has income I_B and sits at point B consuming q_B^d of the good. The price and income for each point is contained in the parentheses after the point's label. (For simplicity, we assume that these individuals are identical in all other respects, including their tastes, so that any differences in quantity demanded are due to differences in price or income.) Suppose that $I_A > I_B$. Note that points A and B cannot be connected to form a demand curve because they are based on different incomes; demand curves are drawn holding all factors other than price constant.

[1] *Ceteris paribus* means "all other things constant."

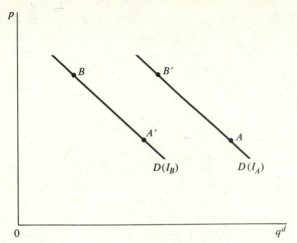

Figure 14.2 Demand curves based on different incomes.

If person A had the smaller income I_B but still faced the price p_A then he or she would be at A', presuming this good is a normal good. If person B had the larger income I_A but still faced the price p_B, then he or she would be at B'. Points B' and A are then points along the same demand curve (income is held constant at I_A), and points B and A' are points along another demand curve (income is held constant at I_B). The difference between the two demand curves is due to the income difference; these two demand curves are shown in Figure 14.2.

The slope of each of these demand curves is β_1, which represents the effect (on average) on quantity demanded of a change in price, holding income constant. The size of the horizontal shift in the demand curve when income changes is determined by β_2.[2]

The demand curves shown in Figure 14.2 are analogous to what we uncover via a multiple regression because they represent the relationship between quantity demanded and price, holding income constant. Simply connecting the two points A and B (the two observations) would be analogous to a two-variable regression (quantity demanded regressed on the single independent variable price) in which income is not held constant.

---

[2] We have actually oversimplified things a bit here in order to facilitate the exposition of the interpretation of the parameters. In fact, the presence of the error term ε suggests that some individuals locate themselves at positions "off the demand curve." Perhaps the reason is that such individuals have different preferences for the good in question.

To further illustrate the usefulness of this framework, let us consider a fairly simple example in which we are interested in determining the extent to which a student's performance in college (measured by his or her grade point average) is associated with two variables—SAT scores and high school grade point average. We might model this situation as

$$\text{CGPA}_i = \beta_0 + \beta_1 \text{SAT}_i + \beta_2 \text{HGPA}_i + \varepsilon_i$$

where CGPA_i is person i's college grade point average, SAT_i is person i's combined verbal and quantitative score on the SAT test, and HGPA_i is person i's high school grade point average. β_1 represents the relationship (on average) between college GPA and SAT scores, holding high school GPA constant. A positive value of β_1 suggests that in comparing students with equal high school GPAs, those with higher SAT scores do on average better in college; the bigger the value of β_1, the larger this difference.

We generally expect students with high SATs to have a tendency to have good high school grades. Presumably, both high SATs and good high school grades are associated with high college GPAs. But β_1 illustrates the link (on average) between CGPA and SAT, holding HGPA constant. In a two-variable regression of CGPA on SAT, however, we would not hold HGPA constant, and the coefficient of SAT would then be likely to overstate the marginal impact of SATs on college performance because it would include the associated effects of HGPA.[3]

In a statistical analysis we are often concerned with separating the effects of variables in an effort to determine their marginal impacts. Multiple regression provides us with a general method for this determination. We now turn to a more general discussion of this framework.

GENERAL SPECIFICATION OF THE MULTIPLE REGRESSION MODEL

In general, we have a single dependent variable Y and a set of k independent variables $X_1, X_2, \ldots, X_k$, and we presume that the

---

[3] Notice that we have implicitly assumed a linear framework in this problem, which may not be appropriate. It is, however, one of the typical assumptions behind a multiple regression, as will be discussed below.

causality goes from the X's to Y but not the other way (the assumptions behind this model will be discussed in more detail later). We express the model as

$$Y_i = \beta_0 + \beta_1 X_{1i} + \beta_2 X_{2i} + \cdots + \beta_k X_{ki} + \varepsilon_i \qquad (14.1)$$

where X_{ji} is the value of the ith observation on the variable X_j. The $(k + 1)$ β's are the parameters of the model. Equation (14.1) suggests that

$$E(Y_i \mid X_{1i}, X_{2i}, \ldots, X_{ki}) = \beta_0 + \beta_1 X_{1i} + \beta_2 X_{2i} + \cdots + \beta_k X_{ki}$$

$$(14.2)$$

where $E(Y_i \mid X_{1i}, X_{2i}, \ldots, X_{ki})$ is the expected value of Y_i conditional on the values of the X's. In general, the interpretation of the β's is as marginal rates of change; for example, β_1 represents the expected change in Y for a one-unit change in X_1, holding X_2 through X_k fixed.

At this point it is helpful to present a graphical interpretation of the model. In Figure 14.3 we present a graph characterizing the problem for the three-variable model

$$Y_i = \beta_0 + \beta_1 X_{1i} + \beta_2 X_{2i} + \varepsilon_i$$

(we restrict our geometric presentation to this case because we cannot graph more than three dimensions). The Y and X_1 axes should be viewed as on the page, and the X_2 axis should be viewed as coming off the page at right angles to both of the other axes. We have marked several observed points; for each point how far toward the top of the page it is gives us the Y coordinate, how far toward the right edge of the page it is gives us the X_1 coordinate, and how far off the page it is gives us the X_2 coordinate. These points might be viewed as specks of dust suspended in the air over the page.

Let us assume for the sake of argument that for each observation all three variables take on nonnegative values; this assumption allows us to restrict our consideration to one of the eight octants of this coordinate system. Let us also presume that $\beta_1 > 0$ and $\beta_2 < 0$. The fact that β_1 exceeds 0 suggests that as X_1 rises (we move farther to the right), the value of Y on average gets larger (the points are on average closer to the top of the page). The fact that β_2 is less than 0 suggests that as X_2 rises (the points are farther off the page), the value of Y gets on average smaller (the points are on average farther away from the top of the page).

The "regression plane" (note that it is not a line) is pictured

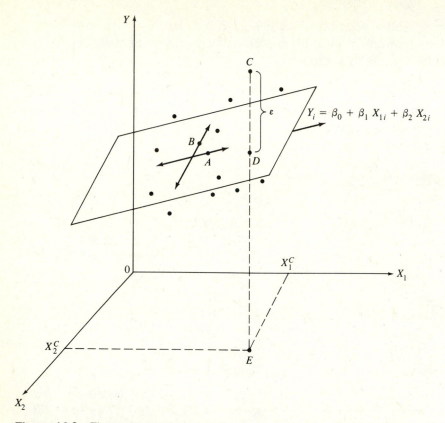

Figure 14.3 The multiple regression plane, three-variable case.

in the graph. The tilt of the plane in the X_1 direction illustrates the relationship on average between changes in X_1 and changes in Y (see the vector containing point A), holding X_2 constant. The tilt of the plane in the X_2 direction illustrates the relationship on average between changes in X_2 and changes in Y (see the vector containing point B), holding X_1 constant. Let us now focus on point C, which is one of our observations.

The height of C represents the actual value of Y that we observe. The values of X_1 and X_2 associated with this point may be determined by projecting C downward to the X_1X_2 plane (point E) and reading the coordinates off the X_1 and X_2 axes (thus X_1^C and X_2^C). The point directly below C on the regression plane (D) has a height equal to what we expect Y to be given X_1^C and X_2^C. The distance between C and D then represents the value of the error term ε.

We can further explore this model by considering a projection of the regression plane into a two-dimensional framework. Let us suppose that X_2 is fixed at X_2^*, and we investigate the relationship between Y and X_1 given that $X_2 = X_2^*$. This situation is depicted in Figure 14.4.

Given that $X_2 = X_2^*$, the intercept term is $\beta_0 + \beta_2 X_2^*$, which is the expected value of Y when $X_1 = 0$. Holding X_2 fixed, each unit change in X_1 is associated with an expected β_1 unit change in Y. This result suggests that the line $Y = (\beta_0 + \beta_2 X_2^*) + \beta_1 X_1$ describes the relationship on average between Y and X_1 given that $X_2 = X_2^*$.

Suppose now that X_2 increases to X_2^{**}. Presuming, as before, that $\beta_2 < 0$, the line would shift downward to

$$Y = (\beta_0 + \beta_2 X_2^{**}) + \beta_1 X_1$$

as shown. Note in both cases that β_1 represents the expected change in Y for a one-unit change in X_1 holding X_2 fixed. That is, β_1 is the common slope of the lines drawn in Figure 14.4, which represents the relationship on average between Y and X_1, holding X_2 fixed at some predetermined level.

As an example, suppose that Y represents aggregate consumption, X_1 represents disposable income, and X_2 represents an interest rate. Clearly $\beta_1 > 0$. It is also reasonable to assume that $\beta_2 < 0$ since we would expect consumption that is financed with borrowed funds to decrease when the interest rate rises. The lines in Figure 14.4 are then essentially two separate consumption functions drawn based on different interest rates; the change in the interest rate causes a shift in the function. Notice the similarity between the statistical

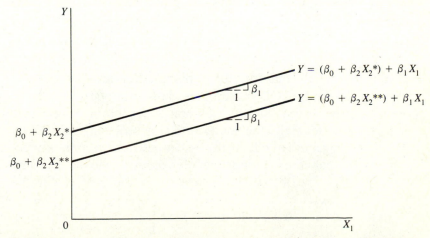

Figure 14.4 A two-dimensional representation of a three-variable relationship.

interpretation of the multiple regression and the way we generally think about shifts in a consumption function in a theoretical context.

The multiple regression framework can be extended to the case of k independent variables, but once $k > 2$ we cannot represent it graphically. Nevertheless, the understanding of the case just described should facilitate an intuitive understanding of the more complicated multidimensional situations we often encounter.

The parameters β_0, β_1, ..., β_k are, of course, generally unknown; they represent a population relationship, but we only observe a sample of observations. The true regression plane of Figure 14.3 must then be estimated. We now turn to a discussion of this problem.

ESTIMATION

The multiple regression model is given by Eq. (14.1). To construct the estimated regression relationship we want to construct an estimate that "fits the data well"; that is, we want to construct the estimate

$$\hat{Y}_i = \hat{\beta}_0 + \hat{\beta}_1 X_{1i} + \hat{\beta}_2 X_{2i} + \cdots + \hat{\beta}_k X_{ki}$$

Following the same general strategy as in the two-variable model, we want to make the residuals

$$\hat{\varepsilon}_i = Y_i - \hat{Y}_i$$

small. The least squares criterion is to minimize

$$\sum_{i=1}^{n} (Y_i - \hat{Y}_i)^2 = \sum_{i=1}^{n} (Y_i - \hat{\beta}_0 - \hat{\beta}_1 X_{1i} - \hat{\beta}_2 X_{2i} - \cdots - \hat{\beta}_k X_{ki})^2$$

The values of the $\hat{\beta}$'s that minimize this function are then our Ordinary Least Squares (OLS) estimates. This multivariate calculus problem can be solved using the general approach of Appendix 1 to Chapter 13. Generally the solution to this problem is expressed using matrix algebra (of which this text does not presume knowledge). The actual computations, which become quite burdensome even with only a few observations, are easily accomplished with the aid of a computer.

In the case of two X's (X_1 and X_2) the formulas are as follows:

$$\hat{\beta}_0 = \bar{Y} - \hat{\beta}_1 \bar{X}_1 - \hat{\beta}_2 \bar{X}_2$$

$$\hat{\beta}_1 = \frac{\widehat{\text{cov}(X_1, Y)}s^2_{X_2} - \widehat{\text{cov}(X_2, Y)}\widehat{\text{cov}(X_1, X_2)}}{s^2_{X_1}s^2_{X_2} - (\widehat{\text{cov}(X_1, X_2)})^2}$$

$$\hat{\beta}_2 = \frac{\widehat{\text{cov}(X_2, Y)}s^2_{X_1} - \widehat{\text{cov}(X_1, Y)}\widehat{\text{cov}(X_1, X_2)}}{s^2_{X_1}s^2_{X_2} - (\widehat{\text{cov}(X_1, X_2)})^2}$$

where $s^2_{X_1}$ and $s^2_{X_2}$ are the sample variances of X_1 and X_2, respectively, and the $\widehat{\text{cov}}$ values are the sample covariances as indicated (see Chapter 8). Note how the parameter estimates depend not only on the relationships between the X's and Y but also on the relationships between the individual X's. (The derivation of the OLS estimates in the three-variable model is discussed in more detail in Appendix 1 to this chapter.)

In this text we do not present general formulas for the OLS estimates (i.e., the $\hat{\beta}$'s); we will presume that they are calculated using statistical software on a computer. A more theoretical treatment of the model would be contained in a more advanced course in econometrics. But most of our intuitive understanding developed in the context of the two-variable regression model carries over into the multiple regression framework.

We now turn to an elaboration of the model and in particular of the assumptions that underlie it. In essence, we extend the Classical Linear Regression Model as developed in Chapter 13 to the case of several X's.

THE CLASSICAL LINEAR REGRESSION MODEL

As we did in the case of the two-variable regression model, we shall now lay out a set of six assumptions that form the Classical Linear Regression Model (CLRM) for the multiple-variable case; we shall also discuss the properties of the OLS estimators under these assumptions. This approach is best seen as an extension of what we earlier developed in Chapter 13 to the case of more than one X variable.

Assumption 1: $Y_i = \beta_0 + \beta_1 X_{1i} + \cdots + \beta_k X_{ki} + \varepsilon_i$

This asssumption is a restatement of the model and of the presumed linear relationship between the X's and Y.

Assumption 2:   The observations on each X_j ($j = 1, \ldots, k$) are a set of fixed numbers and not random variables, and no perfect linear relationship exists between any X and any one or more of the other X's (i.e., there is no *perfect multicollinearity*).

This assumption is really two separate ones, but it is typically stated as one assumption to make neater the parallels between the two- and multiple-variable cases. As in the case of the two-variable model, the assumption that (for each X) the values are determined by a process other than a random process suggests that the values of these explanatory variables are in principle such that the researcher could create another data set containing the exact same X's (presumably by redoing the experiment). Paralleling the discussion in Chapter 13, an alternative assumption is more relevant to the observational studies we typically encounter in economics and leads to the same results (albeit in a more complicated way). This alternative assumption is that each X_j ($j = 1, \ldots, k$) is independent of the error term ε.

The second part of assumption 2 is new; after all, it would be irrelevant in the case in which we had only one X variable. To motivate the need for this assumption, note that if it were violated, it would be implied that the X that can be expressed as a perfect linear function of one or more of the other X's contains no new information contributing to our explanation of the behavior of Y, and it should therefore perhaps be excluded. To demonstrate this point more clearly, let us consider a simple example in the case of two X's; we choose this framework as it lends itself to a simple geometric interpretation. The principle is, however, generalizable.

Suppose that

$$Y_i = \beta_0 + \beta_1 X_{1i} + \beta_2 X_{2i} + \varepsilon_i$$

but

$$X_{2i} = \alpha_0 + \alpha_1 X_{1i}$$

This last expression tells us that, for each observation, X_2 is the same linear function of X_1; that is, a perfect linear relationship exists between the two variables. (There is no error term in this expression because the relationship is perfect, and the correlation coefficient between X_2 and X_1 would be one in absolute value.)

In Figure 14.5, we represent this situation geometrically. The line A in the $X_1 X_2$ plane is a line above which all our observations

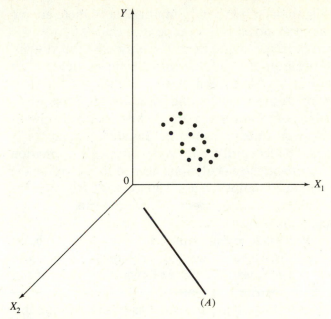

Figure 14.5 Perfect multicollinearity.

lie. Since $X_{2i} = \alpha_0 + \alpha_1 X_{1i}$ for each observation i, the projections of the observations into the $X_1 X_2$ plane (where $Y = 0$) are collinear. That is, if we were to take each observation and project it onto the $X_1 X_2$ plane, each of these projections would be along A. If we had a sheet of glass rising vertically from A, each of our observations (once we incorporate the Y value) would be a speck of dust along that sheet of glass.

We can use these observations to determine how Y on average changes as we move to the southeast (or northwest) along A. But to place a plane, we must know its tilt in *two* directions; a plane can be tilted along its width dimension and along its length dimension. In the problem at hand we would need to be able to determine the tilt of the regression plane not only as we move along A but also as we move along a line perpendicular to A. We cannot determine the latter, however, because we have *no* information as to how Y changes as we move off A. For any plane we could fit to incorporate movement along A, the tilt in the other direction would be completely arbitrary. The regression plane cannot then be fully determined; an infinite number of planes fit the data equally well.

This situation is similar to that of trying to determine two unknowns with only one equation. In general, there must be as many unique dimensions of variability in our data as there are parameters; otherwise the model is not properly specified.[4]

As a somewhat trivial example, if Y represented individuals' weights, X_1 their height in feet, and X_2 their height in inches, then $X_{2i} = 12X_{1i}$ and we have violated assumption 2. Clearly, X_2 contains no new information in this problem and should be omitted. More complicated versions of this problem are, however, possible; if, for example, $X_{1i} = X_{2i} + X_{3i}$, then assumption 2 is still violated.[5]

The problem of perfect multicollinearity causes the estimation procedure to break down; the calculations needed to determine the OLS estimators cannot be completed. But suppose we have *imperfect multicollinearity;* that is, what happens if there is a close, but not perfect, linear relationship between two or more of the X's? In this case, the OLS estimators can be determined, but they are less reliable (they have a larger variance) the closer the relationship is between the two (or more) X's in question. For a further discussion of this question, see an econometrics textbook.

Assumptions 3 through 6 are the same as in the two-variable model. Assumptions 1 and 2 suggest that the sole source of *random* variation in Y_i is due to ε_i, and assumptions 3 through 6 describe ε_i as a random variable. We here merely restate these assumptions without comment.

Assumption 3: $E(\varepsilon_i) = 0$ for each i.

Assumption 4:   var (ε_i) is the same for each i (we shall call this common variance σ^2).

Assumption 5: ε_i's are a set of mutually independent random variables.

Assumption 6: ε_i's are normally distributed.

---

[4] In a formal discussion of this problem, we would find that the process of solving for the OLS estimators breaks down; the OLS estimators are solved for by inverting a matrix that is in fact not invertible.

[5] If however, there is a perfect *nonlinear* relationship between two or more X's, we do not have perfect multicollinearity. In Figure 14.5, in this case, the "vector" A would not really be a vector (which is linear) at all but would rather be some sort of curve. Since the estimated regression relationship is a plane, we would then have all the dimensions of variation necessary to estimate the regression relationship.

From the six CLRM assumptions the general properties of the OLS estimators can be derived. Since this procedure in general requires matrix algebra, we shall here merely summarize the important results. The intuition behind much of this can be gained from our Chapter 13 discussion of the two-variable model.[6]

PROPERTIES OF OLS ESTIMATORS IN THE CLRM

Under the CLRM assumptions, for each $\hat{\beta}_j$ ($j = 0, 1, \ldots, k$) it will be the case that

$$\hat{\beta}_j \sim N[\beta_j, \text{var}(\hat{\beta}_j)]$$

That is, each parameter estimate is a random variable, which is normally distributed with a mean equal to the parameter of interest and a variance that we are not specifying in detail (its general specification requires matrix algebra) but depends on σ^2 [$= \text{var}(\varepsilon)$]. This suggests that each $\hat{\beta}$ is an unbiased estimator and that confidence intervals and hypothesis tests can be accomplished in the general framework used earlier in this text (in particular, using the t distribution since σ^2 will need to be estimated).

Since var ($\hat{\beta}_j$) depends on σ^2 for each j, and since σ^2 is unknown (since we have no observations on ε_i; see Chapter 13 for a discussion of this problem in the two-variable model), we must then form an estimate of σ^2, use this estimate to form an estimated variance of each $\hat{\beta}$, and then base our inferences (confidence intervals and hypothesis tests) upon these estimates and the t distribution. Notice that this procedure is the same one we followed in the two-variable model.

It turns out that if we form[7]

$$s_\varepsilon^2 = \frac{1}{n - k - 1} \sum_{i=1}^{n} (Y_i - \hat{Y}_i)^2 = \frac{1}{n - k - 1} \sum_{i=1}^{n} \hat{\varepsilon}_i^2$$

---

[6] As mentioned in Chapter 13, the above framework is not necessarily the only (or least restrictive) way of specifying the CLRM assumptions, but it is the one from which the results come most directly and easily.

[7] We divide by $(n - k - 1)$ in the s_ε^2 formula because this leads to $E(s_\varepsilon^2) = \sigma^2$ and convenient t distribution results. The intuition behind this result is that $(k + 1)$ observations are needed even to construct the estimated regression relationship, and thus it is only to the extent that n exceeds $(k + 1)$ that we can form s_ε^2; s_ε^2 is then based upon $n - (k + 1) = n - k - 1$ sources of variation. Note that this result is consistent with a degrees of freedom of $(n - k - 1)$ in the t distribution (see below).

(remember that k is the number of X variables and not the total number of variables) and then use this estimated variance of ε in place of the true σ^2 in the formulas for var $(\hat{\beta}_j)$ to form $s^2_{\hat{\beta}_j}$ for $j = 0, 1, \ldots, k$, then

$$\frac{\hat{\beta}_j - \beta_j}{s_{\hat{\beta}_j}} \sim t^{(n-k-1)} \tag{14.3}$$

for each j.[8] Equation (14.3) then becomes the basis for the formation of confidence intervals and hypothesis tests, following the general procedures discussed in the two-variable model. (We often refer to $s_{\hat{\beta}_j}$ as the estimated *standard error* of $\hat{\beta}_j$; statistical programs provide us with this value.)

Under the assumptions of the CLRM, the OLS estimators have the smallest possible variance of any unbiased estimators of the β's.[9] The OLS estimators are then both unbiased and efficient under these assumptions. This result is comparable to our earlier one in the two-variable model.

Let us now illustrate the calculation of a multiple regression and the process of inference within this model in the context of a consumption function example.

EXAMPLE 1: THE CONSUMPTION FUNCTION EXAMPLE REVISITED

In Appendix 3 to Chapter 13 we presented data relevant to the estimation of a consumption function for the United States from 1959 to 1987. In Chapter 13 we modeled consumption in a two-variable context as

$$C_i = \alpha + \beta YD_i + \varepsilon_i$$

where C_i represented consumption in year i and YD_i represented disposable income in year i, and both variables were measured in constant 1982 dollars. Clearly, however, there are factors other than

---

[8] Under the CLRM assumptions, $[(n - k - 1)s_e^2]/\sigma^2 \sim \chi^2_{(n-k-1)}$. This statement generalizes the result in the two-variable model ($k = 1$) and is a basis for Eq. (14.3). (See the appendix to Chapter 11 for a discussion of the relationship between the chi-square and t distributions.)

[9] The normality of the ε's is required for this efficiency result. If the ε's are not normal, then the OLS estimators are minimum-variance *linear* unbiased estimators of the β's.

disposable income that influence consumption; one possibility is that the real value of monetary balances influences consumption in a positive way (this is the so-called "real balance effect").[10] The simplest explanation for this is that real monetary balances are a form of wealth, and we should expect consumption to respond to changes in the stock variable wealth in addition to the flow variable income. In Appendix 3 to Chapter 13 we include the value of the M_1 measure of the money supply (in current dollars) and a price index (the price deflator for consumption with base year 1982) so that the real value of M_1 balances can be calculated as

$$[(M_1/\text{price deflator}) \times 100].$$

We can then form a more sophisticated consumption function

$$C_i = \beta_0 + \beta_1 YD_i + \beta_2 M_i + \varepsilon_i$$

where M_i is the value of M_1 in constant 1982 dollars in year i. We estimated this model and arrived at the following estimated regression relationship (the estimated standard errors are in parentheses under the estimated coefficients):

$$\hat{C}_i = -139.8 + .904 YD_i + .298 M_i$$
$$(58.6) \quad (.015) \qquad (.154)$$

The R^2 for this regression was .997. We interpret these results as follows. We estimate that a one-dollar increase in YD, holding M constant, is associated with (on average) a .904 dollar increase in C. We estimate that a one-dollar increase in M, holding YD constant, is associated with (on average) a .298-dollar increase in C. Of course, the $\hat{\beta}$'s are undoubtedly different from the true β's because of sampling variability. It is thus suggested that we should consider confidence intervals and hypothesis tests to pursue more sophisticated inferences.

We have 29 observations, and k (the number of X's) equals 2. Thus, the degrees of freedom for the t distribution is $n - k - 1 = 26$. We can build 95% confidence intervals for the three β's as follows. In general, the 95% confidence interval for β_j is

$$\hat{\beta}_j \pm t_{.025}^{(n-k-1)} s_{\hat{\beta}_j}$$

---

[10] As mentioned in an earlier example in this chapter, some measure of interest rates is another possible factor influencing consumption. For simplicity we focus here only on real balances.

For β_0,

$$\hat{\beta}_0 \pm t_{.025}^{(26)} s_{\hat{\beta}_0} = -139.8 \pm (2.06)(58.6)$$

$$= -139.8 \pm 120.7 = (-260.5, -19.1)$$

For β_1,

$$\hat{\beta}_1 \pm t_{.025}^{(26)} s_{\hat{\beta}_1} = .904 \pm (2.06)(.015) = .904 \pm .031 = (.873, .935)$$

and for β_2,

$$\hat{\beta}_2 \pm t_{.025}^{(26)} s_{\hat{\beta}_2} = .298 \pm (2.06)(.154) = .298 \pm .317 = (-.019, .615)$$

We now turn to hypothesis testing in this framework. Hypothesis tests follow the same procedure focusing on the prob-value that we used earlier. Equation (14.3) is the distribution theory upon which the prob-value calculation is based. For illustration purposes let us consider hypothesis tests on the two slope parameters β_1 and β_2. β_1 represents the expected change in consumption for a one-dollar change in disposable income holding real balances fixed. We can rule out from the theory the possibility $\beta_1 < 0$; therefore, to verify the theory of consumption we perform a one-sided hypothesis test. Our problem is to find the prob-value for and test the hypothesis

$$H_0: \beta_1 = 0$$

$$H_1: \beta_1 > 0$$

at the .05 level.

As always, the prob-value is the answer to the following question: What is the probability of getting a value of our estimator as extreme as in our sample if H_0 is in fact true? In the problem at hand, this would be

$$\Pr(\hat{\beta}_1 \geq .904) = \Pr\left(\frac{\hat{\beta}_1 - \beta_1}{s_{\hat{\beta}_1}} \geq \frac{.904 - 0}{.015}\right)$$

$$= \Pr(t^{(26)} \geq 60.3)$$

Since $t_{.001}^{(26)} = 3.44$, the prob-value is below .001. Because the prob-value is less than .05, we reject the null hypothesis in favor of the alternative hypothesis at the .05 level, and we conclude that changes in disposable income (holding real balances constant) are on average associated with a change in consumption in the same direction.

For β_2, we can also say that the theory rules out the possibility that $\beta_2 < 0$. In order then to verify the effect of real monetary balances, we find the prob-value for and test the hypothesis

$$H_0: \beta_2 = 0$$

$$H_1: \beta_2 > 0$$

at the .05 level. The prob-value is

$$\Pr(\hat{\beta}_2 \geq .298) = \Pr\left(\frac{\hat{\beta}_2 - \beta_2}{s_{\hat{\beta}_2}} \geq \frac{.298 - 0}{.154}\right)$$

$$= \Pr(t^{(26)} \geq 1.94)$$

and thus

$$.05 > \text{prob-value} > .025$$

Since the prob-value is less than .05, we reject H_0 in favor of H_1 at the .05 level. We therefore conclude that increases in real balances (holding disposable income constant) are associated on average with increases in consumption.[11]

Notice how hypothesis tests and confidence intervals follow the same procedural framework as in the cases we previously considered. Two-sided hypothesis tests have the same relationship to one-sided tests as in the two-variable model considered earlier.

EXAMPLE 2: THE POVERTY RATE EXAMPLE REVISITED

In Chapter 13 we considered a two-variable regression investigating the relationship between the poverty rate (POV) and the unemployment rate (UN) in the 51 United States (and the District of Columbia). In Appendix 4 to Chapter 13 we presented the data on these variables as well as on three other variables: the percent of the population 18 years of age or older that is white (%WHITE), the

[11] In this hypothesis test we conclude that $\beta_2 > 0$. But notice that in our confidence interval for this parameter the value zero was within the interval. This apparent contradiction is not really a contradiction at all but rather is indicative of the distinction between a one-sided and a two-sided hypothesis test. Seeing whether zero is within the confidence interval, which is analogous to a two-sided hypothesis test, does not incorporate the information from the theory that β_2 cannot be negative. The one-sided test, however, does incorporate this information. The different conclusions are then attributable to the different extents to which the theory has been used.

percent of persons 18 years of age or older that have graduated from high school (%HSG), and a measure of monthly welfare payments for a family with no countable income (AFDC). Each of these additional variables potentially affects the poverty rate.

This suggests the regression model

$$POV_i = \beta_0 + \beta_1 \, UN_i + \beta_2 \, \%WHITE_i$$

$$+ \beta_3 \, \%HSG_i + \beta_4 \, AFDC_i + \varepsilon_i$$

An OLS regression on this equation yields

$$\widehat{POV}_i = 33.4 + .302 \, UN_i - .066 \, \%WHITE$$
$$(4.67) \quad (.210) \qquad (.023)$$

$$- .186 \, \%HSG_i - .013 \, AFDC_i$$
$$(.070) \qquad (.004)$$

where the numbers in parentheses are the estimated standard errors of the coefficients.

Let us focus on the coefficient of AFDC. There are competing views as to whether β_4 is negative or positive. One view is that more generous welfare payments (AFDC increases) provide families with the resources they need to achieve a standard of living above the poverty level; this view implies $\beta_4 < 0$. Another view is that more generous welfare payments encourage individuals not to seek work or the training and education associated with a secure labor market status; this view suggests that $\beta_4 > 0$.

To investigate these competing hypotheses we might find the prob-value for and test the hypothesis

$$H_0: \beta_4 = 0$$

$$H_1: \beta_4 \neq 0$$

at the .05 level. (Notice the two-sided H_1 due to the theoretical possility that β_4 could be positive or negative.) The prob-value is

$$Pr(|\hat{\beta}_4 - \beta_4| \geq |-.013 - 0|) = Pr\left(\left|\frac{\hat{\beta}_4 - \beta_4}{s_{\hat{\beta}_4}}\right| \geq \left|\frac{-.013 - 0}{.004}\right|\right)$$

$$= Pr(|t^{(46)}| \geq 3.25)$$

Since the degrees of freedom exceeds 30, we may approximate this as $\Pr(Z \geq 3.25)$; since $\Pr(Z \geq 3.09)$ is .0010 and $\Pr(Z \geq 3.50)$ is .0002, we know that

$$.0010 > \text{prob-value} > .0002$$

Clearly, the prob-value is below .05, and we therefore reject H_0 in favor of H_1 at the .05 level. Our evidence suggests that more generous welfare payments, holding unemployment, percent white, and percent high school graduates constant, are associated with a lower poverty rate.

As a final comment, the R^2 in the above regression was .6595. Since R^2 has the same interpretation in the multiple regression framework as it does in the two-variable regression framework, we then can say that about 66% of the variation in the poverty rate across states is associated with the set of X variables used in the above regression.

PROBLEMS

14.1 Suppose that Y_i is the college GPA of person i, X_{1i} is the combined SAT score of person i, and X_{2i} is the high school GPA of person i. We estimate the regression

$$\hat{Y}_i = \hat{\beta}_0 + \hat{\beta}_1 X_{1i} + \hat{\beta}_2 X_{2i}$$

for a simple random sample of 30 persons, and find the following parameter estimates and estimated standard errors:

Parameter	Estimated value	Standard error
β_0	2.0	.5
β_1	.001	.001
β_2	.4	.15

(a) Build 95% confidence intervals for each parameter.
(b) Find the prob-value for and test the hypothesis

$$H_0: \beta_0 = 0$$

$$H_1: \beta_0 \neq 0$$

at the .05 level.

(c) Find the prob-value for and test the hypothesis

$$H_0: \beta_1 = 0$$

$$H_1: \beta_1 > 0$$

at the .05 level.

(d) Find the prob-value for and test the hypothesis

$$H_0: \beta_2 = 0$$

$$H_1: \beta_2 > 0$$

at the .05 level.

(e) Summarize the implications of your findings as they pertain to an understanding of the relationships under investigation.

14.2 (Requires computer.) Use the data from problem 13.14 to calculate the regression of MIM on PMHS, PURBAN, and MAGE. Carefully and fully interpret your results.

14.3 (Requires computer.) Use the data from problem 13.15 to calculate the regression of NETINV on CFSAL and RINT. Carefully and fully interpret your results.

AN APPLICATION: DUMMY VARIABLES

In this section we present an example to illustrate the inherent flexibility of the multiple regression model. This example involves a particular type of X variable referred to as a "dummy variable." A dummy variable is discrete and is one we create in order to capture information that is not inherently quantitative.

For example, the sex of an individual is a variable that may be associated with his or her earnings. A variable that equals "male" for males and "female" for females is not, however, quantitative. On the other hand, if we were to create a variable "sex" which equals 1 for males and 0 for females we have put (albeit in an arbitrary way) this notion into a quantitative context.

By exploring the possibilities here we can enhance our general understanding of the multiple regression model and in addition provide ourselves with an often useful new tool. Suppose that for a cross section of individuals we observe their income (inc_i), their number of years of schooling (ed_i), and their sex ($sex_i = 1$ for males and $sex_i = 0$ for females). We might then postulate, in a simplified yet illustrative context,

$$inc_i = \beta_0 + \beta_1\ ed_i + \beta_2\ sex_i + \varepsilon_i$$

as our model. If $\beta_2 > 0$, then we have evidence suggesting that males earn on average more than females with equivalent educations, for whatever reason. It is important to note that even if we know that $\beta_2 > 0$, we do not know why this is the case. One possibility is that women are discriminated against in the process of wage determination, but there are other possibilities. Perhaps women on average have less work experience than men because they are more likely to withdraw from the work force temporarily for family-related purposes; this lesser degree of experience (on average) may then lead to lower earnings (on average), even once educational attainment is held constant. To determine why women earn less (on average) than men would require further analysis; see our earlier discussions regarding causality, especially in Chapter 11.

An instructive way to view this model is to note that, for males,

$$inc_i = (\beta_0 + \beta_2) + \beta_1\ ed_i + \varepsilon_i \qquad \text{(males)}$$

whereas, for females,

$$inc_i = \beta_0 + \beta_1\ ed_i + \varepsilon_i \qquad \text{(females)}$$

In Figure 14.6 these two relationships are graphed.

The two regression lines have the same slope (β_1); by assump-

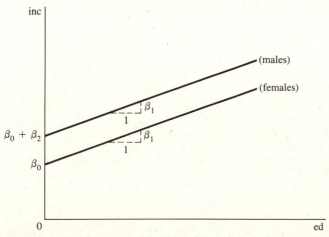

Figure 14.6 Income versus education using a dummy variable for gender, with equal marginal impacts.

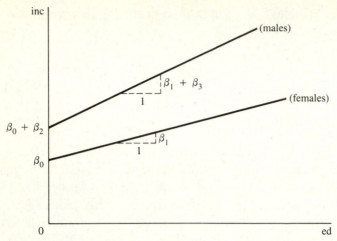

Figure 14.7 Income versus education using a dummy variable for gender, with different marginal impacts.

tion, the average rate of return for education is the same for males and females. But in this model the intercepts are different; there is, if you will, a constant "premium" of β_2 dollars of income for males.

In a more elaborate model, we might have both the slope and the intercept differ across the sexes. In such a model, we might say

$$\text{inc}_i = \beta_0 + \beta_1 \text{ ed}_i + \beta_2 \text{ sex}_i + \beta_3(\text{ed}_i)(\text{sex}_i) + \varepsilon_i$$

Here, our third X variable is the product of the first two. This formulation suggests, for males,

$$\text{inc}_i = (\beta_0 + \beta_2) + (\beta_1 + \beta_3)\text{ed}_i + \varepsilon_i \qquad (\text{males})$$

and, for females,

$$\text{inc}_i = \beta_0 + \beta_1 \text{ ed}_i + \varepsilon_i \qquad (\text{females})$$

The relationship between these two regression lines is shown in Figure 14.7. The intercepts are $\beta_0 + \beta_2$ for males and β_0 for females, and the slopes are $\beta_1 + \beta_3$ for males and β_1 for females. Presuming that β_3 is positive, the rate of return for education is on average higher for males than for females.

Estimation of a dummy variable model follows the same procedures as discussed above; a dummy X is just a special type of X variable.[12] But via this estimation and appropriate hypothesis tests

[12] If a dummy variable is used for Y, many new issues arise, and OLS is no longer an appropriate estimation technique. A discussion of this issue is beyond the scope of this text.

we can determine whether the variable in question (in this example, sex) is associated with the dependent variable even when the other X variables are held constant.

In some situations, an explanatory variable has more than two categories. For example, suppose that we are trying to explain the salary associated with a college graduate's first job in terms of his or her college GPA. But we think that the individual's major may also play a role. Suppose that we classify majors into three groups: humanities, social sciences, and sciences. How might we use a dummy variable framework in this setting?

One thought might be to define a dummy variable that equals 0 if the student was a humanities major, 1 if the student was a social science major, and 2 if the student was a science major. The problem with this approach is that it assumes that the difference between humanities and social sciences (0 to 1) is the same as the difference between social sciences and sciences (1 to 2); there is, however, no reason to believe that such is the case.

To get around this problem, we might define a set of dummy variables as follows: $D_1 = 1$ if a humanities major, 0 otherwise; $D_2 = 1$ if a social sciences major, 0 otherwise; and $D_3 = 1$ if a science major, 0 otherwise. Our regression model would then be

$$Y_i = \beta_0 + \beta_1 X_i + \beta_2 D_{1i} + \beta_3 D_{2i} + \beta_4 D_{3i} + \varepsilon_i$$

where Y_i is the starting salary of graduate i, and X_i is his or her GPA. The problem with this specification, however, is that since $D_{1i} + D_{2i} + D_{3i} = 1$ for each i, we have perfect multicollinearity! This feature implies that the OLS estimation procedure will break down.

The third dummy D_3, of course, contains no information that is unavailable from what we know regarding D_1 and D_2 (i.e., $D_{3i} = 1 - D_{1i} - D_{2i}$ for each i); in order to fully describe the various possibilities in terms of whether one majored in a humanities, social science, or science, only two dummy variables are necessary. Therefore, we should define D_1 and D_2 as before, and then note that if $D_1 = D_2 = 0$, then the graduate was a science major. That is, we should specify

$$Y_i = \beta_0 + \beta_1 X_i + \beta_2 D_{1i} + \beta_3 D_{2i} + \varepsilon_i$$

In this case,

$$Y_i = \beta_0 + \beta_1 X_i + \varepsilon_i \qquad \text{(for science majors)}$$

$$Y_i = (\beta_0 + \beta_2) + \beta_1 X_i + \varepsilon_i \qquad \text{(for humanities majors)}$$

$$Y_i = (\beta_0 + \beta_3) + \beta_1 X_i + \varepsilon_i \qquad \text{(for social science majors)}$$

The values of β_2 and β_3 will then tell us the relationship between the salary situations for the three groups of graduates.

PROBLEMS

14.4 Consider the model, discussed in the text, wherein

$$\text{inc}_i = \beta_0 + \beta_1 \, \text{ed}_i + \beta_2 \, \text{sex}_i + \beta_3(\text{ed}_i)(\text{sex}_i) + \varepsilon_i$$

Using a simple random sample of 25 persons, we found the following parameter estimates and estimated standard errors:

Parameter	Estimated value	Standard error
β_0	5000	1000
β_1	1000	250
β_2	500	200
β_3	300	200

(a) Find the prob-value for and test the hypothesis

$$H_0: \beta_2 = 0$$

$$H_1: \beta_2 \neq 0$$

at the .05 level. What is the implication of your finding?

(b) Find the prob-value for and test the hypothesis

$$H_0: \beta_3 = 0$$

$$H_1: \beta_3 \neq 0$$

at the .05 level. Putting the results of parts (a) and (b) together, summarize your findings regarding the relationship between income, education, and sex.

14.5 Suppose that instead of the model

$$\text{inc}_i = \beta_0 + \beta_1 \, \text{ed}_i + \beta_2 \, \text{sex}_i + \varepsilon_i$$

(sex equals 1 for males and 0 for females) we posited

$$\text{inc}_i = \alpha_0 + \alpha_1 \, \text{ed}_i + \alpha_2 \, \text{sex}_i^* + \varepsilon_i$$

where sex* equals 0 for males and 1 for females. What would you expect concerning the relationship between the parameter estimates and the interpretation of these estimates from OLS regressions of these alternative formulations?

14.6 (Requires computer.) Using the data from problem 13.15, calculate the regression of NETINV on CFSAL, RINT, and a dummy variable equal to 1 if there was a Democratic president and 0 if there was a Republican president. (There was a Democratic president—Carter—for, basically, 1977 through 1980.) Comment on the effects of political party on the investment behavior of firms.

14.7 (Requires computer.) Using the data from problem 13.14, calculate the regression of MIM on PMHS, PURBAN, MAGE, and a set of dummy variables representing regions of the country, as follows: Maine (ME) through Pennsylvania (PA) are Northeast; Ohio (OH) through Kansas (KS) are Midwest; Delaware (DE) through Texas (TX) are South, and Montana (MT) through Hawaii (HI) are West. Comment on the effects of geographical region on the determination of the dependent variable.

SUMMARY REMARKS

In this chapter we have provided an introduction to multiple regression analysis emphasizing the interpretation of the results. We have focused on the simplest form of the problem, which uses OLS in a linear framework. More complicated versions of this framework are at times appropriate, however; in an econometrics course many of these are considered. For now, however, we have a foundation in this important methodology and have discussed the essential interpretive issues.

REVIEW PROBLEMS FOR CHAPTER 14

14.8 Suppose that the true relationship between Y, X_1, and X_2 is

$$Y_i = \beta_0 + \beta_1 X_{1i} + \beta_2 X_{2i}^2 + \varepsilon_i$$

That is, suppose that in the correct specification of the model, X_2 enters in a nonlinear fashion. Not realizing this, you estimate the regression relationship

$$\hat{Y}_i = \hat{\beta}_0 + \hat{\beta}_1 X_{1i} + \hat{\beta}_2 X_{2i}$$

Comment on whether this estimated relationship truly gives us a picture of the relationship between Y and X_2, holding X_1 fixed.

14.9 Suppose that we observe a simple random sample of persons with a disease, and we note their survival time and whether they took a particular drug or not. Discuss how you might investigate the relationship between taking the drug and survival time; in particular, discuss the possible factors you would consider attempting to hold constant and how you might do this. (Note that this study is an observational and not an experimental one.)

*14.10 Suppose that

$$Y_i = \beta_0 + \beta_1 X_{1i} + \beta_2 X_{2i} + \varepsilon_i$$

but that we mistakenly estimate

$$\hat{Y}_i = \hat{\beta}_0 + \hat{\beta}_1 X_{1i}$$

(a) Is $\hat{\beta}_1$ an unbiased estimator of β_1? Fully justify your answer.

(b) What is the relationship between the bias of $\hat{\beta}_1$ as an estimator of β_1 and the correlation coefficient between X_1 and X_2?

*14.11 In a rural area the appraised value of a house depends upon its square footage, whether or not it has central air conditioning, and whether or not it has central heating. An analyst estimates the regression

$$\hat{V}_i = \hat{\beta}_0 + \hat{\beta}_1 F_i + \hat{\beta}_2 A_i + \hat{\beta}_3 H_i$$

where V_i is the appraised value, F_i is the square footage, A_i is a dummy variable equal to 1 if the house has central air conditioning (and 0 otherwise), and H_i is a dummy variable equal to 1 if the house has central heating (and 0 otherwise). It turns out that all houses with central air conditioning have central heating. But of those without central air conditioning, some have central heating and some do not (they are heated by wood stoves, for example).

This analyst is criticized by a colleague who says that the above model is subject to perfect multicollinearity since we know that if $A_i = 1$ then $H_i = 1$, and if $H_i = 0$ then $A_i = 0$.

Comment on this criticism.

APPENDIX 1 TO CHAPTER 14

In this appendix we briefly describe the derivation of the OLS estimators in the three-variable case (two X's and one Y). The problem is to

$$\text{minimize } \Sigma(Y_i - \hat{Y}_i)^2 = \Sigma(Y_i - \hat{\beta}_0 - \hat{\beta}_1 X_{1i} - \hat{\beta}_2 X_{2i})^2$$

by choosing the values of the three $\hat{\beta}$'s. To solve this multivariate calculus problem we take the partial derivatives with respect to each of the choice parameters and set these expressions equal to zero. The partial derivatives are

$$\frac{\partial \Sigma(Y_i - \hat{Y}_i)^2}{\partial \hat{\beta}_0} = -2\Sigma(Y_i - \hat{\beta}_0 - \hat{\beta}_1 X_{1i} - \hat{\beta}_2 X_{2i})$$

$$\frac{\partial \Sigma(Y_i - \hat{Y}_i)^2}{\partial \hat{\beta}_1} = -2\Sigma(Y_i - \hat{\beta}_0 - \hat{\beta}_1 X_{1i} - \hat{\beta}_2 X_{2i}) X_{1i}$$

$$\frac{\partial \Sigma(Y_i - \hat{Y}_i)^2}{\partial \hat{\beta}_2} = -2\Sigma(Y_i - \hat{\beta}_0 - \hat{\beta}_1 X_{1i} - \hat{\beta}_2 X_{2i}) X_{2i}$$

and the first-order conditions are then

$$\Sigma(Y_i - \hat{\beta}_0 - \hat{\beta}_1 X_{1i} - \hat{\beta}_2 X_{2i}) = 0$$

$$\Sigma(Y_i - \hat{\beta}_0 - \hat{\beta}_1 X_{1i} - \hat{\beta}_2 X_{2i}) X_{1i} = 0$$

$$\Sigma(Y_i - \hat{\beta}_0 - \hat{\beta}_1 X_{1i} - \hat{\beta}_2 X_{2i}) X_{2i} = 0$$

These relations are often referred to as the "normal equations." By solving this system of three equations for the three $\hat{\beta}$'s, the OLS estimators are determined. The formulas provided in the text represent the solution to this problem.

To know that these solutions involve a true minimum we must also consider the second-order conditions, but we here omit a discussion of that issue. Intuitively, there is no maximum value of

$$\Sigma(Y_i - \hat{Y}_i)^2$$

because the value of this function can always be increased by moving the estimated regression plane farther away from the observations. Thus, the solution above must be a minimum (remember that the function of interest is quadratic and not of a higher order).

APPENDIX 2 TO CHAPTER 14

In this appendix we discuss some hypothesis tests based on the F distribution in the multiple regression model. For a further discussion see, for example, Jan Kmenta, *Elements of Econometrics,* 2d ed. (New York: Macmillan, 1986), pp. 412–422.

An often interesting test in the multiple regression model

$$Y_i = \beta_0 + \beta_1 X_{1i} + \cdots + \beta_k X_{ki} + \varepsilon_i$$

is to find the prob-value for and test the hypothesis

$H_0: \beta_1 = \beta_2 = \cdots = \beta_k = 0$

$H_1:$ not H_0 (i.e., at least one $\beta_j \neq 0$ ($j = 1, \ldots, k$))

(note that β_0 is not included in H_0). What lies behind this test? If H_0 is true, then the set of k explanatory variables does not help us to explain the variation in the dependent variable Y, and Y_i is simply the constant β_0 plus the random error ε_i. That is, changes in the X's are not associated with changes in Y. On the other hand, if H_1 is true, then we have at least to some extent explained variation in Y, by the one or more X's that have a nonzero coefficient. This hypothesis test then is a way of investigating whether the regression model as a whole makes progress toward the ultimate goal of explaining the behavior of Y. (In this context, when we say "explaining" we do not mean to imply a statement of causality; rather, we mean that the variation in Y is associated with variation in the X's and is then explainable, in a statistical sense, by that variation.) This case is an example of what we refer to as a *joint* hypothesis test; our concern is not with which particular β's are nonzero but with whether any of the (slope) β's are.

If the model were of no help in explaining the variation in Y, then we would expect $R^2 = 0$ (remember R^2 is the fraction of variation in Y explained by the X's); to the extent that R^2 exceeds zero the X's make progress toward explaining the variation in Y. Now the R^2 we achieve through a multiple regression is in fact a "sample R^2"; it is calculated based on the information in a sample as opposed to the population as a whole. That is, it is possible that there is no population relationship between the X's and Y (i.e., the "population R^2" is zero) but that the sample R^2 from our regression is positive. The hypothesis test is then resolved, not by asking "Is $R^2 > 0$?" but rather by asking "Is R^2 big enough?" Is R^2 big enough to indicate a relationship between the set of X's and Y that cannot plausibly be due to random variability? That is, is R^2 statistically significant? The key to addressing this question is the probability distribution of R^2 or some function of R^2.

It turns out that, given the CLRM assumptions,

$$f = \frac{R^2/k}{(1 - R^2)/(n - k - 1)} \sim F^{(k, n-k-1)} \qquad (2A14.1)$$

where k is the number of X's in the model. Note that, ignoring the scale factors, the numerator (R^2) is a measure of variation in Y explained by the X's, and the denominator $(1 - R^2)$ is a measure of variation in Y unexplained by the X's. The larger f is, then, the larger the ratio of explained to unexplained variation, and the more likely that there is in fact a true (population) relationship between the X's and Y and not merely a relationship in our sample due to random variability.

Let us illustrate the use of Eq. (2A14.1) in the context of the regression in problem 14.2, which involved a regression of mean income of males (MIM) on percent of males who are high school graduates (PMHS), percent of persons living in urban areas (PUR-BAN), and median age of males (MAGE). The R^2 in this regression is .5465. To test the hypothesis

$$H_0: \beta_1 = \beta_2 = \beta_3 = 0$$

$$H_1: \text{not } H_0$$

(where β_1, β_2, and β_3 are, respectively, the coefficients of PMHS, PURBAN, and MAGE in a linear regression) our prob-value is

$$\Pr(f \geq f^*)$$

where f^* is the value taken on by f in our regression. Now

$$\Pr(f \geq f^*) = \Pr\left(\frac{R^2/k}{(1 - R^2)/(n - k - 1)} \right.$$

$$\geq \left. \frac{.5465/3}{(1 - .5465)/(51 - 3 - 1)} \right)$$

$$= \Pr(F^{(3,47)} \geq 18.9)$$

Since $F_{.05}^{(3,40)} = 2.84$ we know $F_{.05}^{(3,47)} < 2.84$. Clearly, the prob-value is less than .05. We therefore reject H_0 in favor of H_1 at the .05 level, and we conclude that the X's as a group do help to explain in a significant way the variation in mean male income across states.

In the above example, we were not so much concerned with the relationship between Y and the individual X's, but rather between Y and the X's as a group. In the t tests discussed earlier in this chapter, we were concerned with the individual X's and their relationships to Y, and not the X's as a group. But what if we were interested in a subset of the X's as a group, and their contribution to explaining the variation in Y?

Consider the general model

$$Y_i = \beta_0 + \beta_1 X_{1i} + \cdots + \beta_{k-q}X_{k-q,i}$$
$$+ \beta_{k-q+1}X_{k-q+1,i} + \cdots + \beta_k X_{ki} + \varepsilon_i \quad (2A14.2)$$

and the hypothesis test

$$H_0: \beta_{k-q+1} = \beta_{k-q+2} = \cdots = \beta_k = 0$$

$$H_1: \text{not } H_0 \quad\quad (2A14.3)$$

That is, the null hypothesis is that the coefficients of the last q right-hand-side variables are all zero (the X's are numbered arbitrarily, so the X's of interest can be placed at the end); this result suggests that this subset of q variables does not help to explain in a significant way the variation in Y.

Why might we be interested in such a test? Consider the regression example from earlier in this appendix, wherein we investigated the relationship between the mean income of males (MIM), the percent of males who are high school graduates (PMHS), the percent of the population living in urban areas (PURBAN), and the median age of males (MAGE). Suppose, however, that we also included a set of regional dummies as in problem 14.7. We might be interested in whether the set of regional dummies as a whole helps to explain the variation in the dependent variable—that is, does region matter in some way? The regression without the regional dummies is

$$\text{MIM}_i = \beta_0 + \beta_1 \text{ PMHS}_i + \beta_2 \text{ PURBAN}_i$$
$$+ \beta_3 \text{ MAGE}_i + \varepsilon_i \quad\quad (2A14.4)$$

and the regression with the regional dummies is

$$\text{MIM}_i = \beta_0 + \beta_1 \text{ PMHS}_i + \beta_2 \text{ PURBAN}_i + \beta_3 \text{ MAGE}_i$$
$$+ \beta_4 \text{ NE}_i + \beta_5 \text{ MW}_i + \beta_6 \text{ SO}_i + \varepsilon_i \quad\quad (2A14.5)$$

where NE = 1 if the state is in the Northeast, and 0 otherwise; MW = 1 if the state is in the Midwest, and 0 otherwise; and SO = 1 if the state is in the South, and 0 otherwise. (NE = MW = SO = 0 implies that the state is in the West; see problem 14.7 for information on which states fall into these regions.)

In general, to test the null hypothesis given in Eq. (2A14.3) as pertaining to the model, Eq. (2A14.2), we use the fact that

$$\frac{(R_{UR}^2 - R_R^2)/q}{(1 - R_{UR}^2)/(n - k - 1)} \sim F^{(q, n-k-1)} \quad (2A14.6)$$

R_{UR}^2 is the R^2 from the "unrestricted" regression in which all X's are included; R_R^2 is the R^2 from a "restricted" regression in which the X's associated with the β's in the null hypothesis (2A14.3) are omitted; this second regression imposes the restrictions hypothesized under H_0. The parameter q is the number of β's hypothesized equal to 0 in H_0; $(n - k - 1)$ is the degrees of freedom in the *unrestricted* regression. [In the previous example, the R^2 from a regression of Eq. (2A14.5) would be the unrestricted R^2, and the R^2 from a regression of Eq. (2A14.4) would be the restricted R^2. In this case, $q = 3$ and $n - k - 1 = 44$.]

To understand the use of Eq. (2A14.6), note that if R_{UR}^2 is much larger than R_R^2, then we would reject H_0; we have a significant increase in the explanatory power of the model due to the inclusion of the additional variables. If, however, R_{UR}^2 is only a bit above R_R^2, we do not reject H_0. (We know that $R_{UR}^2 \geq R_R^2$ (why?).) The F distribution tells us when the difference between the two R^2's is large enough to conclude that H_0 is false.

In the problem characterized by Eqs. (2A14.4) and (2A14.5) the null hypothesis is

$$H_0: \beta_4 = \beta_5 = \beta_6 = 0$$

$$H_1: \text{not } H_0$$

From Eq. (2A14.5) the unrestricted R^2 is .5769; the restricted R^2, from Eq. (2A14.4), is .5465. The prob-value is

$$\Pr(f \geq f^*)$$

(where f^* is the value taken on by f in our problem). Now

$$\Pr(f \geq f^*) = \Pr\left[\frac{(R_{UR}^2 - R_R^2)/q}{(1 - R_{UR}^2)/(n - k - 1)}\right.$$

$$\left. \geq \frac{(.5769 - .5465)/3}{(1 - .5769)/(51 - 6 - 1)}\right]$$

$$= \Pr(F^{(3,44)} \geq 1.05)$$

Since $F_{.05}^{(3,40)} = 2.84$ and $F_{.05}^{(3,60)} = 2.76$, we know that $2.84 > F_{.05}^{(3,44)} > 2.76$. Clearly, $\Pr(F^{(3,44)} \geq 1.05) > .05$; that is, the prob-value exceeds .05. We therefore cannot reject H_0 in favor of H_1 at the .05 level. The set of dummy variables for region do not significantly increase our ability to explain variation in the mean income of males across states beyond that already achieved by PMHS, PURBAN, and MAGE.

APPENDIX 3 TO CHAPTER 14
SPECIAL ISSUES IN TIME SERIES DATA

In a time series data set we observe the same unit of observation at different points in time, as contrasted with a cross-sectional data set, in which we observe different units all at the same point in time. There are, however, special issues that typically arise in the context of a time series analysis, and it is the goal of this appendix to provide an introduction to these issues. [Many statistics textbooks provide a more detailed discussion of time series analysis; see, for example, Thomas H. Wonnacott and Ronald J. Wonnacott, *Introductory Statistics for Business and Economics,* 4th ed. (New York: John Wiley & Sons, 1990), chapter 24, or Paul Newbold, *Statistics for Business and Economics,* 3rd ed. (Englewood Cliffs, NJ: Prentice-Hall, 1991), chapter 17.]

Components of a Time Series

Suppose that we are interested in the quarterly sales behavior of a chain of retail stores. We have observed quarterly sales for a 15-year (60-quarter) period, and have plotted these quarterly sales (in millions of dollars) in Figure 3A14.1. (Ignore for the moment the graph entitled "adjusted moving average.") There is clearly a tremendous amount of variation in quarterly sales over time, but we can understand this variation better by categorizing its different sources.

    One source of variation is a *time trend.* That is, there is a clear upward trend over time in the sales of this chain of stores. This observation does not imply that each quarter shows an increase in sales from the previous quarter; however, it does indicate that the long-run trend is upward. To confirm this suspicion we have in fact plotted a *linear trend,* which has an upward slope; this trend line

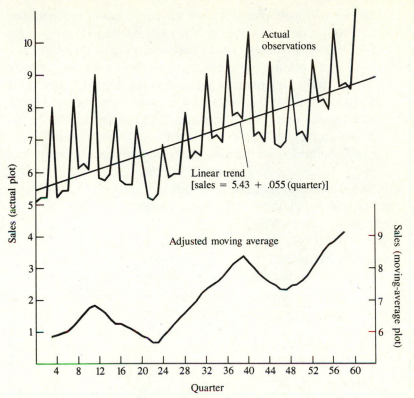

Note: Data for quarters 1 through 60 are as follows: 5.1, 5.2, 5.2, 8.0, 5.2, 5.4, 5.4, 8.2, 6.1, 6.2, 6.1, 9.0, 5.8, 5.7, 5.9, 7.6, 5.7, 5.6, 5.6, 7.4, 5.2, 5.1, 5.3, 6.8, 5.8, 5.9, 5.9, 7.8, 6.4, 6.6., 6.5, 9.0, 7.0, 7.1, 6.9, 9.6, 7.7, 7.8, 7.6, 10.3, 7.1, 7.2, 6.9, 9.4, 6.8, 6.7, 6.9, 8.7, 7.1, 7.2, 6.9, 9.4, 8.1, 8.2, 7.9, 10.4, 8.6, 8.7, 8.5, 11.0.

Figure 3A14.1   A time series of quarterly sales figures.

was determined by calculating the OLS regression of quarterly sales on a time variable (going from 1 to 60 for the 60 quarters of data).

The time trend, though informative, clearly misses much of the variation in the series. (That is, there is substantial variation about the time trend line.) A second source of variation is a recurring pattern in which there is a substantial increase in sales from quarter three to quarter four in each year, with a subsequent decline in the following first quarter. This result is likely due to the Christmas buying season. Such a *seasonal component* of variation is often present in a time series when the period of observation is less than a year.

Time series in which the time periods are less than a year in length are often *seasonally adjusted* so as to remove the seasonal component. Although we do not here consider techniques for this procedure, we will note that a simple way of "smoothing" out the seasonal component is to consider the calculation of a new series based upon a *moving average.* In the problem at hand the procedure would involve calculating the average sale for four consecutive time periods and then seeing how that average changes as we move from quarter to quarter. Since the averages are for four consecutive quarters, each contains one "fourth quarter," and therefore the effects of the Christmas buying season are distributed over time. In Figure 3A14.1 we have added to the plot of the actual data a time series plot of the "adjusted moving average." We have changed the vertical axis for the moving average so as to not juxtapose the two graphs, but the vertical coordinates of the actual observations and the moving average are comparable.

We present here an "adjusted" moving average due to the complication that arises from averaging over an even number of time periods. For time t the moving average is

$$\frac{Y_{t-1} + Y_t + Y_{t+1} + Y_{t+2}}{4}$$

But this is really the moving average for a time "between" t and $t + 1$ (i.e., the center of the interval from $t - 1$ to $t + 2$). Doing the same for time $t - 1$, that is,

$$\frac{Y_{t-2} + Y_{t-1} + Y_t + Y_{t+1}}{4}$$

would give us the comparable figure for a time "between" $t - 1$ and t; averaging these two averages gives us the "adjusted moving average" for time t. Notice that in this averaging process we lose observations at both the start and the end of the time series.

The moving-average series clearly shows the time trend and has smoothed away the seasonal component. But the moving-average series suggests the presence of a third component, the *cyclical* component. Sales rise for a time, then fall for a time, then rise for a time, and so on. The peaks and the troughs are both getting higher (thus the positive long-term trend), but the cycle remains. This finding is, of course, a manifestation of the general business cycle phenomenon; perhaps the down periods are recessions and the up periods are expansions for the economy as a whole. Note that the cyclical component is present in the actual data, but it is not as easily seen as in the smoothed series.

A final component is pure *random variation.* That is, even once we have accounted for the time trend, seasonal factors, and cyclical factors, there still is a random element involved in the determination of most variables. (Note that this is really what we were thinking about when we discussed the error term in the CLRM.)

Implications for Regression Analysis

Time series data pose special problems for regression analysis, as best seen by considering the four sources of variation in a time series discussed above.

The time trend suggests that the variable in question is either related to some other variable that is changing in a similar way over time or, perhaps, that it has a time trend of its own that is not well predicted by the behavior of other variables. In the above example, perhaps consumer income is rising over time, but perhaps also the popularity of this chain is increasing for reasons that are hard to quantify or associate with some other variable.

The seasonal component might be dealt with by using data that are annual, or by using data over shorter time periods that are seasonally adjusted. Alternatively, with the actual observations (which are not seasonally adjusted) we might use some sort of dummy variable framework to deal with the seasonal effect; in the retail sales example, a dummy variable equal to one for the fourth quarter and zero otherwise might be a key part of using an estimated relationship to make a reasonable forecast for future sales.

The cyclical component suggests that the error terms from a regression will be correlated over time. With the time trend regression line shown in the figure, a plot of the residuals over time would show a tendency for there to be strings of positives followed by strings of negatives (which is most clearly seen using the adjusted moving average series). The cyclical component causes the assumption (in the CLRM) of independent error terms to be violated; this problem is common in time series even when there is no problem due to time trend or seasonality, and is known as *serial correlation.* Clearly, in making a forecast we would want to consider where within a cycle we are; if it is more likely than not that we are above the regression line, we want to use that information in making a prediction.

Conclusion

We have here tried to illustrate the complexity of the issues that arise in time series data. There are many different contexts with varying degrees of complexity for addressing these issues; our goal here has been simply to bring attention to them.

DISCUSSION QUESTIONS FOR CHAPTER 14

1. Suppose that the (sample) correlation coefficient between X_1 and Y is positive, but when we estimate the regression

$$\hat{Y}_i = \hat{\beta}_0 + \hat{\beta}_1 X_{1i} + \hat{\beta}_2 X_{2i}$$

we are unable to reject the hypothesis $H_0: \beta_1 = 0$. Explain how this result might occur.

2. Suppose that the correlation coefficient between X and Y over some period is zero. Does this situation imply that changes in X do not cause changes in Y?

3. Discuss the relationship between correlation, two-variable regression, and multiple regression as ways of investigating the relationship between the two variables X and Y.

Epilogue

This text has taken us on a fairly long journey. We began by discussing ways of describing information and making simple inferences. We then studied probability theory in order to improve our understanding of random behavior; this heightened understanding was the key link between the simple (but limited) inferences we began with and the more sophisticated inferences (confidence intervals and hypothesis tests) we eventually developed. Finally, we discussed regression analysis, which builds upon our earlier work and is the most commonly used statistical approach in economics. Throughout this text we have tried to present the basics of statistical analysis as used in economics in a way that emphasizes the intuition behind the material. We have tried to show that statistical analysis can be insightful and informative, and not as hard to understand as is often believed.

WHAT MIGHT COME NEXT?

For a student of economics, the usual next step is to pursue the subject of *econometrics*. In econometrics, we are interested in the application of statistical techniques to economic problems while paying particular attention to what economic theory suggests about the nature of the underlying relationships. One focus is on the likely validity of the assumptions that make up the Classical Linear Regression Model. The properties of Ordinary Least Squares estimators discussed in this text were derived under this set of fairly restrictive assumptions. But what happens to these properties if one or more of these assumptions is violated? In econometric analysis a great deal of attention is paid to determining the properties of estimators under alternative assumptions and what (if any) alternative estimation techniques are available that have more desirable properties.

But before a student moves on to such analyses, he or she should reflect on what we have done in this course to see the usefulness as well as the possible limitations of that work. In particular, it is helpful to recall the assumptions upon which much of our analysis has been based, so that we can be alert for situations in which these assumptions may not be valid.

SOME KEY ASSUMPTIONS WE HAVE MADE

We are often, explicitly or implicitly, making assumptions. When purchasing a new car we make assumptions (consciously or unconsciously) about what our income will be over the life of the car loan. The types of analyses discussed in this text are no different. In particular, there are three major assumptions that have shown up again and again, and it is useful to remind ourselves of them.

First, we have always assumed that the sample we are using to infer the characteristics of the population of interest is a random sample. But in Chapter 1 we suggested that in practice it is often difficult to obtain a sample that is in fact really a random sample. What happens to the quality of our inferences when our sample cannot be well approximated as being random? Are there ways that we should change our process of inference when the random sample assumption is not a good one? We should keep these questions in mind. They are a possible subject for future study.

Second, there has been a key role played by linearity and linear functions in our analysis. The covariance seeks to uncover relationships between random variables which can be expressed in the form of a linear function, the summation (Σ) and expected-value (E) signs are linear operators, the regression analysis we considered is for linear relationships, and so on. But not all relationships are truly linear; do we really believe that even such a relatively simple construct as a demand curve is truly linear? We must then be concerned about what happens when the linearity assumption is no longer a good one. At the very least, we must be aware of where this assumption fits into our analysis.

Finally, in our analysis there has been a key role played by the normal distribution. A main reason for this is the Central Limit Theorem, which suggests that the distribution of the sample mean is at least approximately normal (once the sample size is large enough) even if the parent population is not. (In addition, there are variations of the Central Limit Theorem that suggest the normality of OLS parameter estimates even if the regression equation's error term is not normally distributed.) But even though the sample mean is at least approximately normal once we get a large enough sample, there is no guarantee that the sample we actually have is large enough for this to be a relevant result. What do we do in this type of situation?

These remarks are not meant to imply that the techniques presented in this book are of limited applicability; in fact, these techniques provide good insights into a wide variety of problems considered by economists. Our purpose is rather to remind the reader of the key role played by certain assumptions. When an assumption is untenable, we must search for alternative techniques based on different assumptions. There are methodologies for working with samples that are not random samples; nonlinear methods can be used, and statistics not based upon normal distribution theory are available. These issues might be addressed in more advanced treatments of statistics. Nevertheless, the topics discussed in this text are the basis for most statistical work in economics.

TABLES

Table I SELECTED VALUES OF THE BINOMIAL DISTRIBUTION

(For a set of n independent trials, $p(s) = \dfrac{n!}{s!(n-s)!}\pi^s(1-\pi)^{n-s}$ where s is the number of successes and $\Pr(s) = \pi$ on each individual trial.)

n	s	.05	.10	.15	.20	.25	.30	.35	.40	.45	.50
							π				
1	0	.9500	.9000	.8500	.8000	.7500	.7000	.6500	.6000	.5500	.5000
	1	.0500	.1000	.1500	.2000	.2500	.3000	.3500	.4000	.4500	.5000
2	0	.9025	.8100	.7225	.6400	.5625	.4900	.4225	.3600	.3025	.2500
	1	.0950	.1800	.2550	.3200	.3750	.4200	.4550	.4800	.4950	.5000
	2	.0025	.0100	.0225	.0400	.0625	.0900	.1225	.1600	.2025	.2500
3	0	.8574	.7290	.6141	.5120	.4219	.3430	.2746	.2160	.1664	.1250
	1	.1354	.2430	.3251	.3840	.4219	.4410	.4436	.4320	.4084	.3750
	2	.0071	.0270	.0574	.0960	.1406	.1890	.2389	.2880	.3341	.3750
	3	.0001	.0010	.0034	.0080	.0156	.0270	.0429	.0640	.0911	.1250
4	0	.8145	.6561	.5220	.4096	.3164	.2401	.1785	.1296	.0915	.0625
	1	.1715	.2916	.3685	.4096	.4219	.4116	.3845	.3456	.2995	.2500
	2	.0135	.0486	.0975	.1536	.2109	.2646	.3105	.3456	.3675	.3750
	3	.0005	.0036	.0115	.0256	.0469	.0756	.1115	.1536	.2005	.2500
	4	.0000	.0001	.0005	.0016	.0039	.0081	.0150	.0256	.0410	.0625
5	0	.7738	.5905	.4437	.3277	.2373	.1681	.1160	.0778	.0503	.0312
	1	.2036	.3280	.3915	.4096	.3955	.3602	.3124	.2592	.2059	.1562
	2	.0214	.0729	.1382	.2048	.2637	.3087	.3364	.3456	.3369	.3125
	3	.0011	.0081	.0244	.0512	.0879	.1323	.1811	.2304	.2757	.3125
	4	.0000	.0004	.0022	.0064	.0146	.0284	.0488	.0768	.1128	.1562
	5	.0000	.0000	.0001	.0003	.0010	.0024	.0053	.0102	.0185	.0312

π

n	s	.05	.10	.15	.20	.25	.30	.35	.40	.45	.50
6	0	.7351	.5314	.3771	.2621	.1780	.1176	.0754	.0467	.0277	.0156
	1	.2321	.3543	.3993	.3932	.3560	.3025	.2437	.1866	.1359	.0938
	2	.0305	.0984	.1762	.2458	.2966	.3241	.3280	.3110	.2780	.2344
	3	.0021	.0146	.0415	.0819	.1318	.1852	.2355	.2765	.3032	.3125
	4	.0001	.0012	.0055	.0154	.0330	.0595	.0951	.1382	.1861	.2344
	5	.0000	.0001	.0004	.0015	.0044	.0102	.0205	.0369	.0609	.0938
	6	.0000	.0000	.0000	.0001	.0002	.0007	.0018	.0041	.0083	.0156
7	0	.6983	.4783	.3206	.2097	.1335	.0824	.0490	.0280	.0152	.0078
	1	.2573	.3720	.3960	.3670	.3115	.2471	.1848	.1306	.0872	.0547
	2	.0406	.1240	.2097	.2753	.3115	.3177	.2985	.2613	.2140	.1641
	3	.0036	.0230	.0617	.1147	.1730	.2269	.2679	.2903	.2918	.2734
	4	.0002	.0026	.0109	.0287	.0577	.0972	.1442	.1935	.2388	.2734
	5	.0000	.0002	.0012	.0043	.0115	.0250	.0466	.0774	.1172	.1641
	6	.0000	.0000	.0001	.0004	.0013	.0036	.0084	.0172	.0320	.0547
	7	.0000	.0000	.0000	.0000	.0001	.0002	.0006	.0016	.0037	.0078
8	0	.6634	.4305	.2725	.1678	.1001	.0576	.0319	.0168	.0084	.0039
	1	.2793	.3826	.3847	.3355	.2670	.1977	.1373	.0896	.0548	.0312
	2	.0515	.1488	.2376	.2936	.3115	.2965	.2587	.2090	.1569	.1094
	3	.0054	.0331	.0839	.1468	.2076	.2541	.2786	.2787	.2568	.2188
	4	.0004	.0046	.0185	.0459	.0865	.1361	.1875	.2322	.2627	.2734
	5	.0000	.0004	.0026	.0092	.0231	.0467	.0808	.1239	.1719	.2188
	6	.0000	.0000	.0002	.0011	.0038	.0100	.0217	.0413	.0703	.1094
	7	.0000	.0000	.0000	.0001	.0004	.0012	.0033	.0079	.0164	.0312
	8	.0000	.0000	.0000	.0000	.0000	.0001	.0002	.0007	.0017	.0039

(Continued)

Table I SELECTED VALUES OF THE BINOMIAL DISTRIBUTION (CONTINUED)

π

n	s	.05	.10	.15	.20	.25	.30	.35	.40	.45	.50
9	0	.6302	.3874	.2316	.1342	.0751	.0404	.0207	.0101	.0046	.0020
	1	.2985	.3874	.3679	.3020	.2253	.1556	.1004	.0605	.0339	.0176
	2	.0629	.1722	.2597	.3020	.3003	.2668	.2162	.1612	.1110	.0703
	3	.0077	.0446	.1069	.1762	.2336	.2668	.2716	.2508	.2119	.1641
	4	.0006	.0074	.0283	.0661	.1168	.1715	.2194	.2508	.2600	.2461
	5	.0000	.0008	.0050	.0165	.0389	.0735	.1181	.1672	.2128	.2461
	6	.0000	.0001	.0006	.0028	.0087	.0210	.0424	.0743	.1160	.1641
	7	.0000	.0000	.0000	.0003	.0012	.0039	.0098	.0212	.0407	.0703
	8	.0000	.0000	.0000	.0000	.0001	.0004	.0013	.0035	.0083	.0176
	9	.0000	.0000	.0000	.0000	.0000	.0000	.0001	.0003	.0008	.0020
10	0	.5987	.3487	.1969	.1074	.0563	.0282	.0135	.0060	.0025	.0010
	1	.3151	.3874	.3474	.2684	.1877	.1211	.0725	.0403	.0207	.0098
	2	.0746	.1937	.2759	.3020	.2816	.2335	.1757	.1209	.0763	.0439
	3	.0105	.0574	.1298	.2013	.2503	.2668	.2522	.2150	.1665	.1172
	4	.0010	.0112	.0401	.0881	.1460	.2001	.2377	.2508	.2384	.2051
	5	.0001	.0015	.0085	.0264	.0584	.1029	.1536	.2007	.2340	.2461
	6	.0000	.0001	.0012	.0055	.0162	.0368	.0689	.1115	.1596	.2051
	7	.0000	.0000	.0001	.0008	.0031	.0090	.0212	.0425	.0746	.1172
	8	.0000	.0000	.0000	.0001	.0004	.0014	.0043	.0106	.0229	.0439
	9	.0000	.0000	.0000	.0000	.0000	.0001	.0005	.0016	.0042	.0098
	10	.0000	.0000	.0000	.0000	.0000	.0000	.0000	.0001	.0003	.0010

n	s	.55	.60	.65	.70	.75	.80	.85	.90	.95
1	0	.4500	.4000	.3500	.3000	.2500	.2000	.1500	.1000	.0500
	1	.5500	.6000	.6500	.7000	.7500	.8000	.8500	.9000	.9500
2	0	.2025	.1600	.1225	.0900	.0625	.0400	.0225	.0100	.0025
	1	.4950	.4800	.4550	.4200	.3750	.3200	.2550	.1800	.0950
	2	.3025	.3600	.4225	.4900	.5625	.6400	.7225	.8100	.9025
3	0	.0911	.0640	.0429	.0270	.0156	.0080	.0034	.0010	.0001
	1	.3341	.2880	.2389	.1890	.1406	.0960	.0574	.0270	.0071
	2	.4084	.4320	.4436	.4410	.4219	.3840	.3251	.2430	.1354
	3	.1664	.2160	.2746	.3430	.4219	.5120	.6141	.7290	.8574
4	0	.0410	.0256	.0150	.0081	.0039	.0016	.0005	.0001	.0000
	1	.2005	.1536	.1115	.0756	.0469	.0256	.0115	.0036	.0005
	2	.3675	.3456	.3105	.2646	.2109	.1536	.0975	.0486	.0135
	3	.2995	.3456	.3845	.4116	.4219	.4096	.3685	.2916	.1715
	4	.0915	.1296	.1785	.2401	.3164	.4096	.5220	.6561	.8145
5	0	.0185	.0102	.0053	.0024	.0010	.0003	.0001	.0000	.0000
	1	.1128	.0768	.0488	.0284	.0146	.0064	.0022	.0004	.0000
	2	.2757	.2304	.1811	.1323	.0879	.0512	.0244	.0081	.0011
	3	.3369	.3456	.3364	.3087	.2637	.2048	.1382	.0729	.0214
	4	.2059	.2592	.3124	.3602	.3955	.4096	.3915	.3280	.2036
	5	.0503	.0778	.1160	.1681	.2373	.3277	.4437	.5905	.7738

π

(Continued)

Table I SELECTED VALUES OF THE BINOMIAL DISTRIBUTION (CONTINUED)

π

n	s	.55	.60	.65	.70	.75	.80	.85	.90	.95
6	0	.0083	.0041	.0018	.0007	.0002	.0001	.0000	.0000	.0000
	1	.0609	.0369	.0205	.0102	.0044	.0015	.0004	.0001	.0000
	2	.1861	.1382	.0951	.0595	.0330	.0154	.0055	.0012	.0001
	3	.3032	.2765	.2355	.1852	.1318	.0819	.0415	.0146	.0021
	4	.2780	.3110	.3280	.3241	.2966	.2458	.1762	.0984	.0305
	5	.1359	.1866	.2437	.3025	.3560	.3932	.3993	.3543	.2321
	6	.0277	.0467	.0754	.1176	.1780	.2621	.3771	.5314	.7351
7	0	.0037	.0016	.0006	.0002	.0001	.0000	.0000	.0000	.0000
	1	.0320	.0172	.0084	.0036	.0013	.0004	.0001	.0000	.0000
	2	.1172	.0774	.0466	.0250	.0115	.0043	.0012	.0002	.0000
	3	.2388	.1935	.1442	.0972	.0577	.0287	.0109	.0026	.0002
	4	.2918	.2903	.2679	.2269	.1730	.1147	.0617	.0230	.0036
	5	.2140	.2613	.2985	.3177	.3115	.2753	.2097	.1240	.0406
	6	.0872	.1306	.1848	.2471	.3115	.3670	.3960	.3720	.2573
	7	.0152	.0280	.0490	.0824	.1335	.2097	.3206	.4783	.6983
8	0	.0017	.0007	.0002	.0001	.0000	.0000	.0000	.0000	.0000
	1	.0164	.0079	.0033	.0012	.0004	.0001	.0000	.0000	.0000
	2	.0703	.0413	.0217	.0100	.0038	.0011	.0002	.0000	.0000
	3	.1719	.1239	.0808	.0467	.0231	.0092	.0026	.0004	.0000
	4	.2627	.2322	.1875	.1361	.0865	.0459	.0185	.0046	.0004
	5	.2568	.2787	.2786	.2541	.2076	.1468	.0839	.0331	.0054
	6	.1569	.2090	.2587	.2965	.3115	.2936	.2376	.1488	.0515
	7	.0548	.0896	.1373	.1977	.2670	.3355	.3847	.3826	.2793
	8	.0084	.0168	.0319	.0576	.1001	.1678	.2725	.4305	.6634

| | | | | | | | π | | | |
n	s	.55	.60	.65	.70	.75	.80	.85	.90	.95
9	0	.0008	.0003	.0001	.0000	.0000	.0000	.0000	.0000	.0000
	1	.0083	.0035	.0013	.0004	.0001	.0000	.0000	.0000	.0000
	2	.0407	.0212	.0098	.0039	.0012	.0003	.0000	.0000	.0000
	3	.1160	.0743	.0424	.0210	.0087	.0028	.0006	.0001	.0000
	4	.2128	.1672	.1181	.0735	.0389	.0165	.0050	.0008	.0000
	5	.2600	.2508	.2194	.1715	.1168	.0661	.0283	.0074	.0006
	6	.2119	.2508	.2716	.2668	.2336	.1762	.1069	.0446	.0077
	7	.1110	.1612	.2162	.2668	.3003	.3020	.2597	.1722	.0629
	8	.0339	.0605	.1004	.1556	.2253	.3020	.3679	.3874	.2985
	9	.0046	.0101	.0207	.0404	.0751	.1342	.2316	.3874	.6302
10	0	.0003	.0001	.0000	.0000	.0000	.0000	.0000	.0000	.0000
	1	.0042	.0016	.0005	.0001	.0000	.0000	.0000	.0000	.0000
	2	.0229	.0106	.0043	.0014	.0004	.0001	.0000	.0000	.0000
	3	.0746	.0425	.0212	.0090	.0031	.0008	.0001	.0000	.0000
	4	.1596	.1115	.0689	.0368	.0162	.0055	.0012	.0001	.0000
	5	.2340	.2007	.1536	.1029	.0584	.0264	.0085	.0015	.0001
	6	.2384	.2508	.2377	.2001	.1460	.0881	.0401	.0112	.0010
	7	.1665	.2150	.2522	.2668	.2503	.2013	.1298	.0574	.0105
	8	.0763	.1209	.1757	.2335	.2816	.3020	.2759	.1937	.0746
	9	.0207	.0403	.0725	.1211	.1877	.2684	.3474	.3874	.3151
	10	.0025	.0060	.0135	.0282	.0563	.1074	.1969	.3487	.5987

This table is based upon National Bureau of Standards Applied Mathematics Series No. 6, *Tables of the Binomial Probability Distribution* (Washington, DC: U.S. Government Printing Office, 1950).

Table II PROBABILITIES FOR THE STANDARD NORMAL DISTRIBUTION

$(\Pr(Z \geq a) \text{ when } Z \sim N(0, 1))$

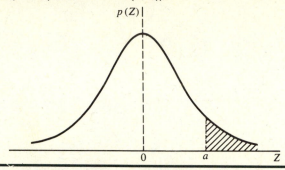

a	.00	.01	.02	.03	.04	.05	.06	.07	.08	.09
0.0	.5000	.4960	.4920	.4880	.4840	.4801	.4761	.4721	.4681	.4641
0.1	.4602	.4562	.4522	.4483	.4443	.4404	.4364	.4325	.4286	.4247
0.2	.4207	.4168	.4129	.4090	.4052	.4013	.3974	.3936	.3897	.3859
0.3	.3821	.3783	.3745	.3707	.3669	.3632	.3594	.3557	.3520	.3483
0.4	.3446	.3409	.3372	.3336	.3300	.3264	.3228	.3192	.3156	.3121
0.5	.3085	.3050	.3015	.2981	.2946	.2912	.2877	.2843	.2810	.2776
0.6	.2743	.2709	.2676	.2643	.2611	.2578	.2546	.2514	.2483	.2451
0.7	.2420	.2389	.2358	.2327	.2296	.2266	.2236	.2206	.2177	.2148
0.8	.2119	.2090	.2061	.2033	.2005	.1977	.1949	.1922	.1894	.1867
0.9	.1841	.1814	.1788	.1762	.1736	.1711	.1685	.1660	.1635	.1611
1.0	.1587	.1562	.1539	.1515	.1492	.1469	.1446	.1423	.1401	.1379
1.1	.1357	.1335	.1314	.1292	.1271	.1251	.1230	.1210	.1190	.1170
1.2	.1151	.1131	.1112	.1093	.1075	.1056	.1038	.1020	.1003	.0985
1.3	.0968	.0951	.0934	.0918	.0901	.0885	.0869	.0853	.0838	.0823
1.4	.0808	.0793	.0778	.0764	.0749	.0735	.0721	.0708	.0694	.0681
1.5	.0668	.0655	.0643	.0630	.0618	.0606	.0594	.0582	.0571	.0559
1.6	.0548	.0537	.0526	.0516	.0505	.0495	.0485	.0475	.0465	.0455
1.7	.0446	.0436	.0427	.0418	.0409	.0401	.0392	.0384	.0375	.0367
1.8	.0359	.0351	.0344	.0336	.0329	.0322	.0314	.0307	.0301	.0294
1.9	.0287	.0281	.0274	.0268	.0262	.0256	.0250	.0244	.0239	.0233
2.0	.0228	.0222	.0217	.0212	.0207	.0202	.0197	.0192	.0188	.0183
2.1	.0179	.0174	.0170	.0166	.0162	.0158	.0154	.0150	.0146	.0143
2.2	.0139	.0136	.0132	.0129	.0125	.0122	.0119	.0116	.0113	.0110
2.3	.0107	.0104	.0102	.0099	.0096	.0094	.0091	.0089	.0087	.0084
2.4	.0082	.0080	.0078	.0075	.0073	.0071	.0069	.0068	.0066	.0064
2.5	.0062	.0060	.0059	.0057	.0055	.0054	.0052	.0051	.0049	.0048
2.6	.0047	.0045	.0044	.0043	.0041	.0040	.0039	.0038	.0037	.0036
2.7	.0035	.0034	.0033	.0032	.0031	.0030	.0029	.0028	.0027	.0026
2.8	.0026	.0025	.0024	.0023	.0023	.0022	.0021	.0021	.0020	.0019
2.9	.0019	.0018	.0018	.0017	.0016	.0016	.0015	.0015	.0014	.0014
3.0	.0013	.0013	.0013	.0012	.0012	.0011	.0011	.0011	.0010	.0010
3.50	.0002326									
4.00	.0000317									
4.50	.0000034									
5.00	.000000287									

The values for $a < 3.5$ are reprinted from Harry H. Kelejian and Wallace E. Oates, *Introduction to Econometrics*, 3rd ed. (New York: Harper & Row, 1989), p. 338. The values for $a \geq 3.5$ are based on E.S. Pearson and H.O. Hartley, eds., *Biometrika Tables for Statisticians*, Vol. 1, 3rd ed. (Cambridge: Cambridge University Press, 1970), table 1, with permission of the *Biometrika* trustees.

Table III SELECTED VALUES OF THE CHI-SQUARE DISTRIBUTION

(That value C^* such that $\Pr(C \geq C^*) = \alpha$, where $C \sim \chi^2_{(df)}$)

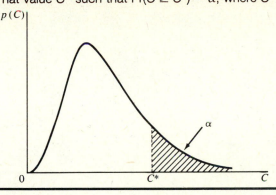

df	.995	.990	.975	.950	.050	.025	.010	.005
1	.00004	.00016	.00098	.00393	3.8415	5.0239	6.6349	7.8794
2	.0100	.0201	.0506	.1026	5.9915	7.3778	9.2103	10.5966
3	.0717	.1148	.2158	.3518	7.8147	9.3484	11.3449	12.8382
4	.2070	.2971	.4844	.7107	9.4877	11.1433	13.2767	14.8603
5	.4117	.5543	.8312	1.1455	11.0705	12.8325	15.0863	16.7496
6	.6757	.8721	1.2373	1.6354	12.5916	14.4494	16.8119	18.5476
7	.9893	1.2390	1.6899	2.1674	14.0671	16.0128	18.4753	20.2777
8	1.3444	1.6465	2.1797	2.7326	15.5073	17.5345	20.0902	21.9550
9	1.7349	2.0879	2.7004	3.3251	16.9190	19.0228	21.6660	23.5894
10	2.1559	2.5582	3.2470	3.9403	18.3070	20.4832	23.2093	25.1882
11	2.6032	3.0535	3.8158	4.5748	19.6751	21.9200	24.7250	26.7568
12	3.0738	3.5706	4.4038	5.2260	21.0261	23.3367	26.2170	28.2995
13	3.5650	4.1069	5.0088	5.8919	22.3620	24.7356	27.6882	29.8195
14	4.0747	4.6604	5.6287	6.5706	23.6848	26.1189	29.1412	31.3194
15	4.6009	5.2294	6.2621	7.2609	24.9958	27.4884	30.5779	32.8013
16	5.1422	5.8122	6.9077	7.9616	26.2962	28.8454	31.9999	34.2672
17	5.6972	6.4078	7.5642	8.6718	27.5871	30.1910	33.4087	35.7185
18	6.2648	7.0149	8.2308	9.3905	28.8693	31.5264	34.8053	37.1565
19	6.8440	7.6327	8.9065	10.1170	30.1435	32.8523	36.1909	38.5823
20	7.4338	8.2604	9.5908	10.8508	31.4104	34.1696	37.5662	39.9968
21	8.0337	8.8972	10.2829	11.5913	32.6706	35.4789	38.9322	41.4011
22	8.6427	9.5425	10.9823	12.3380	33.9244	36.7807	40.2894	42.7957
23	9.2604	10.1957	11.6886	13.0905	35.1725	38.0756	41.6384	44.1813
24	9.8862	10.8564	12.4012	13.8484	36.4150	39.3641	42.9798	45.5585
25	10.5197	11.5240	13.1197	14.6114	37.6525	40.6465	44.3141	46.9279
26	11.1602	12.1981	13.8439	15.3792	38.8851	41.9232	45.6417	48.2899
27	11.8076	12.8785	14.5734	16.1514	40.1133	43.1945	46.9629	49.6449
28	12.4613	13.5647	15.3079	16.9279	41.3371	44.4608	48.2782	50.9934
29	13.1211	14.2565	16.0471	17.7084	42.5570	45.7223	49.5879	52.3356
30	13.7867	14.9535	16.7908	18.4927	43.7730	46.9792	50.8922	53.6720
40	20.7065	22.1643	24.4330	26.5093	55.7585	59.3417	63.6907	66.7660
50	27.9907	29.7067	32.3574	34.7643	67.5048	71.4204	76.1539	79.4900
60	35.5345	37.4849	40.4817	43.1880	79.0819	83.2977	88.3794	91.9517
70	43.2752	45.4417	48.7576	51.7393	90.5312	95.0232	100.425	104.215
80	51.1719	53.5401	57.1532	60.3915	101.879	106.629	112.329	116.321
90	59.1963	61.7541	65.6466	69.1260	113.145	118.136	124.116	128.299
100	67.3276	70.0649	74.2219	77.9295	124.342	129.561	135.807	140.169

Table IV SELECTED VALUES OF THE *t* DISTRIBUTION

(That value t^* such that $Pr(t \geq t^*) = \alpha$, where $t \sim t^{(df)}$)

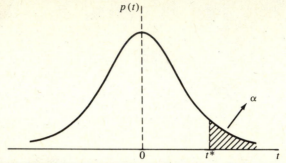

df	.25	.10	.05	.025	.01	.005	.001
1	1.00	3.08	6.31	12.71	31.82	63.66	318.31
2	.82	1.89	2.92	4.30	6.96	9.92	22.33
3	.76	1.64	2.35	3.18	4.54	5.84	10.21
4	.74	1.53	2.13	2.78	3.75	4.60	7.17
5	.73	1.48	2.02	2.57	3.36	4.03	5.89
6	.72	1.44	1.94	2.45	3.14	3.71	5.21
7	.71	1.42	1.90	2.36	3.00	3.50	4.78
8	.71	1.40	1.86	2.31	2.90	3.36	4.50
9	.70	1.38	1.83	2.26	2.82	3.25	4.30
10	.70	1.37	1.81	2.23	2.76	3.17	4.14
11	.70	1.36	1.80	2.20	2.72	3.11	4.02
12	.70	1.36	1.78	2.18	2.68	3.06	3.93
13	.69	1.35	1.77	2.16	2.65	3.01	3.85
14	.69	1.34	1.76	2.14	2.62	2.98	3.79
15	.69	1.34	1.75	2.13	2.60	2.95	3.73
16	.69	1.34	1.75	2.12	2.58	2.92	3.69
17	.69	1.33	1.74	2.11	2.57	2.90	3.65
18	.69	1.33	1.73	2.10	2.55	2.88	3.61
19	.69	1.33	1.73	2.09	2.54	2.86	3.58
20	.69	1.32	1.72	2.09	2.53	2.84	3.55
21	.69	1.32	1.72	2.08	2.52	2.83	3.53
22	.69	1.32	1.72	2.07	2.51	2.82	3.50
23	.68	1.32	1.71	2.07	2.50	2.81	3.48
24	.68	1.32	1.71	2.06	2.49	2.80	3.47
25	.68	1.32	1.71	2.06	2.48	2.79	3.45
26	.68	1.32	1.71	2.06	2.48	2.78	3.44
27	.68	1.31	1.70	2.05	2.47	2.77	3.42
28	.68	1.31	1.70	2.05	2.47	2.76	3.41
29	.68	1.31	1.70	2.04	2.46	2.76	3.40
30	.68	1.31	1.70	2.04	2.46	2.75	3.38
∞	.67	1.28	1.64	1.96	2.33	2.58	3.09

Based on E. S. Pearson and H. O. Hartley, eds., *Biometrika Tables for Statisticians,* Vol. 1, 3rd ed. (Cambridge: Cambridge University Press, 1970), table 12, with permission of the *Biometrika* trustees.

Table V   SELECTED VALUES OF THE F DISTRIBUTION

(That value f^* such that $Pr(F \geq f^*) = .05$, where $F \sim F^{(k_1, k_2)}$)

k_2 \ k_1	1	2	3	4	5	6	7	8	9	10
1	161.4	199.5	215.7	224.6	230.2	234.0	236.8	238.9	240.5	241.9
2	18.51	19.00	19.16	19.25	19.30	19.33	19.35	19.37	19.38	19.40
3	10.13	9.55	9.28	9.12	9.01	8.94	8.89	8.85	8.81	8.79
4	7.71	6.94	6.59	6.39	6.26	6.16	6.09	6.04	6.00	5.96
5	6.61	5.79	5.41	5.19	5.05	4.95	4.88	4.82	4.77	4.74
6	5.99	5.14	4.76	4.53	4.39	4.28	4.21	4.15	4.10	4.06
7	5.59	4.74	4.35	4.12	3.97	3.87	3.79	3.73	3.68	3.64
8	5.32	4.46	4.07	3.84	3.69	3.58	3.50	3.44	3.39	3.35
9	5.12	4.26	3.86	3.63	3.48	3.37	3.29	3.23	3.18	3.14
10	4.96	4.10	3.71	3.48	3.33	3.22	3.14	3.07	3.02	2.98
11	4.84	3.98	3.59	3.36	3.20	3.09	3.01	2.95	2.90	2.85
12	4.75	3.89	3.49	3.26	3.11	3.00	2.91	2.85	2.80	2.75
13	4.67	3.81	3.41	3.18	3.03	2.92	2.83	2.77	2.71	2.67
14	4.60	3.74	3.34	3.11	2.96	2.85	2.76	2.70	2.65	2.60
15	4.54	3.68	3.29	3.06	2.90	2.79	2.71	2.64	2.59	2.54
16	4.49	3.63	3.24	3.01	2.85	2.74	2.66	2.59	2.54	2.49
17	4.45	3.59	3.20	2.96	2.81	2.70	2.61	2.55	2.49	2.45
18	4.41	3.55	3.16	2.93	2.77	2.66	2.58	2.51	2.46	2.41
19	4.38	3.52	3.13	2.90	2.74	2.63	2.54	2.48	2.42	2.38
20	4.35	3.49	3.10	2.87	2.71	2.60	2.51	2.45	2.39	2.35
21	4.32	3.47	3.07	2.84	2.68	2.57	2.49	2.42	2.37	2.32
22	4.30	3.44	3.05	2.82	2.66	2.55	2.46	2.40	2.34	2.30
23	4.28	3.42	3.03	2.80	2.64	2.53	2.44	2.37	2.32	2.27
24	4.26	3.40	3.01	2.78	2.62	2.51	2.42	2.36	2.30	2.25
25	4.24	3.39	2.99	2.76	2.60	2.49	2.40	2.34	2.28	2.24
26	4.23	3.37	2.98	2.74	2.59	2.47	2.39	2.32	2.27	2.22
27	4.21	3.35	2.96	2.73	2.57	2.46	2.37	2.31	2.25	2.20
28	4.20	3.34	2.95	2.71	2.56	2.45	2.36	2.29	2.24	2.19
29	4.18	3.33	2.93	2.70	2.55	2.43	2.35	2.28	2.22	2.18
30	4.17	3.32	2.92	2.69	2.53	2.42	2.33	2.27	2.21	2.16
40	4.08	3.23	2.84	2.61	2.45	2.34	2.25	2.18	2.12	2.08
60	4.00	3.15	2.76	2.53	2.37	2.25	2.17	2.10	2.04	1.99
120	3.92	3.07	2.68	2.45	2.29	2.17	2.09	2.02	1.96	1.91
∞	3.84	3.00	2.60	2.37	2.21	2.10	2.01	1.94	1.88	1.83

k_2	12	15	20	24	30	40	60	120	∞
1	243.9	245.9	248.0	249.1	250.1	251.1	252.2	253.3	254.3
2	19.41	19.43	19.45	19.45	19.46	19.47	19.48	19.49	19.50
3	8.74	8.70	8.66	8.64	8.62	8.59	8.57	8.55	8.53
4	5.91	5.86	5.80	5.77	5.75	5.72	5.69	5.66	5.63
5	4.68	4.62	4.56	4.53	4.50	4.46	4.43	4.40	4.36
6	4.00	3.94	3.87	3.84	3.81	3.77	3.74	3.70	3.67
7	3.57	3.51	3.44	3.41	3.38	3.34	3.30	3.27	3.23
8	3.28	3.22	3.15	3.12	3.08	3.04	3.01	2.97	2.93
9	3.07	3.01	2.94	2.90	2.86	2.83	2.79	2.75	2.71
10	2.91	2.85	2.77	2.74	2.70	2.66	2.62	2.58	2.54
11	2.79	2.72	2.65	2.61	2.57	2.53	2.49	2.45	2.40
12	2.69	2.62	2.54	2.51	2.47	2.43	2.38	2.34	2.30
13	2.60	2.53	2.46	2.42	2.38	2.34	2.30	2.25	2.21
14	2.53	2.46	2.39	2.35	2.31	2.27	2.22	2.18	2.13
15	2.48	2.40	2.33	2.29	2.25	2.20	2.16	2.11	2.07
16	2.42	2.35	2.28	2.24	2.19	2.15	2.11	2.06	2.01
17	2.38	2.31	2.23	2.19	2.15	2.10	2.06	2.01	1.96
18	2.34	2.27	2.19	2.15	2.11	2.06	2.02	1.97	1.92
19	2.31	2.23	2.16	2.11	2.07	2.03	1.98	1.93	1.88
20	2.28	2.20	2.12	2.08	2.04	1.99	1.95	1.90	1.84
21	2.25	2.18	2.10	2.05	2.01	1.96	1.92	1.87	1.81
22	2.23	2.15	2.07	2.03	1.98	1.94	1.89	1.84	1.78
23	2.20	2.13	2.05	2.01	1.96	1.91	1.86	1.81	1.76
24	2.18	2.11	2.03	1.98	1.94	1.89	1.84	1.79	1.73
25	2.16	2.09	2.01	1.96	1.92	1.87	1.82	1.77	1.71
26	2.15	2.07	1.99	1.95	1.90	1.85	1.80	1.75	1.69
27	2.13	2.06	1.97	1.93	1.88	1.84	1.79	1.73	1.67
28	2.12	2.04	1.96	1.91	1.87	1.82	1.77	1.71	1.65
29	2.10	2.03	1.94	1.90	1.85	1.81	1.75	1.70	1.64
30	2.09	2.01	1.93	1.89	1.84	1.79	1.74	1.68	1.62
40	2.00	1.92	1.84	1.79	1.74	1.69	1.64	1.58	1.51
60	1.92	1.84	1.75	1.70	1.65	1.59	1.53	1.47	1.39
120	1.83	1.75	1.66	1.61	1.55	1.50	1.43	1.35	1.25
∞	1.75	1.67	1.57	1.52	1.46	1.39	1.32	1.22	1.00

Short Answers to Odd Numbered Problems

2.1 Discrete: number of wage earners in a family. Continuous: money supply.

2.3 (**a**) true;   (**b**) false;   (**c**) true;   (**d**) true;   (**e**) false;   (**f**) true;   (**g**) false;   (**h**) false;   (**i**) true;   (**j**) false

2.5 $9540

2.7 Y is a random variable since X is a random variable.

CHAPTER 3

3.1

X	$f(X)$
0	3/20
1	6/20
2	8/20
3	2/20
4	1/20

3.3

X	$f(X)$
4000–8999	3/10
9000–13,999	5/10
14,000–18,999	2/10

3.5 mean = 1.6;   variance = 1.04;   median = 2

3.7 mean = 10,275.4;   variance = 11,338,471.64 (Note: Using the relative frequency distribution or histogram would only give an approximation.)

3.9 mean = 2.4;   variance = .84

3.11 1500

3.13 mean = 14;   variance = 144

3.15 (a) $3000;   (b) $1600;   (c) $60,000,000,000

3.17 (a) $\mu_W = \mu_X + \mu_Y$;   (b) $\mu_W = a\mu_X + b\mu_Y + c$; part (a) is the special case where $a = b = 1$ and $c = 0$.

CHAPTER 4

4.1 $\bar{X} = 9$; $s^2 = 8.67$

4.3 $1,590,000

4.5 $\bar{X} = 1.62$; $s^2 = 1.0107$

4.7 18.75 hours

4.9 $\sum (X_i - \bar{X})^2 = \sum (X_i^2 - 2\bar{X}X_i + \bar{X})^2$

$$= \sum X_i^2 - 2\bar{X} \sum X_i + n(\bar{X})^2$$

$$= \sum X_i^2 - 2n(\bar{X})^2 + n(\bar{X})^2$$

$$= \sum X_i^2 - n(\bar{X})^2$$

From this the proof follows directly.

CHAPTER 5

5.1 (a)

x	f(X)
1	1/6
2	1/6
3	1/6
4	1/6
5	1/6
6	1/6

(b) same as relative frequencies

5.3 (a) $2^{10} = 1024$;   (b) 1023/1024

5.5 .37

5.7 (a) .22; (b) .14; (c) .06; (d) .58

5.9 (a) 1/3; (b) 1/7

5.11 (a) $\frac{1}{49} \times \frac{2}{50} \times \frac{3}{51} \times \frac{4}{52}$; (b) $\frac{3}{49} \times \frac{4}{50} \times \frac{3}{51} \times \frac{4}{52}$; (c) $\frac{48}{49} \times \frac{2}{50}$ $\times \frac{3}{51} \times \frac{4}{52}$; (d) $\frac{48}{49} \times \frac{2}{50} \times \frac{3}{51} \times \frac{4}{52} \times 4$; (e) $(\frac{48}{49} \times \frac{2}{50} \times \frac{3}{51}$ $\times \frac{4}{52} \times 4) + (\frac{1}{49} \times \frac{2}{50} \times \frac{3}{51} \times \frac{4}{52})$

5.13 (a) .84, assuming the same success rate as usual.

5.15 8/15 or 53.3%

5.17 .82

5.19 .025

5.21 .316

5.23 (a) .0625; (b) .46875

5.25 A and C are independent.

5.27 (a) $\frac{4}{52} \times \frac{4}{52} \times \frac{1}{52}$; (b) $\frac{4}{52} \times \frac{4}{52} \times \frac{4}{52}$; (c) $\frac{4}{52} \times \frac{4}{52}$ $\times \frac{1}{52}$; (d) $\frac{4}{52} \times \frac{4}{52} \times \frac{1}{52} \times 3$; (e) $\frac{4}{52} \times \frac{4}{52} \times \frac{4}{52} \times 3$

5.29 (a) yes; (b) no; (c) yes

5.31 1/2, as successive tosses are independent.

5.33 Since the lottery picks the numbers randomly, it makes no difference whether you have on your ticket fixed numbers (by your choice) or numbers generated by computer. Thus you should play the cheaper game and let the computer generate your numbers.

5.35 (a) $\frac{4}{52} \times \frac{3}{51} \times \frac{2}{50} \times \frac{1}{49} \times \frac{48}{48} \times 5 \times 13$

(b) $\frac{13}{52} \times \frac{12}{51} \times \frac{11}{50} \times \frac{10}{49} \times \frac{9}{48} \times 4$ (Note: This includes, for simplicity, the chance of a "straight flush," which is not typically viewed as (simply) a flush. See problem (5.38).)

(c) $(\frac{1}{49} \times \frac{48}{48} \times 2) + (\frac{4}{49} \times \frac{3}{48} \times 12) = \frac{240}{2352}$

5.37 If the drug were unrelated to blurry vision, 10 (2%) of those getting the drug would have had vision problems. So 5 people getting the drug who suffer blurry vision wouldn't have had it without the drug; 5/15 is then the fraction of those getting the drug and having blurry vision who have the vision problem due to the drug. (This assumes that there are no errors in the study, i.e. that the 3% and 2% figures are equal to the population parameters.)

5.39 $\frac{1,224,000}{311,875,200}$

5.41 (a) (nc = number correct)
$$Pr(nc = 0) = 2/6; \quad Pr(nc = 1) = 3/6;$$
$$Pr(nc = 2) = 0; \quad Pr(nc = 3) = 1/6.$$
(b) (ncr = number with correct result)
$$Pr(ncr = 0) = 0; \quad Pr(ncr = 1) = 4/6;$$
$$Pr(ncr = 2) = 0; \quad Pr(ncr = 3) = 2/6.$$

5.43 For n spaces, $2/n$.

CHAPTER 6

6.1 $\mu = 1.0 \quad \sigma^2 = 1.0$

6.3

X	p(X)
0	1/8
1	3/8
2	3/8
3	1/8

$$\mu = 3/2, \quad \sigma^2 = 6/8 \text{ so } \sigma = .866$$

6.5 $\sigma^2 = \sum (X - \mu)^2 p(X)$

$$= \sum [(X^2 - 2\mu X + \mu^2)p(X)]$$

$$= \sum [X^2 p(X) - 2\mu X p(X) + \mu^2 p(X)]$$

$$= [\sum X^2 p(X)] - 2\mu \sum X p(X) + \mu^2 \sum p(X)$$

$$= [\sum X^2 p(X)] - 2\mu^2 + \mu^2$$

$$= [\sum X^2 p(X)] - \mu^2$$

6.7 .4444

6.9 .3672

6.11 .0334

6.13 .5077

6.15 (a) In this situation, suppose there are 10 marbles in a hat, 8 of which are red and 2 of which are blue. I pick 5 with replacement. Note $Pr(\text{red}) = .8$ and $Pr(\text{blue}) = .2$. Clearly, the probability of getting exactly 3 reds (and therefore 2 blues) is the same as the probability of getting exactly 2 blues (and therefore 3 reds). The former is $p(3)$ for $\pi = .8$, and the latter is $p(2)$ for $\pi = .2$.

(b)

$$p(s_0 \mid \pi = \pi_0) = \frac{n!}{s_0!(n - s_0)!} (\pi_0)^{s_0} (1 - \pi_0)^{(n-s_0)}$$

$$= \frac{n!}{(n - s_0)!s_0!} (1 - \pi_0)^{(n-s_0)} (\pi_0)^{s_0}$$

Suppose $n - s_0 = s_1$ and $1 - \pi_0 = \pi_1$:

$$p(s_1 \mid \pi = \pi_1) = \frac{n!}{s_1!(n - s_1)!} (\pi_1)^{s_1} (1 - \pi_1)^{(n-s_1)}$$

$$= \frac{n!}{(n - s_0)!s_0!} (1 - \pi_0)^{(n-s_0)} (\pi_0)^{s_0}$$

Clearly, $p(s_1 \mid \pi = \pi_1) = p(n - s_0 \mid \pi = 1 - \pi_0)$. But, from above, $p(s_1 \mid \pi = \pi_1) = p(s_0 \mid \pi = \pi_0)$. Thus, $p(s_0 \mid \pi = \pi_0) = p(n - s_0 \mid \pi = 1 - \pi_0)$.

6.17 (a) $p(Y) = 0$ for $Y < 0$; $p(Y) = 1/12$ for $0 \le Y \le 2$; $p(Y) = 4/12$ for $2 < Y \le 4$; $p(Y) = 1/12$ for $4 < Y \le 6$; $p(Y) = 0$ for $Y > 6$.

6.19 (a) .2061;   **(b)** .3202;   **(c)** .2224;   **(d)** 16.27;   **(e)** .9484

6.21 (a) .0808 or 8.08%;   **(b)** \$21,400

6.23 1.96

6.25 .1810

6.27 .20716

6.29 $\frac{4}{52} \times \frac{3}{51} \times \frac{2}{50} \times \frac{48}{49} \times \frac{47}{48} \times 10 \times 13$ Notes: (1) This does not include the possibility of getting "four of a kind"; (2) This does include the possibility that the remaining 2 cards are of the same denomination, that is, that we have a "full house."

CHAPTER 7

7.1 $E(Y) = 6$

7.3 30

7.5 $\sigma^2 = 35/12$

7.7 $E(1/X) = 294/720 = .408$
$1/E(X) = 2/7 = .286$

7.9 72

CHAPTER 8

8.1 (a)

X \ Y	0	1	p(X)
0	1/8	0	1/8
1	2/8	1/8	3/8
2	1/8	2/8	3/8
3	0	1/8	1/8
p(Y)	4/8	4/8	

(b) Consider $(0, 0)$. Is $p(0, 0) = p_X(0)p_Y(0)$? Is $1/8 = (1/8) \times (4/8)$? No. Since it is then not the case that $p(X, Y) = p(X)p(Y)$ for all (X, Y) we know that X and Y are not independent.

(c) Random variables are independent if knowledge concerning one of them does not influence our evaluation of probabilities regarding the other. In this case, knowledge as to whether $Y = 0$ or $Y = 1$ clearly affects the (conditional) distribution of X. For example, $Pr(X = 3 | Y = 0) = 0$ but $Pr(X = 3 | Y = 1) = 1/4$.

8.3 (a)

X \ Y	0	1	2	p(X)
0	0	0	.04	.04
1	0	.20	.12	.32
2	.25	.30	.09	.64
p(Y)	.25	.50	.25	

(b) Is $p(X, Y) = p(X)p(Y)$ for each (X, Y)? Is $p(0, 0) = p_X(0)p_Y(0)$? No. Thus X and Y are not independent.

(c) If a student is from New York City, he/she may go to an Ivy League school, but may not. If a student is not from New York City, then he/she does attend an Ivy League school.

8.5 (a)

X \ Y	0	1	p(X)
0	.2	.6	.8
1	.05	.15	.2
p(Y)	.25	.75	

(b)

X	p(X | Y = 0)
0	4/5
1	1/5

(c)

Y	p(Y | X = 1)
0	1/4
1	3/4

(d) X and Y are independent, which can be proved by showing that $p(X, Y) = p(X)p(Y)$ for all (X, Y).

8.7

X \ Y	1	2
0	4/9	2/9
1	2/9	1/9

8.9 $E(X) = 1.8$
$\sigma^2 = .56$

8.11 .37

8.13 .26

8.15 (a)

W	p(W)
−.25	.6
.25	.4

(b) −.05; **(c)** −.05; **(d)** They are the same, as they should be, since cov $(X, Y) = E[(X - E(X))(Y - E(Y))]$.

8.17 cov $(X, Y) = E\{[X - E(X)][Y - E(Y)]\}$
$= E[XY - XE(Y) - YE(X) + E(X)E(Y)]$
$= E(XY) - E(X)E(Y) - E(Y)E(X) + E(X)E(Y)$
$= E(XY) - E(X)E(Y)$

8.19 (a) 0
(b) Is $p(X, Y) = p(X)p(Y)$ for each (X, Y)? Is $p(0,0) = p_X(0)p_Y(0)$? No. Thus X and Y are not independent.

(c) $E(X|Y = 0) = 1.5$; $E(X|Y = 1) = 1.5$; $E(X|Y = 2) = 1.5$. $E(Y|X = 0) = 1$; $E(Y|X = 1) = 1$; $E(Y|X = 2) = 1$; $E(Y|X = 3) = 1$. Thus, $E(X|Y)$ is invariant to the value of Y, and $E(Y|X)$ is invariant to the value of X. What we expect $X(Y)$ to be as $Y(X)$ changes does not change. This suggests that the covariance is 0, as seen above. But there is some sort of relationship between X and Y, as seen in part (b); this is just not the type of relationship which the covariance uncovers.

8.21 Since cov $(X, Y) = 0$ we know that $\rho = 0$.

8.23 $\rho = -1$. Intuitively, there are three possible values of (X, Y): $(-1,1)$, $(0,0)$, and $(1,-1)$. For each, $X + Y = 0$ or $Y = -X$; this suggests a perfect (and negative) linear relationship. Thus, $\rho = -1$.

8.25 .822

8.27 $\sum (X_i - \bar{X})(Y_i - \bar{Y})$

$= \sum [X_iY_i - \bar{Y}X_i - \bar{X}Y_i + \bar{X}\bar{Y}]$

$= \sum X_iY_i - \bar{Y} \sum X_i - \bar{X} \sum Y_i + \sum \bar{X}\bar{Y}$

$= \sum X_iY_i - n\bar{X}\bar{Y} - n\bar{X}\bar{Y} + n\bar{X}\bar{Y}$

$= \sum X_iY_i - n\bar{X}\bar{Y}$

Thus,

$$\frac{1}{n-1} \sum (X_i - \bar{X})(Y_i - \bar{Y})$$

$$= \frac{1}{n-1}\left[\left(\sum X_iY_i\right) - n(\bar{X})(\bar{Y})\right]$$

8.29 $E(W) = -7$
$\sigma_W^2 = 109$

8.31 mean = \$23,000; standard deviation = \$6422.62.

8.33 mean = 10; variance = 1.94

8.35 7

8.37 .4364

8.39 .282500;   .826750

CHAPTER 9

9.1 (a)

x	p(X)
1	1/3
2	1/3
3	1/3

$E(X) = 2;$ var $(X) = 2/3$

(b)

$\bar{x}$	$p(\bar{X})$
1	1/9
1.5	2/9
2	3/9
2.5	2/9
3	1/9

$E(\bar{X}) = 2 = E(X);$ var $(\bar{X}) = 1/3 =$ var $(X)/2$

(c)

$\bar{x}$	$p(\bar{X})$
1	1/27
4/3	3/27
5/3	6/27
2	7/27
7/3	6/27
8/3	3/27
3	1/27

$E(\bar{X}) = 2 = E(X);$ var $(\bar{X}) = 2/9 =$ var $(X)/3$

9.3 $E(\bar{X}) = 5;$ var $(\bar{X}) = 16/9$

9.5 (a) .084
(b) .0418
The two answers are different since if the parent population is not normal (and it is not in this case) then the Central Limit Theorem only provides an approximation.

9.7 .9332

9.9 .9429, assuming that the inflation rate is independent from one year to the next (which is unlikely).

9.11 (a) .5987
(b) .8686
(c) The chance of a very low return on the investor's portfolio is decreased by diversification. This diversification

lessens the variation in overall return and increases the chance that the (sample) mean return will be within a relatively narrow range of the population mean of 10%.

9.13 .8749

9.15 .0934

9.17 .0594

9.19 .0084

9.21 $\Psi = 1.96 \dfrac{\sigma}{\sqrt{n}}$, and this is then a likely margin of error.

CHAPTER 10

10.1 $E(\hat{\mu}) = \frac{1}{6} E(X_1) + \frac{1}{6} E(X_2) + \frac{2}{3} E(X_3)$

$\qquad = \frac{1}{6} \mu + \frac{1}{6} \mu + \frac{2}{3} \mu = \mu$

Thus, $\hat{\mu}$ is an unbiased estimator of μ.

10.3 (a) $E(\hat{\mu}_1) = \frac{1}{10} E(X_1) + \frac{1}{10} E(X_2) + \frac{2}{5} E(X_3) + \frac{2}{5} E(X_4)$

$\qquad = \frac{1}{10} \mu + \frac{1}{10} \mu + \frac{2}{5} \mu + \frac{2}{5} \mu = \mu$

$E(\hat{\mu}_2) = \frac{1}{9} E(X_1) + \frac{1}{9} E(X_2) + \frac{1}{9} E(X_3) + \frac{2}{3} E(X_4)$

$\qquad = \frac{1}{9} \mu + \frac{1}{9} \mu + \frac{1}{9} \mu + \frac{2}{3} \mu = \mu$

(b) When comparing unbiased estimators, the one with the smaller variance is better. Let's denote var $(X_i) = \sigma^2$ for each i.

$$\text{var} \, (\hat{\mu}_1) = .34\sigma^2; \text{ var} \, (\hat{\mu}_2) = \frac{39}{81} \sigma^2 = .48\sigma^2.$$

Since var $(\hat{\mu}_1) <$ var $(\hat{\mu}_2)$ for any σ^2, and since both are unbiased, $\hat{\mu}_1$ is a better estimator of μ.

10.5 $E(\hat{\mu}) = \dfrac{n\mu}{n-1} \neq \mu; \quad \text{bias}(\hat{\mu}) = \dfrac{\mu}{n-1}$

10.7 Both are unbiased; when comparing unbiased estimators the one with the smaller variance is better. Let's denote var $(X) = \sigma^2$; thus var $(Y) = .5\sigma^2$.

$$\text{var} \, (\hat{\mu}_1) = \frac{6\sigma^2}{32} \, ; \text{ var} \, (\hat{\mu}_2) = \frac{6\sigma^2}{36} \, .$$

Since both are unbiased we then know that $\hat{\mu}_2$ is a better estimator of μ.

10.9 All are unbiased, so the one we should use is the one with the smallest variance.

$$\text{var}\,(\hat{\mu}_1) = \frac{\sigma^2}{n} \qquad (\text{where } \sigma^2 \text{ is the variance of } X)$$

$$\text{var}\,(\hat{\mu}_2) = \frac{\sigma^2}{n}$$

$$\text{var}\,(\hat{\mu}_3) = .625\,\frac{\sigma^2}{n}$$

$$\text{var}\,(\hat{\mu}_4) = .68\,\frac{\sigma^2}{n}$$

$\hat{\mu}_3$ is then the best as it has the smallest variance.

10.11 $\text{mse}\,(\bar{X}) = \dfrac{\sigma^2}{n}$ where $\sigma^2 = \text{var}\,(X)$

10.13 If $\text{var}\,(X_i) = \sigma^2$ then

$$\text{mse}\,(\hat{\mu}_1) = \frac{20}{64}\,\sigma^2 = .3125\sigma^2$$

$$\text{mse}\,(\hat{\mu}_2) = \frac{4}{25}\,\sigma^2 + \frac{1}{25}\,\mu^2 = .16\sigma^2 + .04\mu^2$$

$$\text{mse}\,(\hat{\mu}_1) > \text{mse}\,(\hat{\mu}_2) \quad \text{if } \sigma^2 > \frac{.04}{.1525}\,\mu^2$$

10.15 $\hat{\mu}_1$ and $\hat{\mu}_2$ are unbiased. The general rule is that if we have a simple random sample of size n from a population with mean μ, then $\hat{\mu} = \Sigma_{i=1}^{n}\, a_i X_i$ is unbiased if $\Sigma_{i=1}^{n}\, a_i = 1$.

10.17 **(a)** All are unbiased.
(b) From most to least efficient, the ranking is $\hat{W}_2$, $\hat{W}_3$, $\hat{W}_4$, $\hat{W}_1$.

10.19 bias $(\hat{A}_1) = \dfrac{\sigma^2}{2}$; bias $(\hat{A}_2) = \sigma^2$; thus bias $(\hat{A}_1)$ is smaller.

Chapter 11

11.1 **(a)** (6.37, 9.63); **(b)** (6.63, 9.37); **(c)** (5.85, 10.15)
11.3 $n = 61.47$, or, rounding up, 62.

11.5 **(a)** (8.16, 11.84); **(b)** (8.47, 11.53); **(c)** (7.50, 12.50)

11.7 (1902,   2098)

11.9 (8730.5,   44826.5) thousand persons.

11.11 (46.53,   53.39)

11.13 (2.69,   7.19) using the exact t value.

11.15 (−53.82,   113.82)

We are 95% confident that the true difference in means is within this interval; the observed difference in sample means could plausibly have been due to random variability when the population means were equal.

11.17 (−.20,   .40)

11.19 (−15585.94,   20839.2)

11.21 (−2.91,   9.85)

11.23 (−1.697,   1.987) using the exact t value.

11.25 Although this conclusion is consistent with the evidence, it is not the only possibility. Perhaps hospitals are more likely to be established in highly infected areas so as to treat those with the illness. Or, perhaps the *true* incidence of the disease is not different in areas with hospitals, but the presence of the medical care causes a higher *reported* incidence of the disease.

11.27 Mary needs a salary in city #2 of above $2300 for her financial situation to improve. A 95% confidence interval for the (population) mean job in city #2 is (2643,   3357). Since the mean job in city #2 then has (with 95% confidence) a salary over $2300, Mary should *expect* her financial situation to improve. (It is of course possible that she will get a job with a below average salary in city #2; the word "expect" is then used in the mathematical sense of an expected value.

11.29 Remember that

$$s_1^2 = \frac{1}{n_1 - 1} \sum_{i=1}^{n_1} (X_{1i} - \bar{X}_1)^2 \quad \text{and}$$

$$s_2^2 = \frac{1}{n_2 - 1} \sum_{i=1}^{n_2} (X_{2i} - \bar{X}_2)^2$$

Substituting these into the formula for the pooled variance yields

$$s_p^2 = \left(\frac{n_1 - 1}{n_1 + n_2 - 2} \right) \left(\frac{1}{n_1 - 1} \sum_{i=1}^{n_1} (X_{1i} - \bar{X}_1)^2 \right)$$

$$+ \left(\frac{n_2 - 1}{n_1 + n_2 - 2} \right) \left(\frac{1}{n_2 - 1} \sum_{i=1}^{n_2} (X_{2i} - \bar{X}_2)^2 \right)$$

$$= \frac{\sum_{i=1}^{n_1} (X_{1i} - \bar{X}_1)^2 + \sum_{i=1}^{n_2} (X_{2i} - \bar{X}_2)^2}{n_1 + n_2 - 2}$$

11.31 (12.70,   14.82)

CHAPTER 12

12.1 $Pr(t^{(24)} \geq 1.25)$;   .25 > prob-value > .10;   cannot reject H_0.

12.3 $Pr(t^{(24)} \geq 2.5)$;   .01 > prob-value > .005;   reject H_0.

12.5 $Pr(t^{(35)} \geq 1.2)$;   prob-value = .1151 (using Z approximation);   cannot reject H_0.

12.7 $Pr(t^{(9)} \leq -3.32)$;   .005 > prob-value > .001;   reject H_0.

12.9 $Pr(t^{(49)} \geq 4.21)$;   .0000317 > prob-value > .0000034 using Z approximation;   prob-value = .0001 using actual t value;   reject H_0.

12.11 $Pr(t^{(18)} \geq 5.66)$;   .001 > prob-value;   reject H_0.

12.13 $Pr(t^{(33)} \geq 1.94)$;   prob-value = .0262 using Z approximation;   reject H_0.

12.15 $Pr(t^{(8)} \geq .36)$;   prob-value > .25;   cannot reject H_0.

12.17 $Pr(t^{(31)} \geq 2.62)$;   prob-value = .013 using actual t value;   prob-value = .0044 using Z approximation;   reject H_0.

12.19 $Pr(|t^{(39)}| \geq 3.69)$;   .0004652 > prob-value > .0000634 using Z approximation;   reject H_0.

12.21 $Pr(|t^{(19)}| \geq 2.87)$;   .01 > prob-value > .002;   reject H_0.

12.23 $Pr(|t^{(48)}| \geq 1.54)$;   prob-value = .13 using actual t value;   prob-value = .1236 using Z approximation;   cannot reject H_0.

12.25 (a) $Pr(t^{(33)} \geq 5.09)$;   prob-value < .000000287 using the Z approximation;   reject H_0.
(b) H_0: $\mu_A - \mu_B = 5$
H_1: $\mu_A - \mu_B > 5$

is the test. Carrying this out, the prob-value is $Pr(t^{(33)} \geq 1.91)$, so prob-value = .0281 using the Z approximation. So we would reject H_0 and conclude that A is "better."

12.27 (a) (o = own a sports car; no = do not own a sports car). The test is

$$H_0: \quad \mu_o - \mu_{no} = 0$$

$$H_1: \quad \mu_o - \mu_{no} > 0$$

Carrying this out, the prob-value is $Pr(t^{(20)} \geq 2.55)$ and .01 > prob-value > .005. We therefore reject H_0, and conclude that owners of sports cars have (on average) higher pulse rates than individuals who do not own sports cars.

(b) No, for example, perhaps those with high pulses are more attracted to sports cars, but the cars actually have no causal effect on pulse.

12.29 The distribution is $\hat{\mu} \sim N(\mu, 6.94)$;   the prob-value is then $Pr(|Z| \geq 2.66) = .0039$. We would then reject H_0.

CHAPTER 13

13.1 $\hat{Y}_i = -15,114.35 + 2714.29 X_i$

13.3 $\hat{Y}_i = 3.036 - .67 X_i$

13.5 The quantity theory of money is $MV = PQ$, where M is the money supply, V is velocity, P is the price level, and Q is the level of output. This implies $M^* + V^* = P^* + Q^*$ where the "*" superscript represents a percent change. A simple monetarist position would view V and Q as constants, in which case $M^* = P^*$ which is consistent with the linearity assumption. But to the extent that V or Q are changing, there are additional variables influencing M^* and P^* which have not been incorporated into the model. (This could lead to there being a nonlinear relationship between M^* and P^* if V^* and Q^* are ignored.)

13.7 (a) $s_\epsilon = 6401$; $\alpha = (-73529.51, \quad 43300.81)$; $\beta = (-1132.48, \quad 6561.06)$

(b) $Pr(t^{(3)} \geq 2.24)$;   .10 > prob-value > .05;   cannot reject H_0.

(c) $Pr(|t^{(3)}| \geq .82)$; $.5 >$ prob-value $> .2$; cannot reject H_0.

13.9 (a) $s_\epsilon = .87$; $\alpha = (-.337, 6.409)$; $\beta = (-4.70, 3.36)$
(b) $Pr(|t^{(4)}| \geq .46)$; prob-value $> .5$; cannot reject H_0.
(c) $Pr(t^{(4)} \geq 2.50)$; $.05 >$ prob-value $> .025$; reject H_0.

13.11 $\sum \hat{\epsilon}_i = \sum (Y_i - \hat{Y}_i)$

$$= \sum (Y_i - \hat{\alpha} - \hat{\beta}X_i)$$

$$= \sum Y_i - n\hat{\alpha} - \hat{\beta} \sum X_i$$

$$= n\bar{Y} - n(\bar{Y} - \hat{\beta}\bar{X}) - \hat{\beta}n\bar{X}$$

$$= n\bar{Y} - n\bar{Y} + n\hat{\beta}\bar{X} - \hat{\beta}n\bar{X}$$

$$= 0$$

13.13 For $X = 13$, $\hat{Y} = 20{,}171.42$. For $X = 8$, $\hat{Y} = 6559.97$. As $X = 13$ is within the range of X's for which we have observations, but $X = 8$ is not, we should be more confident regarding the first prediction and more cautious regarding the second.

13.15 $\widehat{NETINV}_i = 81.492 + .10692 \, CFSAL_i$

$$(8.4625) \quad (.086111)$$

where the numbers in parentheses are the estimated standard errors.

13.17 (a) Since QD_i rises without P_i changing, $\epsilon_i > 0$.
(b) We should expect P_i to rise, as a demand increase causes price to rise.
(c) This suggests that P_i and ϵ_i are correlated, and that this is due to two-way causality. Assumption 2 of the CLRM is then violated.

13.19 In the regression model $Y_i = \alpha + \beta X_i$, X_i "helps" explain the variation in Y_i if and only if $\beta \neq 0$. If we can reject H_0: $\beta = 0$ then that suggests that there is a statistically significant relationship between X and Y. But this is effectively what R^2 seeks to uncover! (Note that the R^2 we calculate is really a sample R^2 since it is based only on the data in our sample. Algebraically

$$R^2 = \frac{\sum (\hat{Y}_i - \bar{Y})^2}{\sum (Y_i - \bar{Y})^2}$$

Since the denominator is always positive, let's focus on the numerator.

$$\sum (\hat{Y}_i - \bar{Y})^2 = \sum (\hat{\alpha} + \hat{\beta}X_i - \hat{\alpha} - \hat{\beta}\bar{X})^2$$
$$= \hat{\beta}^2 \sum (X_i - \bar{X})^2$$

(remember that $\bar{Y} = \hat{\alpha} + \hat{\beta}\bar{X}$). Since $\sum (X_i - \bar{X})^2 > 0$, we then see that $R^2 > 0$ if and only if $\hat{\beta} \neq 0$, and by extension, R^2 is significantly above 0 if and only if $\hat{\beta}$ is significantly different from 0.

13.21 (a) $E(\hat{\beta}) = \beta \dfrac{\sum (X_i - \bar{X})X_i^2}{\sum (X_i - \bar{X})^2} \neq \beta$

(b) Define $W_i = X_i^2$. Then the correct specification is $Y_i = \alpha + \beta W_i + \epsilon_i$. If

$$\hat{\beta}^* = \frac{\sum (W_i - \bar{W})Y_i}{\sum (W_i - \bar{W})^2}$$

then $\hat{\beta}^*$ will be an unbiased estimator of β. The proof is the same as that for the classical linear regression model presented in the text.

CHAPTER 14

14.1 (a) $\beta_0 = (.975, \ 3.025)$; $\beta_1 = (-.00105, \ .00305)$; $\beta_2 = (.0925, \ .7075)$
(b) $Pr(|t^{(27)}| \geq 4)$; $.002 >$ prob-value;   reject H_0
(c) $Pr(t^{(27)} \geq 1)$; $.25 >$ prob-value $> .10$;   cannot reject H_0
(d) $Pr(t^{(27)} \geq 2.67)$; $.01 >$ prob-value $> .005$;   reject H_0
(e) We estimate that a one point increase in SAT is associated with (on average) a .001 point increase in college GPA, holding high school GPA constant. This effect is not, however, statistically significant. We estimate that a one point increase in high school GPA is associated with (on average) a .4 point increase in college GPA,

holding SAT score constant. This relationship is statistically significant.

14.3

$$\widehat{\text{NETINV}_i} = 86.936 + .16711 \text{ CFSAL}_i - 3.0876 \text{ RINT}_i$$
$$(7.9041) \quad (.081091) \quad\quad (1.3558)$$

The coefficient of CFSAL is not statistically significant at the .05 level in a two-sided test (prob-value $= Pr(|t^{(15)}| \geq 2.06)$; $.10 >$ prob-value $> .05$). The coefficient of RINT is statistically significant at the .05 level in a two-sided test (prob-value $= Pr(|t^{(15)}| \geq 2.28)$; $.05 >$ prob-value $> .02$). We then estimate that a one billion dollar increase in CFSAL is associated with, on average, a .16711 billion dollar increase in NETINV when RINT is held constant. This relationship, however, is not statistically significant (note, however, that if our hypothesis test had been a one-sided test, or a test at the 10% level, then we would have concluded that this coefficient was statistically significant). We also estimate that a one percentage point increase in RINT is associated with, on average, a 3.0876 billion dollar decrease in NETINV, holding CFSAL constant. This relationship is statistically significant.

14.5 We should expect $\hat{\beta}_1 = \hat{\alpha}_1$. Also, we should expect

$$\hat{\alpha}_0 = \hat{\beta}_0 + \hat{\beta}_2$$
$$\hat{\beta}_0 = \hat{\alpha}_0 + \hat{\alpha}_2$$

Combining these last two expressions suggests $\hat{\alpha}_2 = -\hat{\beta}_2$. From an interpretive point of view, everything should be the same.

14.7 The dummy variables are as follows:

$$\text{NE}_i = 1 \text{ if Northeast, 0 otherwise;}$$

$$\text{MW}_i = 1 \text{ if Midwest, 0 otherwise;}$$

$$\text{SO}_i = 1 \text{ if South, 0 otherwise}$$

$$(\text{NE}_i = \text{MW}_i = \text{SO}_i = 0 \text{ implies West})$$

The estimated regression is

$$\widehat{MIM_i} = -1291.2 + 198.12 \ PMHS_i + 49.379 \ PURBAN_i$$

$$(5392.3) \quad (53.947) \qquad (14.264)$$

$$- \ 42.210 \ MAGE_i + 245.68 \ NE_i + 755.72 \ MW_i$$

$$(151.58) \qquad\qquad (723.40) \qquad (607.99)$$

$$+ \ 1267.9 \ SO_i$$

$$(862.67)$$

where the numbers in parentheses are the standard errors. In a test of

$$H_0: \quad \beta = 0$$

$$H_1: \quad \beta \neq 0$$

(at the .05 level) for each of the coefficients of the dummy variables we would fail to reject H_0. These results suggest no significant effect on mean male income of region, holding PMHS and PURBAN constant. (See appendix 2 to this chapter for a further discussion and investigation of this issue.)

14.9 The other variables we'd try to hold constant would be factors influencing survival time. For example, various measures of health (e.g. blood pressure, weight, etc.) or proxies for such (e.g. income, location of residence, etc.) might be included in a multiple regression to attempt this.

14.11 The relationship between A and H is as follows:
heat and air: $H = A = 1$
heat but no air: $H = 1$ and $A = 0$
no heat and no air: $H = A = 0$
This is not a perfect linear relationship; note that $H_i + A_i$ can equal 0, 1, or 2. Knowing that $A = 0$, for example, does not tell me the value of H. Similarly, knowing that $H = 1$ does not tell me the value of A. So the relationship still contains an element of randomness. Although there is imperfect multicollinearity, we do not have the problem of perfect multicollinearity.

INDEX